ILLUSTRATED PHYSIOLOGY

CHURCHILL LIVINGSTONE

Medical Division of Longman Group Limited

Distributed in the United States of America by
Churchill Livingstone Inc., 19 West 44th Street, New
York, N.Y. 10036 and by associated companies, branches
and representatives throughout the world.

First Edition	. . . .	1963
Italian translation	. . . .	1966
Second Edition	. . . .	1970
Danish translation	. . . . . .	1973
Third Edition	. . . .	1975
Reprinted	. . . .	1976
Reprinted	. . . .	1977
Reprinted	. . . .	1979
Reprinted	. . . .	1981

ISBN 0 443 01293 8

Printed in Hong Kong by
Wing Tai Cheung Printing Co. Ltd.

ILLUSTRATED PHYSIOLOGY

BY

ANN B. McNAUGHT

M.B., Ch.B., Ph.D.
Lecturer in Physiology, University of Glasgow

AND

ROBIN CALLANDER

FFPh MMAA AIMBI
Medical Illustration Unit, University of Glasgow

THIRD EDITION

CHURCHILL LIVINGSTONE

EDINBURGH LONDON AND NEW YORK

1975

THIS BOOK
IS
AFFECTIONATELY
DEDICATED

To my husband, James A. Gilmour

Ann B. McNaught

To my wife, Elizabeth

Robin Callander

PREFACE
FIRST EDITION

This book grew originally from the need to provide visual aids for the large number of students in this department who come to the study of human physiology with no background of mammalian anatomy and often without any conventional training in either the biological or the physical and chemical sciences. Such students include post-graduates studying for the Diploma or Degree in Education; under-graduates working for Degrees in Science; laboratory technicians taking courses for the Ordinary and Higher National Certificates in Biology; and the increasing number of medical auxiliaries — physio-therapists, occupational therapists, radiographers, cardiographers, dietitians, almoners and social workers — many of whom are required to study to quite advanced levels at least regional parts of the subject.

Many medical, dental and pharmacy students, as well as nurses in training, have been good enough to indicate that they too find diagrams which summarize the salient points of each topic valuable aids to learning or revision. It is our hope that some of these groups may find our book helpful.

Each page is complete in itself and has been designed to oppose its neighbour. It is hoped that this will facilitate the choice of those pages thought suitable for any one course while making it easy to omit those which are too detailed for immediate consideration.

A book of this sort is largely derivative and it is impossible to acknowledge our wider debt. We wish to record our gratitude, however, to Professor R.C. Garry for his generous permission to borrow freely from the large collection of teaching diagrams built up over the years by himself and his staff; to Dr H.S.D. Garven and Dr G. Leaf for permission to use some of their own teaching material; and to Messrs Ciba Pharmaceutical

Products Inc. from whose fine book of Medical Illustrations by
Dr Frank H. Netter the diagram on the Cranial Nerves has been
modified.

We are indebted to the following colleagues and friends who
read parts of the original draft and offered helpful criticism:-
Dr H.S.D. Garven, Dr J.S. Gillespie, Mr J.A. Gilmour, Dr M. Holmes,
Dr B.R. Mackenna, Mr T. McClurg Anderson, Dr I.A. Boyd,
Dr R.Y. Thomson, Dr J.B. deV. Weir.

We should like to express our gratitude to Mr Charles Macmillan
and Mr James Parker of Messrs. E. & S. Livingstone Ltd. for their
unfailing courtesy and encouragement, and to Mrs Elizabeth
Callander for help in preparing the index.

ANN B. MCNAUGHT

ROBIN CALLANDER.

January, 1963.   *Institute of Physiology,*
                                                       The University of Glasgow.

THIRD EDITION

We continue to be encouraged by the favourable reception
accorded "Illustrated Physiology" and are again indebted to
reviewers and colleagues for suggested improvements. In the
1965 revision we recorded our thanks to Professor E.A. Dawes,
Reckitt Professor of Biochemistry in the University of Hull, and
to Drs. T.D.M. Roberts and N.C. Spurway of the Institute of
Physiology, University of Glasgow. On this occasion we
wish to add the names of Dr. G.S. Dawes, Director of the
Nuffield Institute for Medical Research, the University of
Oxford, and Drs. J. Womersley, G. Stenhouse, H.C. McKirdy,
M. Gladden and K. Lindsay of this department.

A.B. McN.
R.C.

1974

CONTENTS

1. INTRODUCTION: The TISSUES 3

The Amoeba 3
The Phenomena of Life . . . 4
The Paramecium 5
The Cell 6
Cell Division (mitosis) . . . 7

Differentiation of Animal Cells . 8
Epithelia 9,10
Connective Tissues . . . 11-13
Muscular Tissues 14
Nervous Tissues 15-17

Cell Division (meiosis) . . 18
Development of Individual . 19
"Master" Tissues 20
"Vegetative" Tissues . . . 21
The Body Systems . . . 22

2. NUTRITION and METABOLISM-The SOURCES, RELEASE and USES of ENERGY 23

Basic Elements in Protoplasm . 25
Carbohydrates 26
Fats 27
Proteins 28
Nucleic Acids 29
Source of Energy:
        Photosynthesis . 30
Carbon Cycle 31

Nitrogen Cycle 32
Nutrition 33
Energy-Giving Foods . . . 34
Body-Building Foods . . . 35
Protective Foods . . 36,37
Digestion 38
Protein Metabolism . . . 39
Carbohydrate Metabolism . 40

Fat Metabolism 41
Release of Energy 42
Heat Balance 43
Maintenance of Body
        Temperature . . 44,45
Growth 46,47
Energy Requirements . 48,49
Balanced Diet 50

3. DIGESTIVE SYSTEM 51

Digestive System 52
Progress of Food along
        Alimentary Canal . . 53
Digestion in Mouth 54
Control of Salivary Secretion 55
Oesophagus 56
Swallowing 57
Stomach 58
Gastric Juice 59

Movements of Stomach . . 60
Vomiting 61
Pancreas 62
Pancreatic Juice 63
Liver and Gall Bladder . . 64
Expulsion of Bile 65
Small Intestine 66
Basic Pattern of Gut Wall . 67
Intestinal Juice 68

Movements of Small Intestine . 69
Absorption in Small Intestine . 70
Transport of Absorbed
        Foodstuffs . . 71
Large Intestine 72
Movements of Large Intestine 73
Innervation of Gut Wall . . 74
Nervous Control of Gut
        Movements . . 75

4. TRANSPORT SYSTEM- The HEART, BLOOD VESSELS and BODY FLUIDS: 77
HAEMOPOIETIC SYSTEM

Cardiovascular System . . . 78
General Course of Circulation 79
Heart 80-82
Cardiac Cycle 83
Heart Sounds 84
Origin and Conduction of
        the Heart Beat . . 85
Electrocardiogram 86
Nervous Regulation of
        Action of Heart . 87
Cardiac Reflexes 88,89
Cardiac Output 90
Blood Vessels 91

Blood Pressure 92
Measurement of Arterial B.P. 93
Elastic Arteries 94
Nervous Regulation of Muscular
        Arteries and Arterioles . 95
Reflex and Chemical Regulation
        of Arteriolar Tone . 96,97
Capillaries 98
Veins 99
Blood Flow 100
Pulmonary Circulation . . 101
Distribution of Water and
        Electrolytes in Body Fluids . 102

Water Balance 103
Blood 104
Blood Coagulation . . . 105
Factors Required for
        Normal Haemopoiesis. 106
Haemopoiesis 107
Blood Groups . . . 108,109
Rhesus Factor 110
Inheritance of
        Rhesus Factor . . 111
Lymphatic System . . . 112
Spleen 113
Cerebrospinal Fluid . . 114

5. RESPIRATORY SYSTEM 115

Respiratory System . . . 117
Air Conducting Passages . 118
Lungs: Respiratory Surfaces 119
Thorax 120
Mechanism of Breathing . 121
Artificial Respiration . . 122

Capacity of Lungs 123
Composition of Respired Air 124
Movement of Respiratory Gases 125
Dissociation of Oxygen
        from Haemoglobin 126
Uptake and Release of CO_2 127

Carriage & Transfer of O_2,CO_2 128,129
Nervous Control of
        Respiratory Movements. 130
Chemical Regulⁿ of Respiration.131
Voluntary and Reflex Factors
        in Regulation of Respiration 132

6. EXCRETORY SYSTEM 133

Excretory System . . . 134
Kidney 135
Formation of Urine-*Filtration*. 136
Formation of Urine- *Concent?* 137
Formation of Urine-*Mechanism of Water Reabsorption*. 138
"Clearance" of Inulin in Nephron.139
Urea "Clearance" 140
Diodone "Clearance" 141
Maintenance of Acid-Base Balance . . 142
Maintenance of Acid-Base Balance 143
Regulation of Water Balance.144-5
Urinary Bladder and Ureters 146
Storage & Expulsion of Urine 147
Urine 148

7. ENDOCRINE SYSTEM 149

Endocrine System 150
Thyroid 151
Underactivity of Thyroid . . 152
Overactivity of Thyroid . . 153
Parathyroids 154
Underactivity of Parathyroids 155
Overactivity of Parathyroids 156
Suprarenal Glands . . . 157
Underactivity of Suprarenal Cortex 158
Overactivity of Suprarenal Cortex 159
Suprarenal Medulla . . . 160
Adrenaline 161
Development of Pituitary. 162
Anterior Pituitary . . . 163
Underactivity of Anterior Pituitary. 164
Overactivity of Pituitary Eosinophil Cells.165
Overactivity of Pituitary Basophil Cells . 166
Panhypopituitarism . . 167
Posterior Pituitary . . 168
Oxytocin 169
Antidiuretic Hormone . 170
Underactivity of Post. Pituitary 171
Aldosterone and ADH: *Maintenance of Blood Volume* 172
Pancreas: Islets of Langerhans 173
Thymus 174

8. REPRODUCTIVE SYSTEM 175

Male Reproductive System . . 176
Testis 177
Male Secondary Sex Organs . 178
Puberty in Male 179
Female Reproductive System . 180
Adult Pelvic Sex Organs in Ordinary Female Cycle . 181
Ovary in Ordinary Adult Cycle 182
Ovary in Pregnancy . . . 183
Puberty in Female . . . 184
Ovarian Hormones . . . 185
Uterus and Uterine Tubes 186,187
Uterine Tubes in cycle ending in Pregnancy . 188
Uterus 189,190,192,194-196
Placenta 191
Foetal Circulation 193
Mammary Glands . . 197-199
Menopause 200
Relationship between Ant. Pituitary, Ovarian and Endometrial Cycles 201

9. "MASTER" TISSUES: CENTRAL NERVOUS SYSTEM, LOCOMOTOR SYSTEM 203

Nervous System 204
Development of Nervous System. 205
Cerebrum 206
Horizontal Section through Brain. 207
Vertical Section through Brain. 208
Coronal Section through Brain. 209
Cranial Nerves 210
Spinal Cord 211
Synapse 212
Nerve Impulse 213
Reflex Action 214
Stretch Reflexes . . . 215
Spinal Reflexes 216
"Edifice" of the C.N.S. . . 217
Reflex Action 218
Arrangement of Neurones . 219
Sense Organs 220
Smell 221
Taste 222
Pathways and Centres for Taste 223
Eye 224
Protection of the Eye . . 225
Muscles of Eye 226
Control of Eye Movements . 227
Iris, Lens and Ciliary Body . 228
Action of Lens 229
Fundus Oculi 230
Retina 231
Mechanism of Vision . . 232,233
Visual Pathways to the Brain. 234
Stereoscopic Vision . . . 235
Light Reflex 236
Ear 237
Cochlea 238
Mechanism of Hearing . . 239
Auditory Pathways to Brain. 240
Special Proprioceptors . . 241
Organ of Equilibrium: Mechanism of Action . 242
Vestibular Pathways to Brain. 243
General Proprioceptors . . 244
Proprioceptor Pathways to Brain. 245
Cutaneous Sensation . . 246
Sensory Pathways from Skin of Face . . 247
Pain and Temperature Pathways from Trunk and Limbs 248
Touch and Pressure Pathways from Trunk and Limbs . 249
Sensory Cortex 250
Motor Cortex 251
Motor Pathways to Head and Neck.252
Motor Pathways to Trunk and Limbs. 253
Motor Unit 254
Final Common Pathway . 255
Extrapyramidal System . 256
Cerebellum . . 257,258,259
Control of Muscle Movement . 260
Locomotor System . . . 261
Skeletal Muscles . . 262,263
Muscular Movements . . 264
Reciprocal Innervation . . 265
Skeletal Muscle/Contraction. 266
Autonomic Nervous System. 268,269
Autonomic Reflex 270
Chemical Transmission at Nerve Endings . . 271

CHAPTER 1

INTRODUCTION:

The TISSUES

The AMOEBA

      All living things are made of PROTOPLASM. Protoplasm exists in MICROSCOPIC UNITS called CELLS. The SIMPLEST living creatures consist of ONE CELL. The AMOEBA (which lives in pond water) exemplifies the BASIC STRUCTURE of all animal cells and shows the PHENOMENA which distinguish living from non-living things.

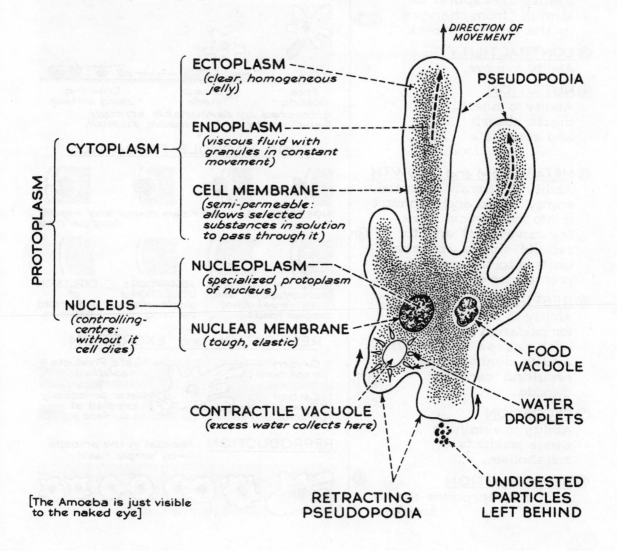

ECTOPLASM
(clear, homogeneous jelly)

ENDOPLASM
(viscous fluid with granules in constant movement)

CELL MEMBRANE
(semi-permeable: allows selected substances in solution to pass through it)

CYTOPLASM

NUCLEOPLASM
(specialized protoplasm of nucleus)

NUCLEAR MEMBRANE
(tough, elastic)

NUCLEUS
(controlling-centre: without it cell dies)

PROTOPLASM

DIRECTION OF MOVEMENT

PSEUDOPODIA

FOOD VACUOLE

WATER DROPLETS

CONTRACTILE VACUOLE
(excess water collects here)

RETRACTING PSEUDOPODIA

UNDIGESTED PARTICLES LEFT BEHIND

[The Amoeba is just visible to the naked eye]

3

The PHENOMENA which characterize all living things ---  are shown by the AMOEBA

① ORGANIZATION
Autoregulation — inherent ability to control all life processes.

② IRRITABILITY
Ability to respond to stimuli (from changes in the environment).

③ CONTRACTILITY
Ability to move.

④ NUTRITION
Ability to ingest, digest, absorb and assimilate food.

⑤ METABOLISM and GROWTH
Ability to liberate potential energy of food and to convert it into mechanical work (*e.g. movement*) and to rebuild simple absorbed units into the complex protoplasm of the living cell.

⑥ RESPIRATION
Ability to take in oxygen for oxidation of food with release of energy; and to eliminate the resulting carbon dioxide.

⑦ EXCRETION
Ability to eliminate waste products of metabolism.

⑧ REPRODUCTION
Ability to reproduce the species.

PSEUDOPODIA FORMATION

Small elevation arises on surface

Cytoplasm streams forward

forming long pseudopodium

②,③

Free floating

Contact made

Crawling along surface

Attracted by favourable stimuli.
Repelled by unfavourable stimuli.

FOOD VACUOLE FORMATION

④

INGESTION: *Amoeba flows round and engulfs food particle.*

SECRETION of Enzymes (*chemical agents*) which DIGEST (*break down*) complex foods.

ABSORPTION & UTILIZATION of simple units by the living cell.

EXPULSION of undigested particles.

RESPIRATION and EXCRETION

⑥,⑦

Oxygen (*in solution*)

Carbon Dioxide (*in solution*)

Waste Products (*in solution*)

Water periodically expelled at surface

REPRODUCTION
Asexual in the amoeba — by simple fission.

⑧

The PARAMECIUM

The Paramecium (another ONE-CELLED fresh water creature) shows:—

MODIFICATION and localization of STRUCTURE --------- for --------- SPECIALIZATION and localization of CERTAIN FUNCTIONS

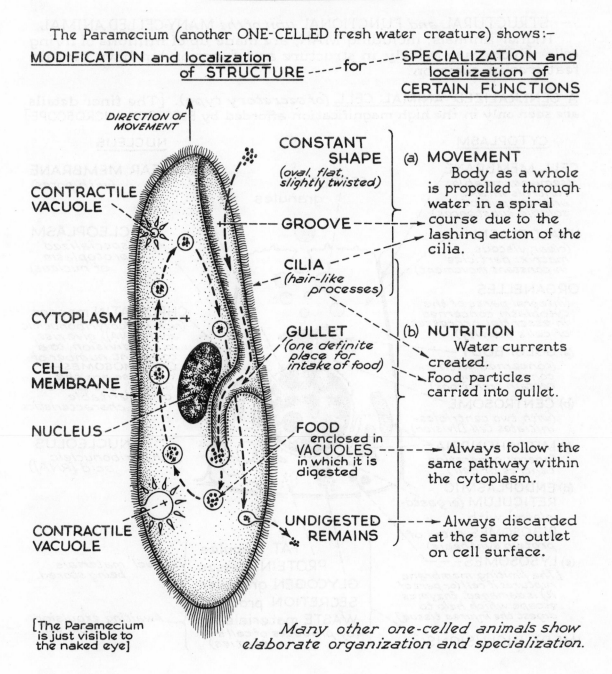

DIRECTION OF MOVEMENT

CONTRACTILE VACUOLE

CONSTANT SHAPE
(oval, flat, slightly twisted)

(a) MOVEMENT
Body as a whole is propelled through water in a spiral course due to the lashing action of the cilia.

GROOVE

CILIA
(hair-like processes)

CYTOPLASM

GULLET
(one definite place for intake of food)

(b) NUTRITION
Water currents created.
Food particles carried into gullet.

CELL MEMBRANE

NUCLEUS

FOOD
enclosed in VACUOLES in which it is digested

Always follow the same pathway within the cytoplasm.

CONTRACTILE VACUOLE

UNDIGESTED REMAINS

Always discarded at the same outlet on cell surface.

[The Paramecium is just visible to the naked eye]

Many other one-celled animals show elaborate organization and specialization.

The CELL

— STRUCTURAL *and* FUNCTIONAL *unit of the* MANY-CELLED ANIMAL.
  Higher animals, including MAN, are made up of millions of living cells which vary widely in structure and function but have certain features in common.

A GENERALIZED ANIMAL CELL *(of secretory type)*. [The finer details are seen only in the high magnification afforded by ELECTRON MICROSCOPE.]

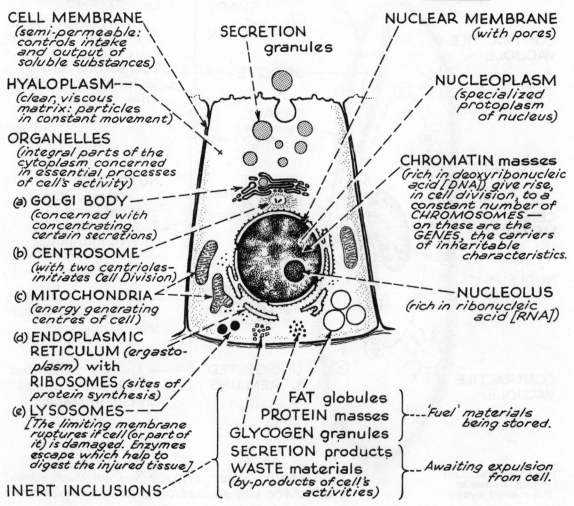

CYTOPLASM

CELL MEMBRANE
(semi-permeable: controls intake and output of soluble substances)

HYALOPLASM
(clear, viscous matrix: particles in constant movement)

ORGANELLES
(integral parts of the cytoplasm concerned in essential processes of cell's activity)

(a) **GOLGI BODY**
(concerned with concentrating certain secretions)

(b) **CENTROSOME**
(with two centrioles- initiates Cell Division)

(c) **MITOCHONDRIA**
(energy generating centres of cell)

(d) **ENDOPLASMIC RETICULUM** *(ergasto- plasm)* with **RIBOSOMES** *(sites of protein synthesis)*

(e) **LYSOSOMES**
[The limiting membrane ruptures if cell (or part of it) is damaged. Enzymes escape which help to digest the injured tissue]

INERT INCLUSIONS

SECRETION granules

FAT globules
PROTEIN masses
GLYCOGEN granules
} ---'Fuel' materials being stored.

SECRETION products
WASTE materials *(by-products of cell's activities)*
} -- Awaiting expulsion from cell.

NUCLEUS

NUCLEAR MEMBRANE *(with pores)*

NUCLEOPLASM *(specialized protoplasm of nucleus)*

CHROMATIN masses *(rich in deoxyribonucleic acid [DNA]) give rise, in cell division, to a constant number of* **CHROMOSOMES** *— on these are the* **GENES**, *the carriers of inheritable characteristics.*

NUCLEOLUS *(rich in ribonucleic acid [RNA])*

CELL DIVISION (MITOSIS)

All cells arise from the division of pre-existing cells. In MITOSIS there is an exact QUALITATIVE division of the NUCLEUS and a less exact QUANTITATIVE division of the CYTOPLASM.

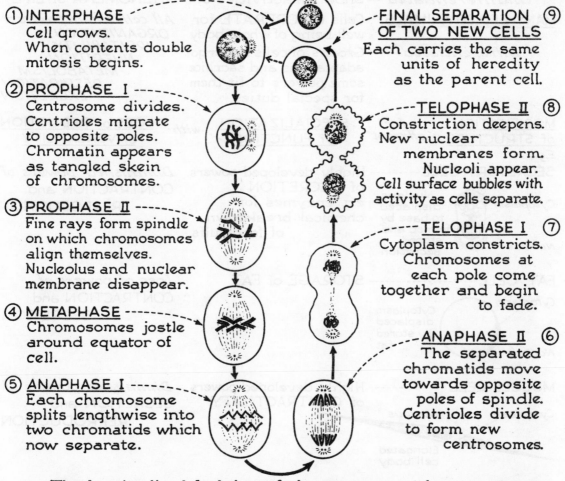

① INTERPHASE
Cell grows.
When contents double mitosis begins.

② PROPHASE I
Centrosome divides.
Centrioles migrate to opposite poles.
Chromatin appears as tangled skein of chromosomes.

③ PROPHASE II
Fine rays form spindle on which chromosomes align themselves.
Nucleolus and nuclear membrane disappear.

④ METAPHASE
Chromosomes jostle around equator of cell.

⑤ ANAPHASE I
Each chromosome splits lengthwise into two chromatids which now separate.

FINAL SEPARATION ⑨
OF TWO NEW CELLS
Each carries the same units of heredity as the parent cell.

TELOPHASE II ⑧
Constriction deepens.
New nuclear membranes form.
Nucleoli appear.
Cell surface bubbles with activity as cells separate.

TELOPHASE I ⑦
Cytoplasm constricts.
Chromosomes at each pole come together and begin to fade.

ANAPHASE II ⑥
The separated chromatids move towards opposite poles of spindle.
Centrioles divide to form new centrosomes.

The longitudinal halving of chromosomes and genes ensures that each new cell receives the same hereditary factors as the original cell.

The number of chromosomes is constant for any one species.

The cells of the human body (somatic cells) carry 23 pairs — i.e. 46 chromosomes.

[For clarity only 4 chromosomes (2 pairs) are shown in these diagrams]

DIFFERENTIATION of ANIMAL CELLS

SPECIALIZATION distinguishes multicellular creatures from more primitive forms of life.

ONE-CELLED ANIMALS — Capable of INDEPENDENT existence —
Undifferentiated — Show all activities or ⋯ PHENOMENA of LIFE

MANY-CELLED ANIMALS — Cells CO-OPERATE for well-being of whole body. — *All cells retain powers of*
ORGANIZATION
IRRITABILITY
NUTRITION
METABOLISM
RESPIRATION
EXCRETION

Differentiated — Groups of cells undergo adaptations and sacrifice some powers to fit them for special duties.

MODIFICATION of STRUCTURE ⋯ *for efficient* ⋯ **SPECIALIZATION of FUNCTION** ⋯ *with* ⋯ **LOSS or REDUCTION of VERSATILITY**
E.g.

SECRETORY CELL ⋯⋯ Highly developed powers of SECRETION e.g. enzymes for chemical breakdown of foodstuffs. — *Diminished powers of* CONTRACTION and REPRODUCTION

Cytoplasm — *Nucleus displaced to base by formed and stored secretion*
Nucleus

FAT CELL ⋯⋯⋯⋯⋯⋯ STORAGE of FAT — *Loss of powers of* CONTRACTION and SECRETION

Cytoplasm — *Cytoplasm displaced by stored fat*
Nucleus

MUSCLE CELL ⋯⋯⋯ Highly developed powers of CONTRACTILITY — *Diminished powers of* SECRETION and REPRODUCTION

Cytoplasm — *Nucleus*
Elongated cell body

NERVE CELL ⋯⋯⋯⋯ Highly developed powers of IRRITABILITY — *Loss of powers of* REPRODUCTION
Cytoplasm

Nucleus x 500 — *Cytoplasm drawn out into long branching processes* — (response to stimuli and transmission of impulses over long distances) — *i.e. if nerve cell is destroyed no regeneration is possible.*

ORGANIZATION OF TISSUES

Different cell types are not mixed haphazardly in the body.
Cells which are alike are arranged together to form TISSUES.
There are *four* main types of tissue:— 1. EPITHELIA or LINING,
2. CONNECTIVE or SUPPORTING, 3. MUSCULAR, 4. NERVOUS.

EPITHELIA

STRUCTURAL MODIFICATIONS	SITE	SPECIALIZED FUNCTIONS
Sheets of cells with minimum intercellular substance		*Line all internal and external surfaces of body*

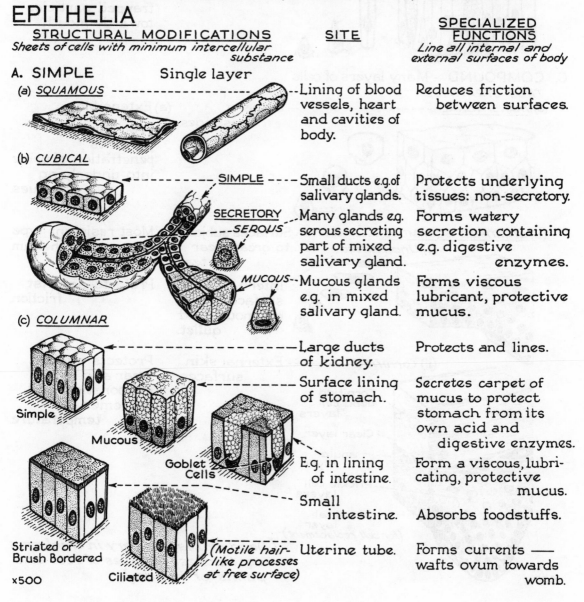

A. SIMPLE   Single layer

(a) *SQUAMOUS* — Lining of blood vessels, heart and cavities of body. — Reduces friction between surfaces.

(b) *CUBICAL*

SIMPLE — Small ducts e.g. of salivary glands. — Protects underlying tissues: non-secretory.

SECRETORY — *SEROUS* — Many glands e.g. serous secreting part of mixed salivary gland. — Forms watery secretion containing e.g. digestive enzymes.

MUCOUS — Mucous glands e.g. in mixed salivary gland. — Forms viscous lubricant, protective mucus.

(c) *COLUMNAR*

Simple — Large ducts of kidney. — Protects and lines.

Mucous — Surface lining of stomach. — Secretes carpet of mucus to protect stomach from its own acid and digestive enzymes.

Goblet Cells — E.g. in lining of intestine. — Form a viscous, lubricating, protective mucus.

— Small intestine. — Absorbs foodstuffs.

Striated or Brush Bordered

Ciliated (Motile hair-like processes at free surface) — Uterine tube. — Forms currents — wafts ovum towards womb.

x500

9

B. PSEUDOSTRATIFIED COLUMNAR
CILIATED *(with Goblet cells)* ------- Respiratory passages.

(a) Protective.

(b) Cilia form currents to move mucus-trapped particles towards mouth.

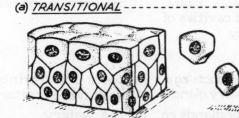

C. COMPOUND - Many layers of cells
TRUE STRATIFIED
(a) *TRANSITIONAL* ----------------------- Urinary passages.

(a) Extensible.

(b) Protective — prevents penetration of urine into underlying tissues.

(b) *STRATIFIED SQUAMOUS* --------------
(i) *Uncornified* ----- Surfaces subjected to great wear and tear.

Most resistant type of epithelium.

Internal surfaces e.g. mouth and gullet.

Protects against friction.

(ii) *Cornified* ---------- External skin surfaces.

Protects against wear and tear, evaporation and extremes of temperature.

Cornified dead cell layers

Clear layer

-anular layer

Prickle cell layers

Germinal layer
(for cell replacement)

×500

[Dead cell layers vary in thickness — e.g. thickest on soles of feet and palms of hands.]

CONNECTIVE TISSUES

STRUCTURAL MODIFICATIONS	SPECIALIZED FUNCTIONS
Cells plus large amount of intercellular matrix and extracellular elements.	*Form framework, connecting, supporting and packing tissues of the body.*

1. CELLS (floating free) in a FLUID MATRIX *(without fibrils until it clots).*

   BLOOD   Lymph

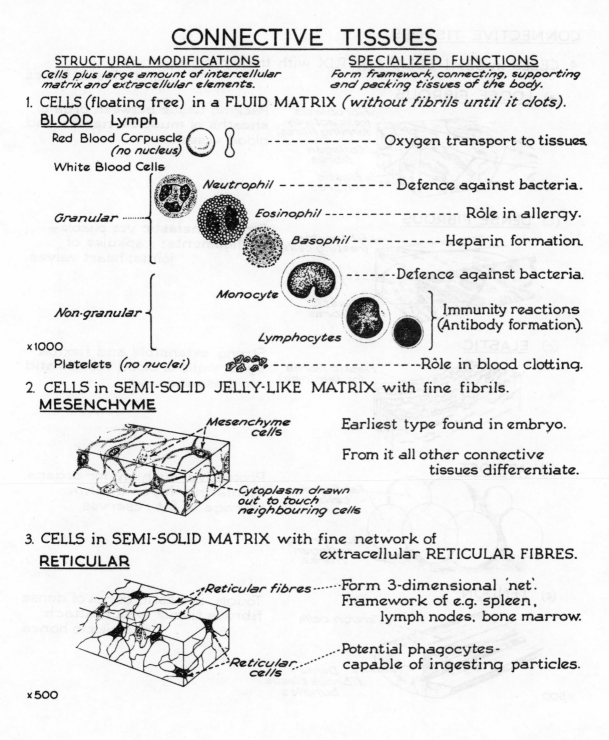

    Red Blood Corpuscle *(no nucleus)* — — — — — — — — — Oxygen transport to tissues.

    White Blood Cells

                              Neutrophil — — — — — — — — Defence against bacteria.

   Granular . . . .   *Eosinophil* — — — — — — — — — Rôle in allergy.

                               Basophil — — — — — — Heparin formation.

                                   — — — — — — Defence against bacteria.

                Monocyte

   Non-granular — — — — — — — — — — — — — — — — Immunity reactions (Antibody formation).

x1000   *Lymphocytes*

   Platelets *(no nuclei)* — — — — — — — — — — — — Rôle in blood clotting.

2. CELLS in SEMI-SOLID JELLY-LIKE MATRIX with fine fibrils.

   MESENCHYME

                    Mesenchyme cells   Earliest type found in embryo.

                                      From it all other connective tissues differentiate.

            -- *Cytoplasm drawn out to touch neighbouring cells*

3. CELLS in SEMI-SOLID MATRIX with fine network of extracellular RETICULAR FIBRES.

   RETICULAR

            Reticular fibres . . . . . Form 3-dimensional 'net'.
                              Framework of e.g. spleen, lymph nodes, bone marrow.

                            Potential phagocytes- capable of ingesting particles.

           Reticular cells

x500

CONNECTIVE TISSUES

4. CELLS in SEMI-SOLID MATRIX with thicker collagenous and elastic fibres.

(a) LOOSE FIBROUS

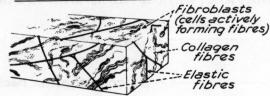

Fibroblasts
(cells actively
forming fibres)

Collagen
fibres

Elastic
fibres

"Packing" tissue between organs:
sheaths of muscles, nerves and
blood vessels.

(b) DENSE FIBROUS

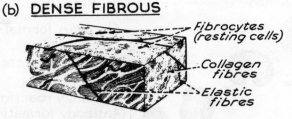

Fibrocytes
(resting cells)

Collagen
fibres

Elastic
fibres

Strong, inelastic yet pliable—
e.g. ligaments; capsules of
joints; heart valves.

(c) ELASTIC

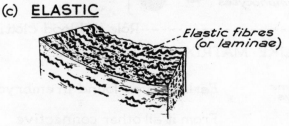

Elastic fibres
(or laminae)

Strong, extensible and flexible—
e.g. in walls of blood vessels and
air passages.

(d) ADIPOSE

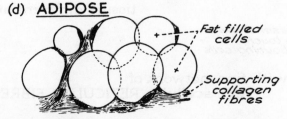

Fat filled
cells

Supporting
collagen
fibres

Protective "cushion" for organs.
Insulating layer in skin.
Storage of fat reserves.

(e) TENDON

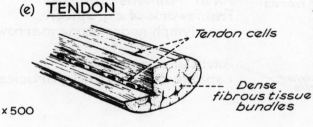

Tendon cells

Dense
fibrous tissue
bundles

Tough, inelastic cords of dense
fibrous tissue which attach
muscles to bones.

x 500

CONNECTIVE TISSUES

5. CELLS in SOLID ELASTIC MATRIX with fibres.

(a) HYALINE CARTILAGE

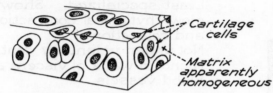

- Cartilage cells
- Matrix apparently homogeneous

Firm yet resilient
e.g. in air passages; ends of bones at joints.
Transition stage in bone
formation.

(b) WHITE FIBRO-CARTILAGE

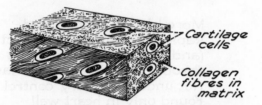

- Cartilage cells
- Collagen fibres in matrix

Tough. Resistant to stretching. Acts as shock absorber between vertebrae.

(c) ELASTIC or YELLOW FIBRO-CARTILAGE

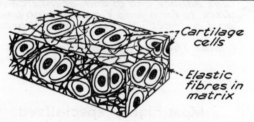

- Cartilage cells
- Elastic fibres in matrix

×500

More flexible, resilient —
e.g. in larynx and ear.

6. CELLS in SOLID RIGID MATRIX impregnated with Calcium and Magnesium Salts.

BONE

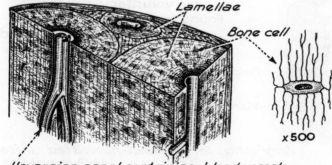

- Lamellae
- Bone cell

×500

Haversian canal containing blood vessels.

×100

Forms rigid framework of body.

MUSCULAR TISSUES

All have ELONGATED cells with special development of CONTRACTILITY.

1. SMOOTH, UNSTRIPED, VISCERAL or INVOLUNTARY muscle

Nucleus

Least specialized.   Shows slow rhythmical contraction and relaxation.
Not under voluntary control.
Found in walls of viscera and blood vessels.

2. CARDIAC or HEART muscle

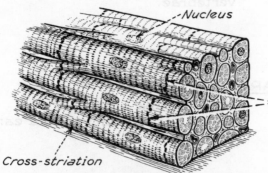

Nucleus

Cross-striation

More highly specialized. Rapid rhythmical contraction (and relaxation) spreads through whole muscle mass.
Not under voluntary control.
Found only in heart wall.

Cells adhere, end to end, at intercalated discs to form long 'fibres' which branch and connect with adjacent 'fibres'.

(Note: There is no protoplasmic continuity between cells.)

3. STRIATED, STRIPED, SKELETAL or VOLUNTARY muscle

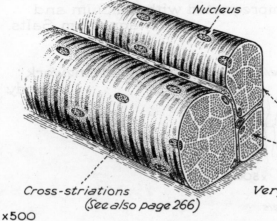

Nucleus

Most highly specialized.
Very rapid, powerful contractions of individual fibres.
Under voluntary control.
Found in e.g. muscles of trunk, limbs, head.

Thick covering membrane (sarcolemma)

Many myofibrils embedded in sarcoplasm

Cross-striations
(See also page 266)

Very long, large multi-nucleated units. No branching.

x500

14

NERVOUS TISSUES

Nervous tissue is divided into:-

(a) **NEURONES** or *True* nerve cells specialized in
IRRITABILITY
CONDUCTION
INTEGRATION

(b) **ACCESSORY** or *Supporting* cells
Not RECEPTIVE
Not CONDUCTING

MOTOR — pass messages from brain and spinal cord to effector organs (muscles and glands)
ASSOCIATION — relay messages between neurones.
SENSORY — receive and pass messages from environment to brain and spinal cord.
NEUROGLIA in Central Nervous System *(Brain and Spinal Cord)*
SHEATH *(Schwann)* cells on peripheral nerve fibres, i.e. outside C.N.S.
SATELLITE cells in ganglia of peripheral nervous system.

- -

TYPICAL NERVE CELL *(Motor)*

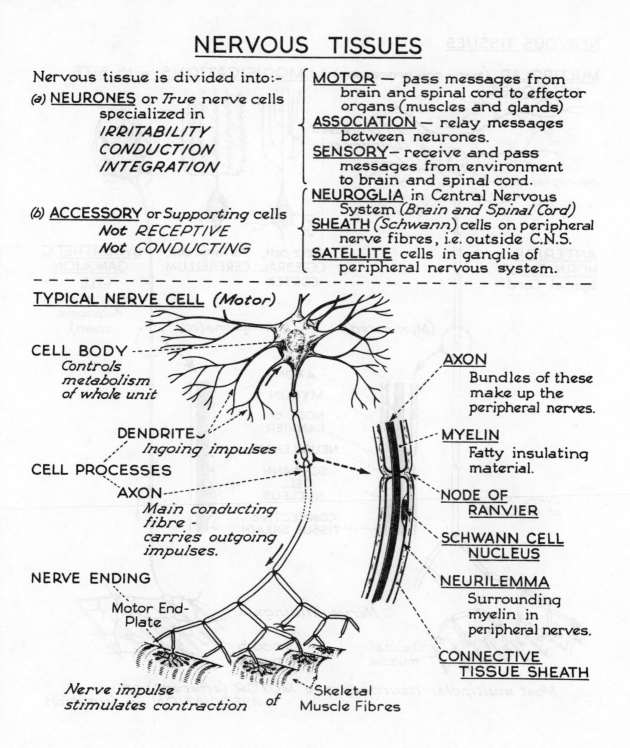

CELL BODY
Controls metabolism of whole unit

DENDRITES
Ingoing impulses

CELL PROCESSES

AXON
Main conducting fibre - carries outgoing impulses.

NERVE ENDING

Motor End-Plate

Nerve impulse stimulates contraction of
Skeletal Muscle Fibres

AXON
Bundles of these make up the peripheral nerves.

MYELIN
Fatty insulating material.

NODE OF RANVIER

SCHWANN CELL NUCLEUS

NEURILEMMA
Surrounding myelin in peripheral nerves.

CONNECTIVE TISSUE SHEATH

NERVOUS TISSUES

MULTIPOLAR (many cell processes) NEURONES

MODIFICATIONS with SITE

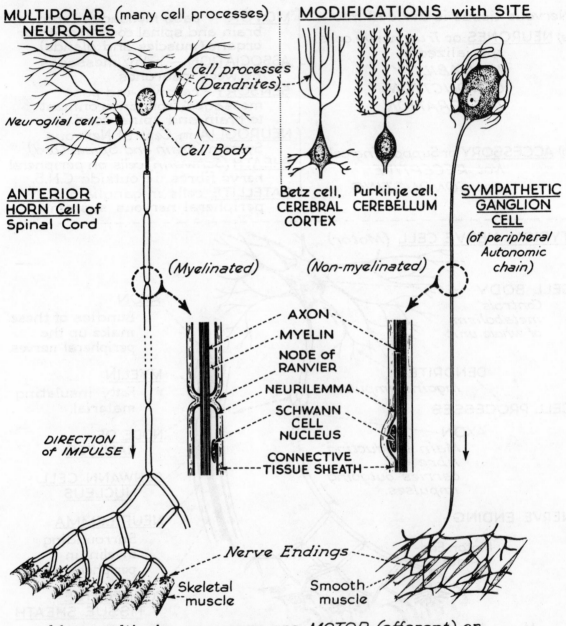

Neuroglial cell

Cell processes (Dendrites)

Cell Body

ANTERIOR HORN Cell of Spinal Cord

Betz cell, CEREBRAL CORTEX

Purkinje cell, CEREBELLUM

SYMPATHETIC GANGLION CELL (of peripheral Autonomic chain)

(Myelinated)

(Non-myelinated)

DIRECTION of IMPULSE

AXON
MYELIN
NODE of RANVIER
NEURILEMMA
SCHWANN CELL NUCLEUS
CONNECTIVE TISSUE SHEATH

Nerve Endings

Skeletal muscle

Smooth muscle

Most multipolar neurones are MOTOR (efferent) or ASSOCIATION in function.

NERVOUS TISSUES

UNIPOLAR (one main process leaves cell body)
NEURONE

BIPOLAR (two main processes)
NEURONE

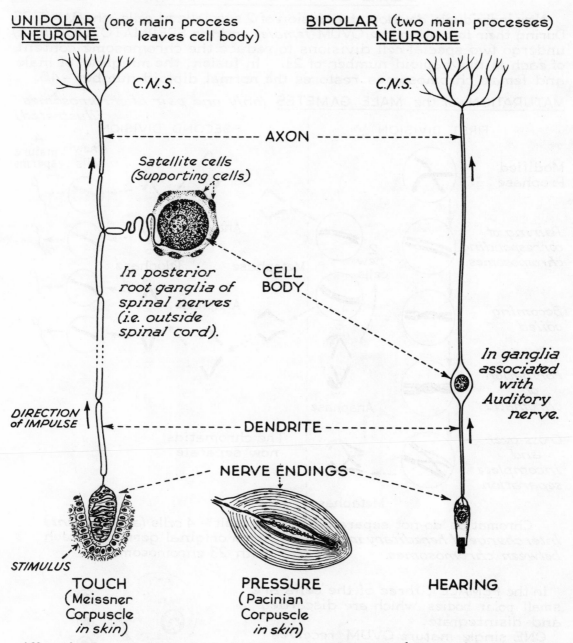

C.N.S.

C.N.S.

AXON

Satellite cells
(Supporting cells)

In posterior
root ganglia of
spinal nerves
(i.e. outside
spinal cord).

CELL
BODY

In ganglia
associated
with
Auditory
nerve.

DIRECTION
of IMPULSE

DENDRITE

NERVE ENDINGS

STIMULUS

TOUCH
(Meissner
Corpuscle
in skin)

PRESSURE
(Pacinian
Corpuscle
in skin)

HEARING

All unipolar and bipolar neurones are SENSORY (afferent) in function.

CELL DIVISION (MEIOSIS)

New individuals develop after fusion of 2 specialized cells—the GAMETES. During their formation the OVUM (*female*) and the SPERMATOZOON (*male*) undergo two special cell divisions to reduce the chromosome content of each to the haploid number of 23. In fusion, the mingling of male and female chromosomes restores the normal diploid number—46.

MATURATION of the MALE GAMETES *(only one pair of chromosomes illustrated)*

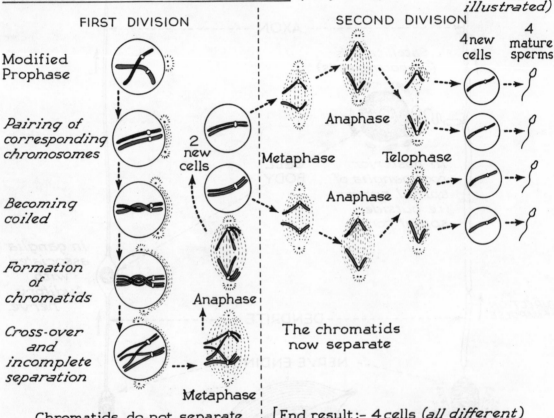

FIRST DIVISION

Modified Prophase

Pairing of corresponding chromosomes

2 new cells

Becoming coiled

Formation of chromatids

Cross-over and incomplete separation

Anaphase

Metaphase

SECOND DIVISION

Metaphase

Anaphase

Anaphase

Telophase

4 new cells

4 mature sperms

The chromatids now separate

Chromatids do not separate
Interchange of hereditary material between chromosomes.

[End result:— 4 cells (*all different*) from original germ cell - each with 23 chromosomes]

In the **FEMALE**, three of the 'cells' are small polar bodies which are discarded and disintegrate.

ONE single mature OVUM receives most of the cytoplasm.

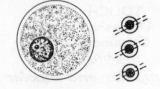

DEVELOPMENT of the INDIVIDUAL

All tissues of the human body are derived from the single cell—the FERTILIZED OVUM.

FERTILIZATION | CLEAVAGE
2 cells   4 cells   8 cells   Ball of 16 cells

FERTILIZATION
Fusion of ovum and spermatozoon

CLEAVAGE
Repeated mitotic divisions: each cell receives equal number of *Maternal* and *Paternal* chromosomes.

WOMB   Ectoderm   Connecting stalk   Mesoderm
Amniotic cavity
Embryonic germ disc
Endoderm
Yolk sac
Extraembryonic coelom
Blastocyst-hollow sphere

BLASTULATION and IMPLANTATION in womb

DIFFERENTIATION of tissues

Primitive foregut   Amniotic cavity   Connecting stalk
Head end
Yolk sac
Extraembryonic coelom

WOMB
Amniotic cavity
Embryo
Mesoderm of umbilical cord
Yolk sac
Extraembryonic coelom
Cavity of womb

EMBRYONIC GERM DISC
ECTODERM gives rise to ----- Epithelia of *External* surfaces.
----- Nervous tissues.
MESODERM gives rise to ----- Muscular tissues.
----- Connective tissues.
----- Urogenital system.
----- Lining of body cavities and blood vessels.
ENDODERM gives rise to ----- Epithelia of most *Internal* surfaces.
----- Some glands (e.g. liver, pancreas, thyroid).

"MASTER" TISSUES

It is an aid to understanding the general Design of the body to think of the systems as grouped into "MASTER" Tissues and "VEGETATIVE" Systems.

The "MASTER" Tissues are those which specialize in receiving stimuli from the environment and reacting to them —
NERVOUS TISSUE and SKELETAL MUSCLE.

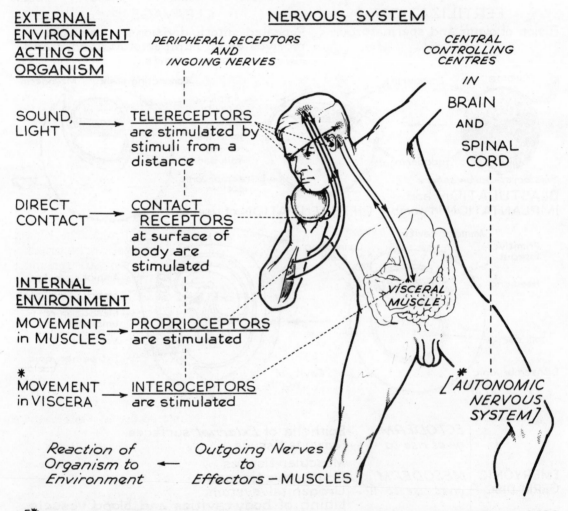

EXTERNAL
ENVIRONMENT
ACTING ON
ORGANISM

NERVOUS SYSTEM

*PERIPHERAL RECEPTORS
AND
INGOING NERVES*

*CENTRAL
CONTROLLING
CENTRES*

IN

BRAIN

AND

SPINAL
CORD

SOUND,
LIGHT → TELERECEPTORS
are stimulated by
stimuli from a
distance

DIRECT
CONTACT → CONTACT
RECEPTORS
at surface of
body are
stimulated

INTERNAL
ENVIRONMENT

MOVEMENT
in MUSCLES → PROPRIOCEPTORS
are stimulated

*
MOVEMENT
in VISCERA → INTEROCEPTORS
are stimulated

*VISCERAL
MUSCLE*

*[AUTONOMIC
NERVOUS
SYSTEM]*

*Reaction of
Organism to
Environment* ← *Outgoing Nerves
to
Effectors* – MUSCLES

*[*Note that a special part of the Nervous System — the AUTONOMIC
NERVOUS SYSTEM — regulates the activities of the "Vegetative" Systems.]*

20

"VEGETATIVE" SYSTEMS

The VEGETATIVE Systems are those which serve the basic utilities of life. They deal with the source of energy – FOOD – breaking it down and transporting it around the body to make it available to the cells for Growth and Repair, for Reproduction and for Movement, etc., and get rid of Waste. Much of their activity goes on below the level of consciousness.

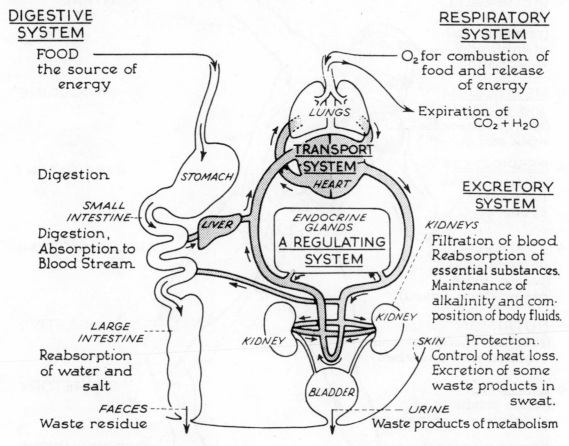

DIGESTIVE SYSTEM

FOOD
the source of
energy

Digestion

SMALL INTESTINE

Digestion,
Absorption to
Blood Stream

LARGE INTESTINE

Reabsorption
of water and
salt

FAECES
Waste residue

STOMACH

LIVER

TRANSPORT SYSTEM

HEART

ENDOCRINE GLANDS
A REGULATING SYSTEM

KIDNEY

KIDNEY

BLADDER

RESPIRATORY SYSTEM

O_2 for combustion of
food and release
of energy

Expiration of
$CO_2 + H_2O$

LUNGS

EXCRETORY SYSTEM

KIDNEYS
Filtration of blood.
Reabsorption of
essential substances.
Maintenance of
alkalinity and com-
position of body fluids.

SKIN   Protection.
Control of heat loss.
Excretion of some
waste products in
sweat.

URINE
Waste products of metabolism

The activities of these systems are regulated partly through a section of the MASTER tissues - the AUTONOMIC NERVOUS SYSTEM. Its complementary divisions - PARASYMPATHETIC and SYMPATHETIC - carry opposite or antagonistic messages to viscera. These impulses are rapidly adjusted and balanced by the BRAIN CENTRES to integrate the actions of the vegetative systems to the constantly changing needs of the whole body.

The BODY SYSTEMS

The tissues are arranged to form ORGANS.
Organs are grouped into SYSTEMS.

The ESSENTIAL LIFE ----are delegated to-------- SEPARATE
PROCESSES SYSTEMS

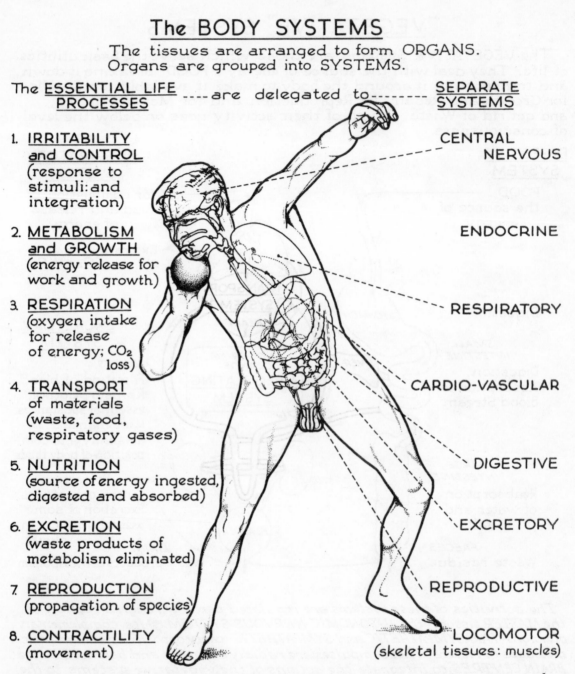

1. IRRITABILITY
and CONTROL
(response to
stimuli: and
integration)

CENTRAL
NERVOUS

2. METABOLISM
and GROWTH
(energy release for
work and growth)

ENDOCRINE

3. RESPIRATION
(oxygen intake
for release
of energy; CO_2
loss)

RESPIRATORY

4. TRANSPORT
of materials
(waste, food,
respiratory gases)

CARDIO-VASCULAR

5. NUTRITION
(source of energy ingested,
digested and absorbed)

DIGESTIVE

6. EXCRETION
(waste products of
metabolism eliminated)

EXCRETORY

7. REPRODUCTION
(propagation of species)

REPRODUCTIVE

8. CONTRACTILITY
(movement)

LOCOMOTOR
(skeletal tissues: muscles)

*The systems do not work independently. The body works as a whole.
Health and well-being depend on the co-ordinated effort of every part.*

22

CHAPTER 2

NUTRITION and METABOLISM

The SOURCES, RELEASE and USES of ENERGY

BASIC CONSTITUENTS of PROTOPLASM

Protoplasm is made up of certain
ELEMENTS ——— present mainly in ——— CHEMICAL COMBINATION

Chemical Symbol	ELEMENTS	e.g. in a 70 kg man average amounts (grams)			
H	– HYDROGEN	6,580	A large amount of the H and O is present as WATER	C, H & O combine chemically to form CARBOHYDRATES and FATS (chief sources of *ENERGY* in living protoplasm)	C, H, O and N combine to form PROTEINS (main *BUILDING* constituents of all protoplasm)
O	– OXYGEN	43,550			
C	– **CARBON**	12,590			
N	– NITROGEN	1,815			

These make up most of the Body Weight

Ca	– CALCIUM	1,700	Important constituents of blood and of hard tissues – e.g. bones and teeth.	
P	– PHOSPHORUS	680		
Cl	– CHLORINE	115	Important constituents of body fluids.	
Na	– SODIUM	70		
K	– POTASSIUM	70	Important constituent of all cells.	
S	– SULPHUR	100		
Mg	– MAGNESIUM	42	Important for activity of Brain, Nerves and Muscles.	
Fe	– IRON	7	Important component of red blood cells.	

These eight make up much of the remaining Body Weight

Trace elements *make up the last few grams or so.*
These include:- Manganese, copper, iodine, zinc, cobalt, molybdenum, nickel, aluminium, chromium, titanium, silicon, rubidium, lithium, arsenic, fluorine, bromine, selenium, boron, barium and strontium.

NOTE:- *Very few of these inorganic substances are found free in protoplasm. Most are in chemical combination. Apart from water, the chief constituents are present as compounds of CARBON, i.e. they are ORGANIC SUBSTANCES.*

25

CARBOHYDRATES

CARBOHYDRATES are found in Plant and Animal Protoplasm. The simpler carbohydrates are known as SUGARS; the more complex as STARCH. Starch is made up of hundreds of units of sugar tied chemically together. They consist of atoms of C, H and O.

GLUCOSE - $C_6H_{12}O_6$ - is one of the simplest SUGARS — a **MONOSACCHARIDE**
[Found in plants and animals]   *The arrangement of atoms in the molecule may be represented thus:-*

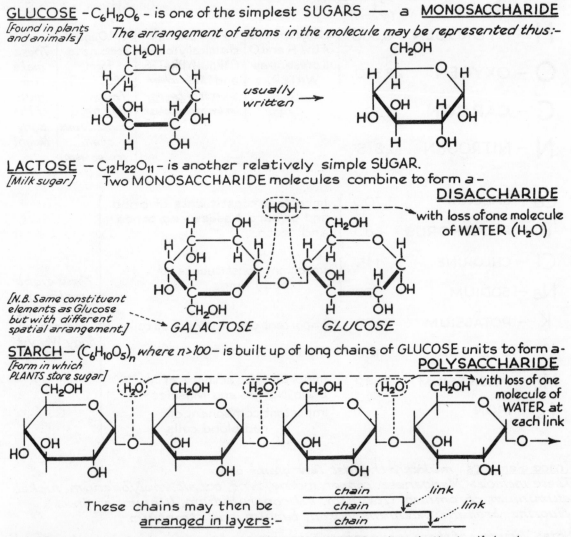

usually written →

LACTOSE — $C_{12}H_{22}O_{11}$ - is another relatively simple SUGAR.
[Milk sugar]   Two MONOSACCHARIDE molecules combine to form a –
DISACCHARIDE
HOH - - - - → with loss of one molecule of WATER (H_2O)

[N.B. Same constituent elements as Glucose but with different spatial arrangement.]

GALACTOSE   *GLUCOSE*

STARCH — $(C_6H_{10}O_5)_n$ *where n>100* – is built up of long chains of GLUCOSE units to form a-
[Form in which PLANTS store sugar]   **POLYSACCHARIDE**
with loss of one molecule of WATER at each link

These chains may then be arranged in layers:-

chain   *link*
chain   *link*
chain

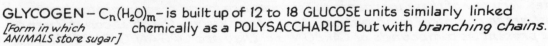

GLYCOGEN – $C_n(H_2O)_m$ - is built up of 12 to 18 GLUCOSE units similarly linked
[Form in which ANIMALS store sugar]   chemically as a POLYSACCHARIDE but with *branching chains*.

FATS

FATS are present in Plant and Animal Tissues. The molecule of fat is made up of smaller molecules of FATTY ACIDS linked chemically with a molecule of GLYCEROL. They consist of atoms of the elements C, H and O.

SIMPLE LIPIDS

GLYCEROL can react with 1 molecule of a FATTY ACID to produce a MONOGLYCERIDE

e.g. BUTYRIC ACID —— " —— MONOBUTYRIN

CH_2OH
$CHOH$ + HO $OCCH_2CH_2CH_3$ ⟶ CH_2OH
CH_2OH $CHOH$
                       $CH_2OOCCH_2CH_2CH_3$

with the loss of one molecule of WATER (H_2O)

GLYCEROL can react with 2 molecules of FATTY ACID to form a —— DIGLYCERIDE

e.g. PALMITIC ACID —— " —— DIPALMITIN

CH_2OH
$CHOH$ + HO $OC CH_2CH_2CH_2CH_2CH_2CH_2CH_2CH_2 CH_2 CH_2CH_2CH_2CH_2CH_2CH_3$
CH_2OH HO $OC CH_2CH_2CH_2CH_2CH_2CH_2CH_2CH_2 CH_2 CH_2CH_2CH_2CH_2CH_2CH_3$

with the loss of two molecules of H_2O

In its most common form FAT is made up of one molecule of GLYCEROL linked with 3 molecules of FATTY ACID to form a —— TRIGLYCERIDE

e.g. STEARIC ACID —— " —— TRISTEARIN

CH_2OH HO $OCCH_2CH_2CH_2CH_2CH_2CH_2CH_2CH_2CH_2CH_2CH_2CH_2CH_2CH_2CH_2CH_2CH_3$
$CHOH$ + HO $OCCH_2CH_2CH_2CH_2CH_2CH_2CH_2CH_2CH_2CH_2CH_2CH_2CH_2CH_2CH_2CH_3$
CH_2OH HO $OCCH_2CH_2CH_2CH_2CH_2CH_2CH_2CH_2CH_2CH_2CH_2CH_2CH_2CH_2CH_2CH_3$

with the loss of three molecules of H_2O

There are many other naturally occurring FATTY ACIDS. GLYCEROL can combine with 3 molecules of the same or 3 molecules of different FATTY ACIDS to form TRIGLYCERIDES (NEUTRAL FATS).

More complex fat-like substances include

PHOSPHATIDES — made up of GLYCEROL, FATTY ACIDS, PHOSPHORIC ACID and a NITROGEN base. They are widely distributed in animal tissues.

CEREBROSIDES — are even more complex compounds combining SUGAR in their structure.

COMPOUND LIPIDS

Other fat-like substances (e.g. Cholesterol, Bile Salts and many Hormones) are known as STEROIDS

They all share this PARENT STRUCTURE

PROTEINS

PROTEIN is the chief organic constituent of all Protoplasm—Plant or Animal. The protein molecule is made up of smaller units called AMINO-ACIDS. These consist of atoms of C,H,O and N with sometimes S and P.

GLYCINE ———————— is the simplest ————— AMINO-ACID
The arrangement of its atoms may be represented like this:-

[The name 'amino' comes from this NH_2 group] → H-N-C-C-OH [The name 'acid' comes from this COOH group]

Both groups are found in all the 20 or so kinds of amino-acid found in nature.

ALANINE ———————— is another simple ——————— AMINO-ACID

H-N-C—C-OH
   (CH_3) O

All amino-acids share the same core- -N-C-C- . Only the group in this position differs in each.

GLYCINE and
ALANINE can combine easily to form a ——— DIPEPTIDE

(HOH) - - - - - - - - - → with the loss of one molecule of WATER (H_2O).

H-N-C-C-N-C-C-OH
   (H) O   H (CH_3) O

AMINO-ACIDS can link in this way many times to form a POLYPEPTIDE

H-N-C-C-N-C-C-N-C-C-N-C-C-OH
   (R) O   (R) O   (R) O   (R) O   (R) O

→ each link causing loss of one molecule of WATER (H_2O)

Any of the 20 Amino-acids may be linked like this. Long chains with 50 to 400 molecules of amino-acids may make up a molecule of PROTEIN.
Note:- *The -N-C-C- cores of the amino acids form the backbone of the protein chain.*

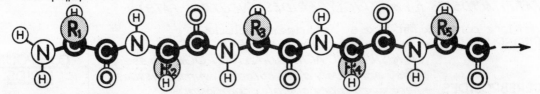

There are thousands of different kinds of Protein but only about 20 different amino-acids. Differences between proteins depend on the amino-acids present and on their number and arrangement.

The various amino-acids can be present in any proportion, arranged in any order, so that the resulting PROTEIN has its own sequence of groups ® branching off along the sides — this sequence gives it its own special properties which distinguish it from other proteins. The number of possible variations accounts for the great variety of living species, each with a protoplasm specific to itself.

NUCLEIC ACIDS

NUCLEOPROTEINS *(Protein attached to NUCLEIC ACID)* are amongst the most important constituents of Cells and their Nuclei.

NUCLEIC ACIDS are important in controlling heredity. They have a complex chain-like structure built up from MONONUCLEOTIDES —
e.g. Deoxyadenylic acid.

A
NUCLEOTIDE
contains

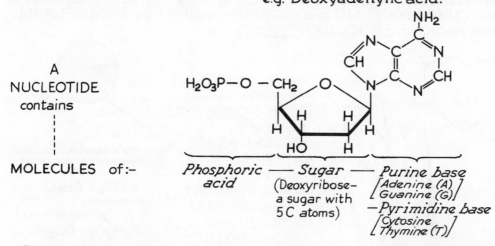

MOLECULES of :—

Phosphoric — Sugar — Purine base
acid (Deoxyribose- [*Adenine (A)*]
a sugar with [*Guanine (G)*]
5 C atoms) —*Pyrimidine base*
[*Cytosine*]
[*Thymine (T)*]

These mononucleotides may condense together to form <u>POLYNUCLEOTIDES</u>.

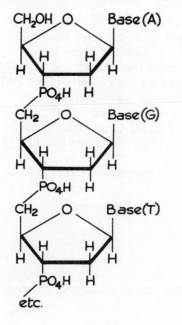

etc.

Such a chain of nucleotides *(in the DNA molecule in Chromosomes)*, by the specific arrangement of its bases, carries information from one generation of cells to the next.

[The information thus carried in a polynucleotide chain may determine many genetic factors, e.g. whether a child will have brown or blue eyes.]

29

SOURCE of ENERGY: PHOTOSYNTHESIS

The essential life processes or the phenomena which characterize life depend on the use of ENERGY. The SUN is the SOURCE of ENERGY for ALL LIVING THINGS.

Only GREEN PLANTS can TRAP and STORE the sun's energy and build from relatively simple substances the energy-rich and body-building compounds — *CARBOHYDRATES, FATS and PROTEINS* — required by PROTOPLASM. This process is called PHOTOSYNTHESIS.

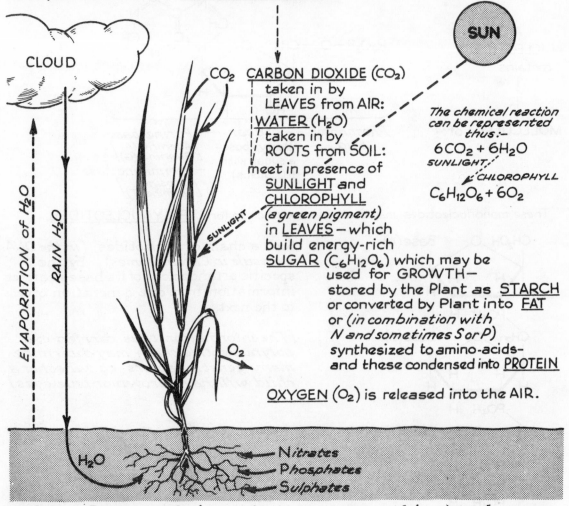

CLOUD

CO_2

SUN

CARBON DIOXIDE (CO_2)
  taken in by
  LEAVES from AIR:
WATER (H_2O)
  taken in by
  ROOTS from SOIL:
meet in presence of
SUNLIGHT and
CHLOROPHYLL
(*a green pigment*)
in LEAVES — which
build energy-rich
SUGAR ($C_6H_{12}O_6$) which may be
  used for GROWTH—
  stored by the Plant as STARCH
  or converted by Plant into FAT
  or (*in combination with
  N and sometimes S or P*)
  synthesized to amino-acids-
  and these condensed into PROTEIN

OXYGEN (O_2) is released into the AIR.

The chemical reaction can be represented thus:-

$$6CO_2 + 6H_2O$$
SUNLIGHT
CHLOROPHYLL
$$C_6H_{12}O_6 + 6O_2$$

EVAPORATION of H_2O

RAIN H_2O

SUNLIGHT

O_2

H_2O

Nitrates
Phosphates
Sulphates

When Plants or their products are eaten this stored energy becomes available to animals and man.

CARBON 'CYCLE' in NATURE

Animal bodies are unable to build *Proteins, Carbohydrates* or *Fats* from the raw materials. They must be built up for them by *Plants*.

CARBON is the basic building unit of all these compounds.

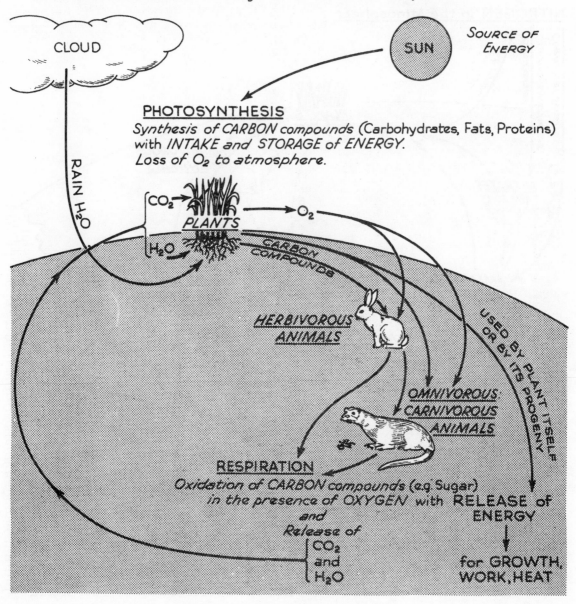

CLOUD

SUN

SOURCE OF ENERGY

PHOTOSYNTHESIS
Synthesis of CARBON compounds (Carbohydrates, Fats, Proteins) with INTAKE and STORAGE of ENERGY.
Loss of O_2 to atmosphere.

RAIN H_2O

CO_2

PLANTS

O_2

H_2O

CARBON COMPOUNDS

HERBIVOROUS ANIMALS

USED BY PLANT ITSELF OR BY ITS PROGENY

OMNIVOROUS CARNIVOROUS ANIMALS

RESPIRATION
Oxidation of CARBON compounds (e.g. Sugar) in the presence of OXYGEN with RELEASE of ENERGY and Release of
CO_2
and
H_2O

for GROWTH, WORK, HEAT

NITROGEN 'CYCLE' in NATURE

Although animals are surrounded by NITROGEN in the air they can only use the NITROGEN trapped by Plants.

NITROGEN in the Atmosphere

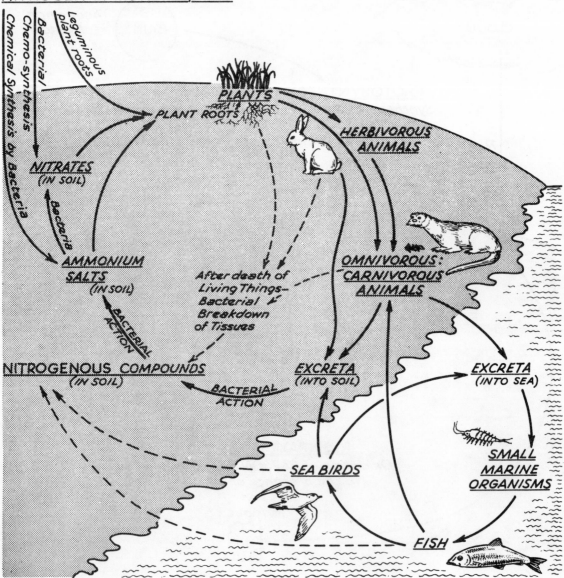

NUTRITION

Man eats as <u>FOOD</u> the substances which PLANTS
(and, through them, ANIMALS) have made.
These are broken down in man's body
into simpler chemical units which provide

BUILDING and PROTECTIVE MATERIALS

Man requires more or less the same
elements as plants. (Some, such as
the minerals iodine, sodium or iron,
he assimilates in INORGANIC FORM.)

Most must be built up for him
by plants:-
ORGANICALLY COMBINED
Carbon, Nitrogen, and Sulphur, etc.
ESSENTIAL AMINO ACIDS

ESSENTIAL FATTY ACIDS
Certain VITAMINS

These are
<u>USED to</u>

<u>BUILD, MAINTAIN or REPAIR</u>
PROTOPLASM

- -

BODY BUILDING REQUIREMENTS
of the INDIVIDUAL determine

<u>QUALITY</u>
of DIET

ENERGY

*Stored originally
by plants*

<u>RELEASED</u> in man's cells
by
<u>OXIDATION</u>

*When food is 'burned' it gives up
its stored energy*

PROTEINS	yield 4·1 kilocalories	*Units*	
CARBOHYDRATES	yield 4·1 kilocalories	*of Energy*	
FATS	yield 9·2 kilocalories	*per gram.*	

Most of this appears as <u>HEAT</u>
and is <u>USED for</u>
<u>KEEPING BODY WARM;</u>
some is used for <u>WORK</u> *of Cells*

- -

ENERGY REQUIREMENTS
of the INDIVIDUAL determine

<u>QUANTITY</u>
of DIET

For a
BALANCED DIET
TOTAL INTAKE
of
Essential Constituents and Energy Units
must balance...
...Amounts *stored* plus amounts *lost* from body plus amounts *used*
as Work or Heat.

ENERGY-GIVING FOODS

All the main foodstuffs yield ENERGY – the energy originally trapped by Plants. CARBOHYDRATES and FATS are the chief energy-giving foods. PROTEINS can give energy but are mainly used for building and repairing protoplasm.

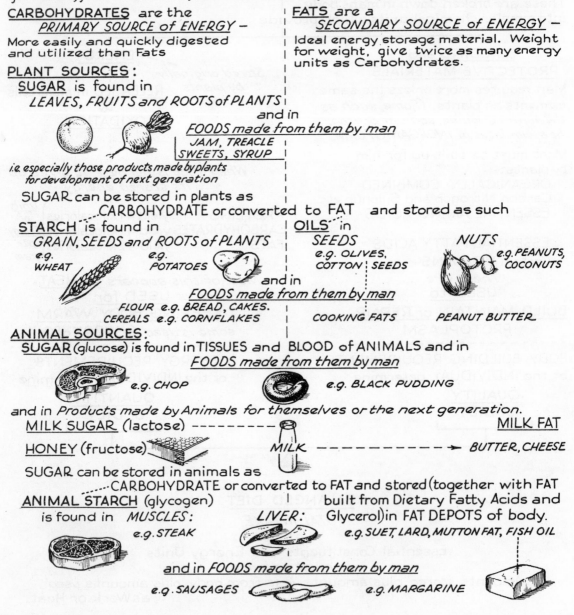

CARBOHYDRATES are the
PRIMARY SOURCE of ENERGY –
More easily and quickly digested
and utilized than Fats

FATS are a
SECONDARY SOURCE of ENERGY –
Ideal energy storage material. Weight
for weight, give twice as many energy
units as Carbohydrates.

PLANT SOURCES :
SUGAR is found in
LEAVES, FRUITS and ROOTS of PLANTS
and in
FOODS made from them by man
JAM, TREACLE
SWEETS, SYRUP
*i.e. especially those products made by plants
for development of next generation*

SUGAR can be stored in plants as
CARBOHYDRATE or converted to FAT and stored as such

STARCH is found in
GRAIN, SEEDS and ROOTS of PLANTS
e.g.
WHEAT
e.g.
POTATOES

OILS in
SEEDS
e.g. OLIVES,
COTTON SEEDS

NUTS
e.g. PEANUTS,
COCONUTS

and in
FOODS made from them by man
FLOUR e.g. BREAD, CAKES.
CEREALS e.g. CORNFLAKES

COOKING FATS

PEANUT BUTTER

ANIMAL SOURCES :
SUGAR (glucose) is found in TISSUES and BLOOD of ANIMALS and in
FOODS made from them by man
e.g. CHOP
e.g. BLACK PUDDING

and in *Products made by Animals for themselves or the next generation.*
MILK SUGAR (lactose) – – – – – – – –
HONEY (fructose)
MILK – – – – – – – – – – ➝
MILK FAT
BUTTER, CHEESE

SUGAR can be stored in animals as
CARBOHYDRATE or converted to FAT and stored (together with FAT
ANIMAL STARCH (glycogen)   built from Dietary Fatty Acids and
is found in *MUSCLES*:   *LIVER*:   Glycerol) in FAT DEPOTS of body.
e.g. STEAK   e.g. SUET, LARD, MUTTON FAT, FISH OIL

and in *FOODS made from them by man*
e.g. SAUSAGES   e.g. MARGARINE

BODY-BUILDING FOODS

PROTEIN is the chief body-building food. Because it is the chief constituent of protoplasm, tissues of Plants and Animals are the richest sources.

PLANT SOURCES

Protoplasm of Plant Tissues and — *Stores made by Plants — or Foods made for the next generation from them by man. in seeds and roots*

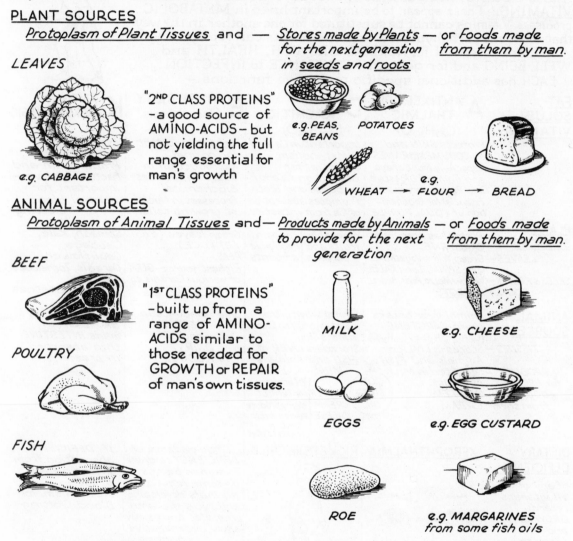

LEAVES

"2ND CLASS PROTEINS" – a good source of AMINO-ACIDS – but not yielding the full range essential for man's growth

e.g. CABBAGE

e.g. PEAS, BEANS   POTATOES

WHEAT → *e.g.* FLOUR → BREAD

ANIMAL SOURCES

Protoplasm of Animal Tissues and — *Products made by Animals — or Foods made to provide for the next generation from them by man.*

BEEF

"1ST CLASS PROTEINS" – built up from a range of AMINO-ACIDS similar to those needed for GROWTH or REPAIR of man's own tissues.

POULTRY

MILK

e.g. CHEESE

EGGS

e.g. EGG CUSTARD

FISH

ROE

e.g. MARGARINES *from some fish oils*

These foods between them also contain other important body-building elements:– e.g. CALCIUM and PHOSPHORUS for making BONES and TEETH hard. IRON for building HAEMOGLOBIN– the OXYGEN CARRYING substance in RED BLOOD CELLS. IODINE for building the THYROID HORMONE.

PROTECTIVE FOODS — I

Even when provided with BODY-BUILDING and ENERGY-GIVING FOODS, human tissues cannot build or repair their protoplasm or release the energy in the absence of PROTECTIVE substances called VITAMINS. These appear to be important links in METABOLIC processes. Vitamins cannot be substituted for one another in the way that different Carbohydrates, Proteins or Fats can replace others.

ALL are essential for NORMAL GROWTH, HEALTH and WELL-BEING and for general RESISTANCE to INFECTION.

EACH has additional specific protective functions:—

FAT-SOLUBLE VITAMINS	A ANTIXEROPH-THALMIC ($C_{20}H_{30}O$)	D ANTI-RACHITIC ($C_{28}H_{44}O$)	E ANTISTERILITY (Tocopherols $\alpha\beta\gamma$)	K ANTI-HAEMORRHAGIC
	Protects SKIN and MUCOUS MEMBRANES (especially front of eye and lining of Digestive and Respiratory tracts). Essential for Regeneration of VISUAL PURPLE.	Important in Ca and P Metabolism. Essential for deposition Ca and P in Bones and Teeth. Promotes absorption of Ca from intestine.	Perhaps important for normal Reproduction in adult. Influences Biochemical Processes in Nuclei of growing cells.	Essential for production of Prothrombin and Factor VII in Liver — important for normal blood clotting.
PLANT SOURCES	Present as precursor CAROTENE.	Vegetables) contain fruits and)negligible cereals) amounts.	Green leaves (e.g. LETTUCE) PEAS. Richest source - GERM of various Cereals eg. WHEAT GERM OIL.	Spinach, Kale, Cabbage, Cauliflower, Cereals, Tomatoes Carrots, Potatoes.
LEAVES —	Green, Yellow Vegetables (esp. SPINACH and KALE).			
SEEDS & FRUITS —	Yellow Maize, Peas, Beans.			
ROOTS —	CARROTS.			
ANIMAL SOURCES	Animal liver breaks down CAROTENE ($C_{40}H_{56}+2H_2O \rightarrow 2C_{20}H_{30}O$)	Formed when ultra-violet rays fall on Sterols in Man's Skin. Stored in LIVER of FISH and Animals	Small amounts in MEAT and DAIRY PRODUCE	Synthesized by BACTERIA in man's INTESTINE then absorbed in presence of BILE.
TISSUES —	Stored in LIVER of Animals and FISH			
PRODUCTS for PROGENY	Secreted in MILK EGG YOLK	MILK EGGS		
FOODS made from them	BUTTER CREAM	COD LIVER OIL BUTTER CREAM CHEESE	Vitamin content of dairy produce varies with plant in animal's diet.	
DIETARY DEFICIENCY	XEROPHTHALMIA	RICKETS in CHILD	Little evidence of DEFICIENCY syndrome in human beings. (In some female animals developing embryos die, and, in male, testes may show degenerative changes.)	If DEFICIENT absorption from intestine → upset in mechanism for Blood Clotting
of any or all vitamins retards GROWTH and leads to SEVERE DISEASE and eventually DEATH		OSTEOMALACIA in ADULT		

PROTECTIVE FOODS – II

WATER-SOLUBLE VITAMINS

B Complex

Includes

B_1 (Aneurin Thiamine) ANTINEURITIC

B_2 RIBOFLAVIN

NICOTINIC ACID

} Each forms part of a COENZYME SYSTEM – concerned with RELEASE of ENERGY from FOOD-STUFFS.

B_6 PYRIDOXINE – Coenzyme in metabolism of certain amino acids.

PANTOTHENIC ACID

BIOTIN – Tissue enzyme in Carbohydrate metabolism.

CHOLINE / INOSITOL } Lipotrophic factors – probably promote utilization of fats.

PARA-AMINOBENZOIC ACID – No known rôle in man.

FOLIC ACID – An **ANTIANAEMIA** factor.

B_{12} COBALAMINE - **ANTIPERNICIOUS ANAEMIA** factor – Involved in maturation of red blood cells

Together form essential parts of tissue ENZYME SYSTEMS – concerned in metabolism and Release of Energy by the cells

C ANTISCORBUTIC $(C_6H_8O_6)$

Probably acts as a Respiratory Catalyst in cells. Essential for formation and maintenance of INTERCELLULAR CEMENT and CONNECTIVE TISSUE. (Especially necessary for healthy state of blood vessels)

PLANT SOURCES

Common natural sources for most members of B Complex:-
PULSES e.g. GREEN PEAS (B_1, B_2)
SEEDS and <u>OUTER COATS of GRAIN</u> e.g. rice, wheat (B_1, B_2, Nic. A)
NUTS e.g. PEANUTS (B_1)
YEAST and Yeast Extracts (B_1, B_2, Nicotinic Acid)

Green vegetables, e.g. Parsley. Citrus fruits, e.g. oranges, lemons, grapefruits, limes. Tomatoes, <u>Rosehips</u>, Blackcurrants, Red & Green Peppers. Peas. Turnips, Potatoes.

ANIMAL SOURCES

B_2 and Nicotinic Acid are <u>Synthesized by BACTERIA</u> in the human intestine. Found in <u>LIVER</u> (B_1, B_2, Nicotinic Acid, B_{12}, etc.), BACON and LEAN MEAT, MEAT Extract, MILK (B_2, Nicotinic Acid)
EGGS (B_2)
CHEESE (B_2)

Stored in body – High concentration in Adrenal Glands. Found in Meat, Liver. Secreted in MILK.

DIETARY DEFICIENCY

Deficiency in Pyridoxine, Pantothenic Acid, Biotin Choline Para-amino benzoic A. not observed in man.

B_1 BERI-BERI

Gross disturbance of function of NERVOUS TISSUE with great MUSCULAR PARALYSIS and WEAKNESS and disturbance in SENSATION. OEDEMA. Gastro-intestinal upsets. HEART FAILURE

B_2 RIBOFLAVINOSIS

Roughening of skin. Cornea becomes cloudy. Cracks and fissures around lips & tongue.

NICOTINIC ACID PELLAGRA

Roughening and reddening of skin. Tongue red and sore. In severe cases – gastro-intestinal upsets and mental derangement

B_{12} PERNICIOUS ANAEMIA

Members of B Complex frequently absent together giving grave disturbances of chemical reactions in all tissues.

SCURVY

Intercellular cement breaks down. Capillary walls leak → haemorrhages into tissues - e.g. gums swell and bleed easily. Wounds heal slowly.

DIGESTION

The organic substances of man's food are chemically <u>similar</u> to those which his body will form from them.   In detail they <u>differ</u>.

Conversion of <u>FOOD SUBSTANCE</u> ——— to ——— <u>BODY SUBSTANCE</u>

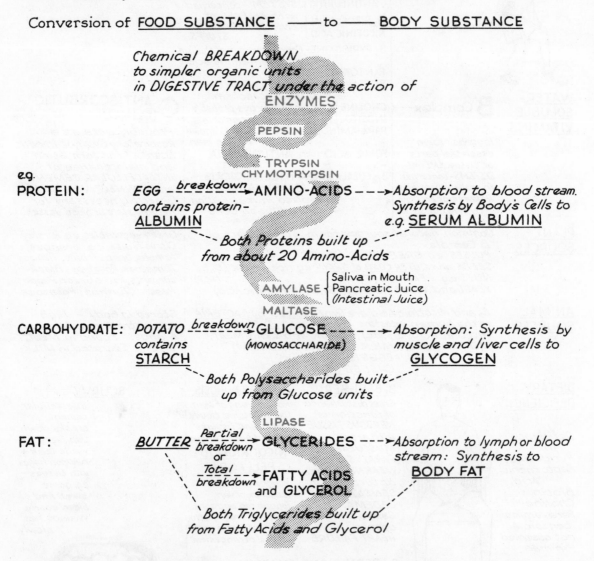

Chemical BREAKDOWN
to simpler organic units
in DIGESTIVE TRACT under the action of
ENZYMES

PEPSIN

TRYPSIN
CHYMOTRYPSIN

eg.
PROTEIN: EGG —*breakdown*→ AMINO-ACIDS ---→ *Absorption to blood stream.*
contains protein-   *Synthesis by Body's Cells to*
<u>ALBUMIN</u>   *e.g.* <u>SERUM ALBUMIN</u>

---*Both Proteins built up* ---
from about 20 Amino-Acids

AMYLASE ⎰ Saliva in Mouth
        ⎱ Pancreatic Juice
        (Intestinal Juice)

MALTASE

CARBOHYDRATE: *POTATO* —*breakdown*→ GLUCOSE --- --→ *Absorption: Synthesis by*
contains   (MONOSACCHARIDE)   *muscle and liver cells to*
<u>STARCH</u>   <u>GLYCOGEN</u>

---*Both Polysaccharides built*---
up from Glucose units

LIPASE

FAT: *BUTTER* —*Partial*--→ GLYCERIDES ---→ *Absorption to lymph or blood*
              breakdown   *stream: Synthesis to*
              or   <u>BODY FAT</u>
              Total---→ FATTY ACIDS
              breakdown   and GLYCEROL

Both Triglycerides built up
from Fatty Acids and Glycerol

DIGESTION *is brought about by* <u>SPECIFIC ENZYMES</u> *themselves made of Protein — each acts as a CATALYST for speeding up one particular chemical breakdown without effect on any others.*

PROTEIN METABOLISM

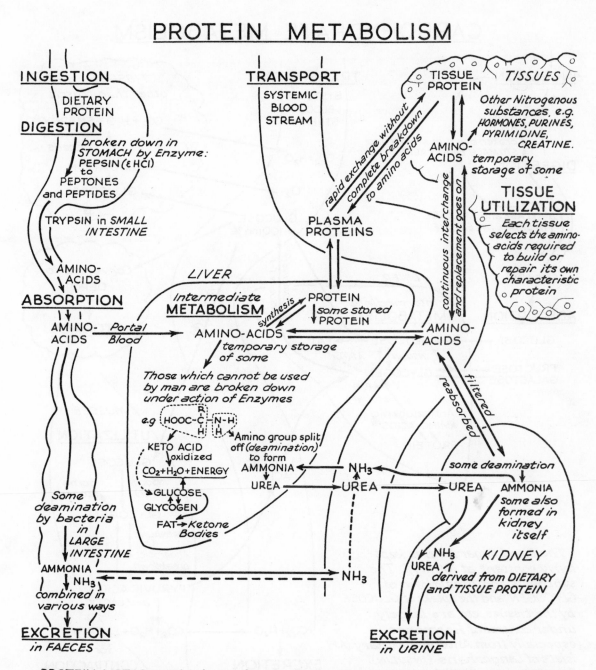

PROTEIN METABOLISM *is under control of Hormones – chiefly the* GROWTH HORMONE *(SOMATOTROPHIN) of the* ANTERIOR PITUITARY.

CARBOHYDRATE METABOLISM

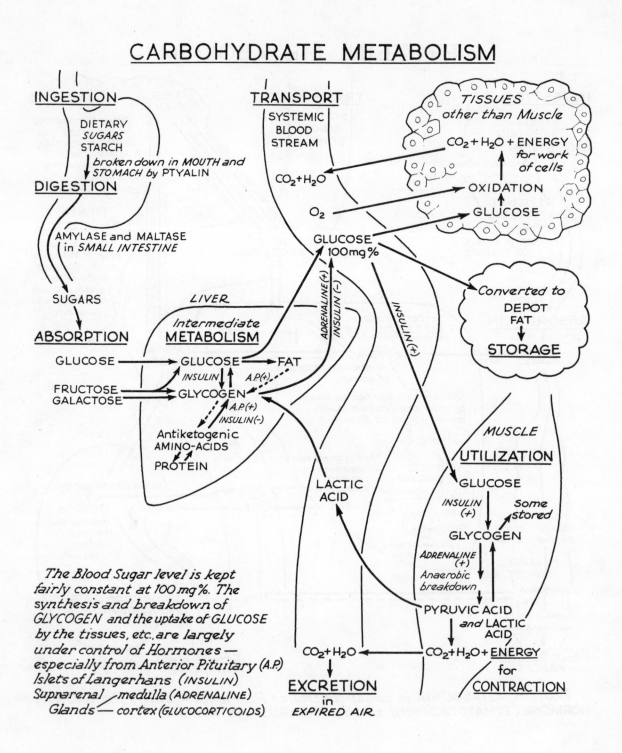

INGESTION
DIETARY *SUGARS* STARCH
broken down in MOUTH and STOMACH by PTYALIN

DIGESTION
AMYLASE and MALTASE *in SMALL INTESTINE*

SUGARS

ABSORPTION
GLUCOSE
FRUCTOSE
GALACTOSE

LIVER
Intermediate METABOLISM
GLUCOSE → FAT
INSULIN
GLYCOGEN
A.P.(+)
A.P.(+)
INSULIN(-)
Antiketogenic AMINO-ACIDS
PROTEIN

TRANSPORT
SYSTEMIC BLOOD STREAM
$CO_2 + H_2O$
O_2
GLUCOSE 100 mg %

ADRENALINE(+)
INSULIN(-)
INSULIN(+)

LACTIC ACID

TISSUES other than Muscle
$CO_2 + H_2O + ENERGY$ *for work of cells*
OXIDATION
GLUCOSE

Converted to DEPOT FAT
STORAGE

MUSCLE
UTILIZATION
GLUCOSE
INSULIN (+)
Some stored
GLYCOGEN
ADRENALINE (+)
Anaerobic breakdown
PYRUVIC ACID *and* LACTIC ACID
$CO_2 + H_2O + ENERGY$ *for* CONTRACTION

$CO_2 + H_2O$ ←
$CO_2 + H_2O$
EXCRETION *in EXPIRED AIR*

The Blood Sugar level is kept fairly constant at 100 mg%. The synthesis and breakdown of GLYCOGEN and the uptake of GLUCOSE by the tissues, etc., are largely under control of Hormones — especially from Anterior Pituitary (A.P.) Islets of Langerhans (INSULIN) Suprarenal medulla (ADRENALINE) Glands — cortex (GLUCOCORTICOIDS)

40

FAT METABOLISM

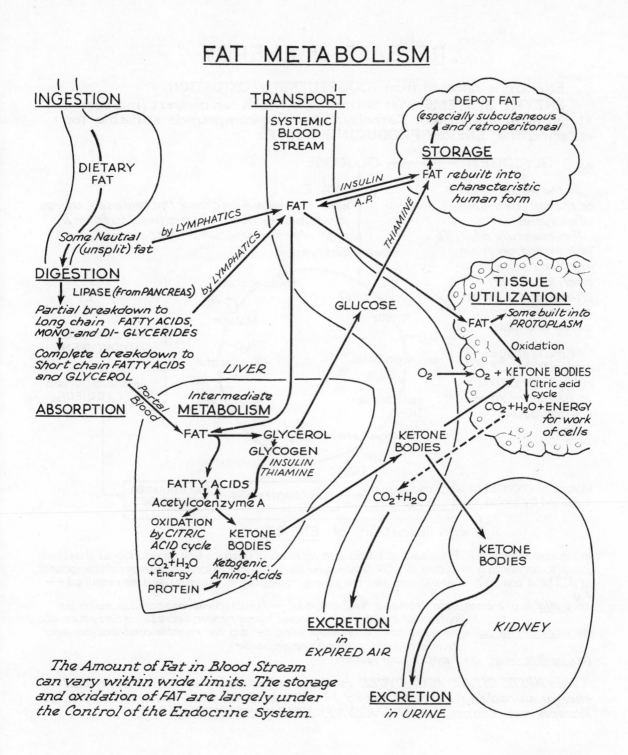

INGESTION

TRANSPORT
SYSTEMIC
BLOOD
STREAM

DEPOT FAT
*(especially subcutaneous
and retroperitoneal*

STORAGE
FAT *rebuilt into
characteristic
human form*

DIETARY
FAT

INSULIN
A.P.

*Some Neutral
(unsplit) fat*

by LYMPHATICS

FAT

THIAMINE

by LYMPHATICS

DIGESTION
LIPASE *(from PANCREAS)*

*Partial breakdown to
Long chain FATTY ACIDS,
MONO- and DI- GLYCERIDES*

GLUCOSE

TISSUE
UTILIZATION
FAT *Some built into
PROTOPLASM*

Oxidation

*Complete breakdown to
Short chain FATTY ACIDS
and GLYCEROL*

LIVER

O_2 → O_2 + KETONE BODIES
Citric acid
cycle
$CO_2 + H_2O$ + ENERGY
*for work
of cells*

ABSORPTION
*Portal
Blood*

Intermediate
METABOLISM

FAT → GLYCEROL
GLYCOGEN
*INSULIN
THIAMINE*

KETONE
BODIES

FATTY ACIDS
Acetylcoenzyme A

$CO_2 + H_2O$

OXIDATION
*by CITRIC
ACID* cycle

KETONE
BODIES

$CO_2 + H_2O$
+ Energy

*Ketogenic
Amino-Acids*

KETONE
BODIES

PROTEIN

EXCRETION
*in
EXPIRED AIR*

KIDNEY

*The Amount of Fat in Blood Stream
can vary within wide limits. The storage
and oxidation of FAT are largely under
the Control of the Endocrine System.*

EXCRETION
in URINE

RELEASE of ENERGY

ENERGY is released from FOODSTUFFS by OXIDATION.

ENZYME SYSTEMS exist within cells which can convert (in a series of steps) Fats, Proteins or Carbohydrates into compounds suitable for entering the "ENERGY-PRODUCING CYCLE".

e.g   GLYCOGEN ⟶ GLUCOSE

In the MITOCHONDRIA of the cell is a system of enzymes.

The molecule of "FUEL" is handed on from enzyme to enzyme – each of which changes it a little before passing it on to the next.

At most stages some HYDROGEN or CARBON DIOXIDE is liberated from the molecule.

PYRUVATE

when Pyruvate is oxidized the released energy is not lost as heat – but is used to form a high energy compound—Acetylcoenzyme A.

The liberated HYDROGEN is transported by CARRIERS

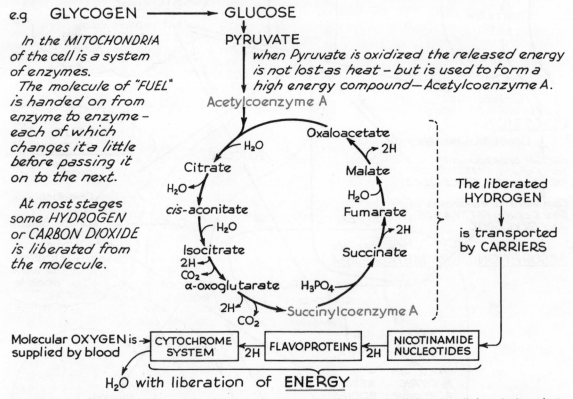

Molecular OXYGEN is supplied by blood

H$_2$O with liberation of **ENERGY**

In the body this ENERGY, instead of being lost as heat, is used to link an additional phosphate group to an existing molecule of ADP (*adenosine diphosphate*) to form the energy-rich compound ATP These compounds are stored and the energy 'trapped' in them is used as required —
e.g.
to *build* more complex STORAGE, STRUCTURAL or FUNCTIONAL materials such as
        Protein of new tissue in growth and repair, Secretions, Enzymes, etc.
for *work of the cell* — in ways not fully understood — e.g. for muscle contraction and
        transmission of nerve impulse.
to *liberate heat* to keep body warm.

This KREBS CITRIC-ACID CYCLE is the chief pathway by which potential energy stored in food is released by cells to make their own energy-rich Phosphorus compound–ADENOSINE TRIPHOSPHATE (ATP).

HEAT BALANCE

ENERGY is released in cells by OXIDATION. It is used for work and to keep body warm. The Body Temperature is kept relatively constant (*with a slight fluctuation throughout the 24 hours*) in spite of wide variations in Environmental Temperature and Heat Production.

HEAT PRODUCTION ———— *must balance* ———— HEAT LOSS

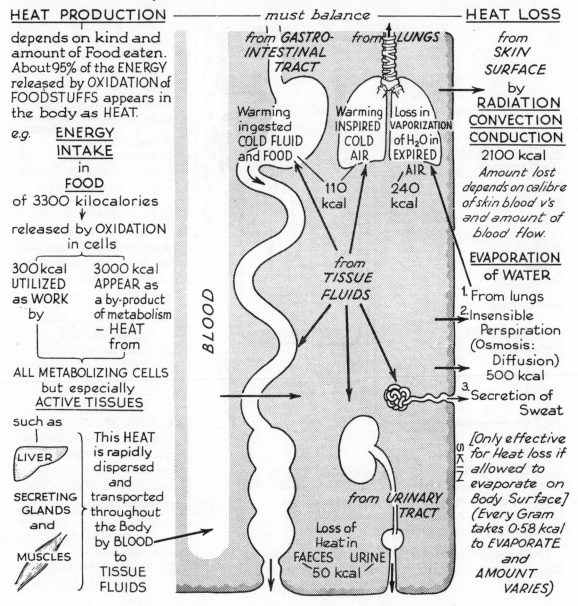

HEAT PRODUCTION

depends on kind and amount of Food eaten. About 95% of the ENERGY released by OXIDATION of FOODSTUFFS appears in the body as HEAT.

e.g. **ENERGY INTAKE** in **FOOD**

of 3300 kilocalories

↓

released by OXIDATION in cells

| 300 kcal UTILIZED as WORK by | 3000 kcal APPEAR as a by-product of metabolism – HEAT from |

ALL METABOLIZING CELLS but especially ACTIVE TISSUES

such as

LIVER

SECRETING GLANDS and

MUSCLES

This HEAT is rapidly dispersed and transported throughout the Body by BLOOD to TISSUE FLUIDS

from GASTRO-INTESTINAL TRACT

Warming ingested COLD FLUID and FOOD

from LUNGS

Warming INSPIRED COLD AIR

110 kcal

Loss in VAPORIZATION of H_2O in EXPIRED AIR

240 kcal

BLOOD

from TISSUE FLUIDS

from URINARY TRACT

Loss of Heat in FAECES   URINE
50 kcal

SKIN

HEAT LOSS

from SKIN SURFACE by RADIATION CONVECTION CONDUCTION

2100 kcal

Amount lost depends on calibre of skin blood v's and amount of blood flow.

EVAPORATION of WATER

1. From lungs
2. Insensible Perspiration (Osmosis: Diffusion) 500 kcal
3. Secretion of Sweat

[Only effective for Heat loss if allowed to evaporate on Body Surface] (Every Gram takes 0.58 kcal to EVAPORATE and AMOUNT VARIES)

MAINTENANCE of BODY TEMPERATURE

Any tendency for the BODY TEMPERATURE to *RISE*
as by

INCREASED HEAT PRODUCTION is balanced by **INCREASED HEAT LOSS**

1. by increased cellular OXIDATION of FOODSTUFFS as occurs e.g. with MUSCULAR ACTIVITY

EXTRA HEAT is dispersed quickly by Blood Stream

Rise in Blood Temperature affects HYPOTHALAMUS

2.
HOT ENVIRONMENTAL TEMPERATURE
i.e. above BODY TEMPERATURE (37°C)
[Body would tend to gain heat by RADIATION, CONVECTION and CONDUCTION.]

stimulates

HEAT SENSITIVE nerve endings in SKIN

Reduced SYMPATHETIC VASOCONSTRICTOR tone to

SECRETOMOTOR

INGOING NERVE IMPULSES

OUTGOING NERVE IMPULSES

BLOOD VESSELS DILATED

FAT

1. SKIN BLOOD VESSELS DILATE

more blood to skin surface → increased heat loss from skin by –
RADIATION
CONVECTION
CONDUCTION
} (Cannot occur if air temperature is above body's)

Increased by voluntary ingestion of cold foods and fluids and by use of fans.

2. SWEAT GLANDS SECRETE

Increased heat loss by evaporation from skin surface (unless atmosphere is already water saturated as e.g. in Tropics)

3. DIMINISHED HEAT INSULATION
by voluntary reduction of CLOTHING worn

4. DIMINISHED HEAT PRODUCTION
SKELETAL MUSCLE 'tone' reduced – and often voluntary relaxation → less work done → less heat produced

5. REDUCTION of 'ENERGY INTAKE'
by voluntary restriction of PROTEIN in DIET

[The increase in activity during the day probably accounts for the gradual Physiological rise in body temperature from about 96.5°F (35.8°C) in the early morning to about 99.2°F (37.3°C) in the late afternoon.

Unless exercise is very strenuous or environment is very hot and humid these measures → RESTORE BODY TEMPERATURE to NORMAL

MAINTENANCE of BODY TEMPERATURE

Any tendency for the BODY TEMPERATURE to *FALL*
as by

├ - DIMINISHED HEAT PRODUCTION is balanced by DIMINISHED HEAT LOSS

1. by decreased cellular
   OXIDATIONS as during
   MUSCULAR INACTIVITY
   (e.g. during sleep)

LESS HEAT
transported by Blood Stream

2. or
COLD ENVIRONMENTAL
TEMPERATURE
*i.e. below CRITICAL
TEMPERATURE
(about 30°C. for
naked body)*

Drop in Blood
Temperature
↓ *affects*
HYPOTHALAMUS

stimulates

COLD SENSITIVE
nerve endings
in SKIN

INGOING
NERVE
IMPULSES

*Increased
SYMPATHETIC
VASOCONSTRICTOR
tone to ↗*

SUPPRESSION

OUTGOING
NERVE IMPULSES

*Suprarenal
Medulla*

Adrenaline

Smooth
Muscle
and
Blood
Vessels

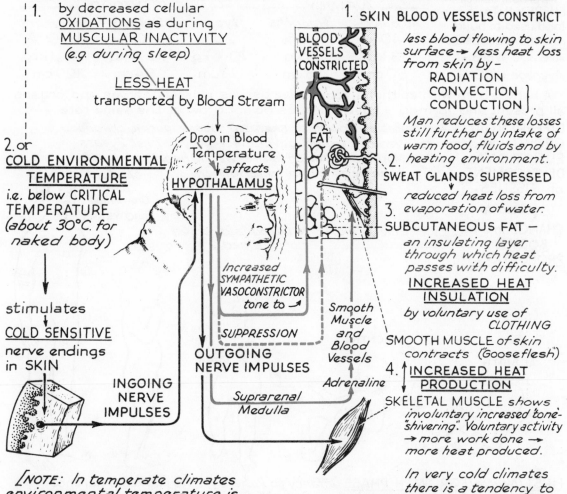

BLOOD
VESSELS
CONSTRICTED

FAT

1. SKIN BLOOD VESSELS CONSTRICT
   *less blood flowing to skin
   surface → less heat loss
   from skin by –*
   RADIATION
   CONVECTION
   CONDUCTION }
   *Man reduces these losses
   still further by intake of
   warm food, fluids and by
   heating environment.*

2. SWEAT GLANDS SUPRESSED
   *reduced heat loss from
   evaporation of water.*

3. SUBCUTANEOUS FAT –
   *an insulating layer
   through which heat
   passes with difficulty.*
   INCREASED HEAT
   INSULATION
   *by voluntary use of
   CLOTHING*
   SMOOTH MUSCLE *of skin
   contracts ("Gooseflesh")*

4. ↑ INCREASED HEAT
   PRODUCTION
   SKELETAL MUSCLE *shows
   involuntary increased 'tone'
   "shivering". Voluntary activity
   → more work done →
   more heat produced.*

   *In very cold climates
   there is a tendency to*

5. INCREASE 'ENERGY INTAKE'
   *by voluntary increase in
   DIETARY PROTEIN which has
   stimulating effect on metabolism.*

*[NOTE: In temperate climates
environmental temperature is
usually lower than body temperature
so that there is a continuous loss of
heat from body surface.]*

Unless environmental temperature
is very low these measures tend to ── RESTORE BODY TEMPERATURE
to NORMAL

GROWTH

Each individual grows, by repeated cell divisions, from a single cell to a total of 30 trillion or more cells. Growth is most rapid before birth and during 1ST year of life.

The proportion of ENERGY INTAKE in food used to build and maintain tissue —	*INFANCY*		*CHILDHOOD*		
	At Birth	*3 months*	*1 year*	*2 years*	*9-11 years*
	40%	40%	20%	20%	4-10%
Average WEIGHT —	3 kg	5 kg	10·4 kg	12·4 kg	27·1 kg
Average HEIGHT —	50 cm	58 cm	73 cm	84 cm	129 cm

A baby is born with epithelia, connective tissues, muscles, nerves and organs all present and formed — but all tissues do not grow at the same rate.

DIFFERENTIAL GROWTH and FUNCTIONAL DEVELOPMENT of TISSUES

e.g. Rapid growth of skeletal tissue during childhood. Nervous tissue develops rapidly in first 2 years. Most rapid growth is first at the head then legs begin to lengthen.

↓ *lead to*

CHANGE in BODY PROPORTIONS

Chiefly PROTEIN being laid down →
or retained

Lymphoid Tissue Growth Spurt
10 YEARS

2 YEARS

1 YEAR

3 MONTHS

At BIRTH

FIRST *(Neutral)* GROWTH PHASE *(Infancy to Puberty).* No marked difference between sexes – Regulated by Growth-Stimulating Hormone of Anterior Pituitary – stimulates Growth of all Tissues and Organs.

FACTORS INFLUENCING NORMAL GROWTH :

HEREDITY: *To large extent rate of growth/sequence of events is predetermined. Inherited factors control pattern and limitations of growth.*

ENVIRONMENT: *Nutrition: For optimal growth the body requires an adequate and balanced diet.*

ENDOCRINE GLANDS: *All play part: mental as well as physical development is influenced.*

GROWTH

GROWTH SPURT 11-12 years	ADOLESCENCE →		ADULTHOOD →	
	GIRL	BOY	WOMAN	MAN
	13 years	15 years	17 years	19 years
	10-15%	10-15%	4%	4%
	42·8 kg	51·1 kg	54·8 kg	65 kg
	152 cm	163 cm	162 cm	173 cm

Growth and development of reproductive organs

More FAT being deposited now

A part from PROTEIN replacement in REPAIR, weight increase is now by deposition of FAT.

SEXUAL GROWTH PHASE *(Puberty to Maturity)*

Marked difference between sexes is initiated by Gonad-Stimulating Hormones of Anterior Pituitary acting on Sex Glands and stimulating their production of Sex Hormones. These hormones are largely responsible for development of secondary sex characteristics and development of reproductive organs.

REPRODUCTIVE PHASE

--- These hormones maintain secondary sex characteristics and reproductive ability during reproductive phase of adult life.

ENERGY REQUIREMENTS *MALE*

DAILY FOOD INTAKE must supply **TOTAL ENERGY REQUIREMENTS** for **1. ACTIVITY**

SPECIAL – *Individual requirements vary with TYPE of WORK or PLAY and, from minute to minute, on the INTENSITY of WORK and frequency and length of REST PAUSES.*

EVERYDAY ACTIVITIES– *such as SITTING, STANDING, WALKING, ETC.*

2. SPECIFIC DYNAMIC ACTION of FOOD (S.D.A.)

The mere taking of food stimulates metabolism of cells so that heat production increases (30% by protein, 4-5% by carbohydrate and fat).
Must allow 10% above Basal Requirements on average mixed diet.

3. BASAL METABOLISM

(measured when body is at rest)
Energy Expenditure of cells doing 'vegetative' processes of living – e.g. tasks involved in RESPIRATION, CIRCULATION, DIGESTION, EXCRETION, SECRETION, SYNTHESIS of special substances, KEEPING BODY TEMPERATURE at 37°C., GROWTH and REPAIR.

Average 70kg 25 year old ADULT (surface area 1·8 square metres) | ADOLESCENT 16 years | CHILD 8 years

HEAVY WORK MODERATE LIGHT (Sedentary)

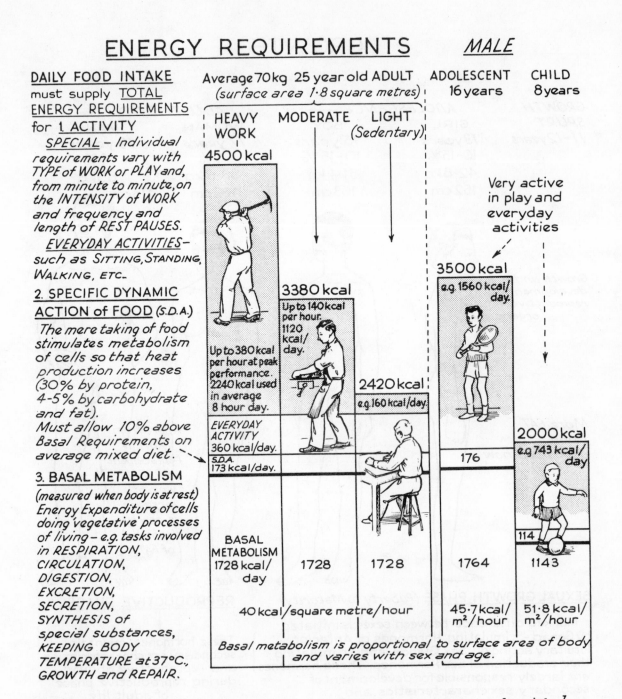

HEAVY WORK 4500 kcal
Up to 380 kcal per hour at peak performance. 2240 kcal used in average 8 hour day.
EVERYDAY ACTIVITY 360 kcal/day.
S.D.A. 173 kcal/day.
BASAL METABOLISM 1728 kcal/day

MODERATE 3380 kcal
Up to 140 kcal per hour. 1120 kcal/day.
1728

LIGHT (Sedentary) 2420 kcal
e.g. 160 kcal/day.
1728

40 kcal/square metre/hour

ADOLESCENT 16 years
Very active in play and everyday activities
3500 kcal
e.g. 1560 kcal/day.
176
1764
45·7 kcal/m²/hour

CHILD 8 years
2000 kcal
e.g. 743 kcal/day
114
1143
51·8 kcal/m²/hour

Basal metabolism is proportional to surface area of body and varies with sex and age.

[All figures are approximate and intended only as a general guide.]

ENERGY REQUIREMENTS *FEMALE*

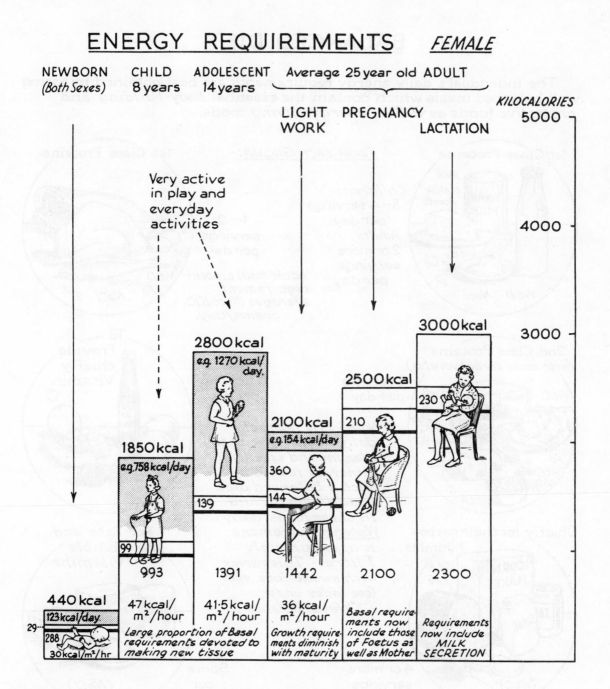

NEWBORN (Both Sexes) · CHILD 8 years · ADOLESCENT 14 years · Average 25 year old ADULT · LIGHT WORK · PREGNANCY · LACTATION

KILOCALORIES — 5000, 4000, 3000

Very active in play and everyday activities

Newborn: 440 kcal — 123 kcal/day — 288 — 30 kcal/m² /hr — 29

Child 8 years: 1850 kcal — e.g. 758 kcal/day — 99 — 993 — 47 kcal/ m²/hour — Large proportion of Basal requirements devoted to making new tissue

Adolescent 14 years: 2800 kcal — e.g. 1270 kcal/day — 139 — 1391 — 41.5 kcal/ m²/hour — Growth requirements diminish with maturity

Light Work: 2100 kcal — e.g. 154 kcal/day — 360 — 144 — 1442 — 36 kcal/ m²/hour — Growth requirements diminish with maturity

Pregnancy: 2500 kcal — 210 — 2100 — Basal requirements now include those of Foetus as well as Mother

Lactation: 3000 kcal — 230 — 2300 — Requirements now include MILK SECRETION

[Proportion needed for growth diminishes in both sexes with age.]

BALANCED DIET

The individual's daily energy requirements are best obtained by eating well-balanced meals which contain the essential *body-building* and *protective* foods as well as *energy-giving* foods.

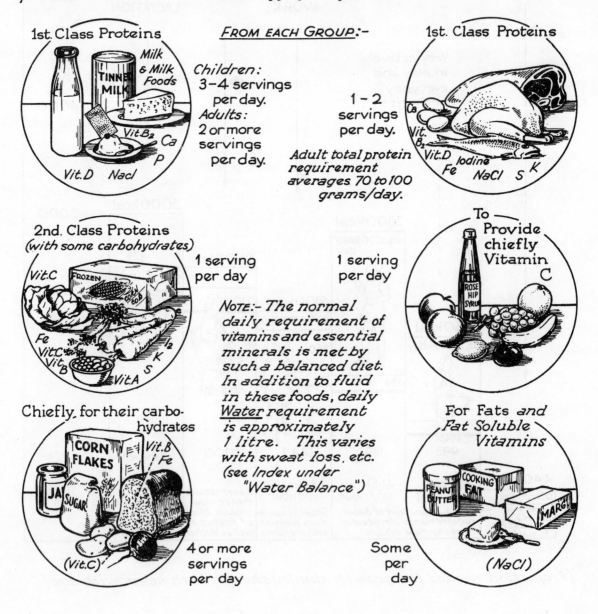

1st. Class Proteins

Milk & Milk Foods

TINNED MILK

Vit.B₂ — $Vit.B_2$ Ca P

Vit.D   Nacl

FROM EACH GROUP:-

Children:
3–4 servings per day.
Adults:
2 or more servings per day.

1 – 2 servings per day.

Adult total protein requirement averages 70 to 100 grams/day.

1st. Class Proteins

Ca $Vit.B_2$ $Vit.D$ Iodine   Fe   NaCl   S   K

2nd. Class Proteins
(with some carbohydrates)

Vit.C   FROZEN COD

Fe   VitC   Vit.B I_2 K   S   Vit.A

1 serving per day

1 serving per day

NOTE:- *The normal daily requirement of vitamins and essential minerals is met by such a balanced diet. In addition to fluid in these foods, daily* Water *requirement is approximately 1 litre. This varies with sweat loss, etc. (see Index under "Water Balance")*

To Provide chiefly Vitamin C

ROSE HIP SYRUP

Chiefly, for their carbohydrates

CORN FLAKES   Vit.B   Fe

JAM   SUGAR

(Vit.C)

4 or more servings per day

For Fats *and* Fat Soluble *Vitamins*

PEANUT BUTTER   COOKING FAT   MARG

(NaCl)

Some per day

CHAPTER 3

DIGESTIVE SYSTEM

DIGESTIVE SYSTEM

The ALIMENTARY CANAL and ASSOCIATED GLANDS } Special System for dealing with FOOD and FLUIDS

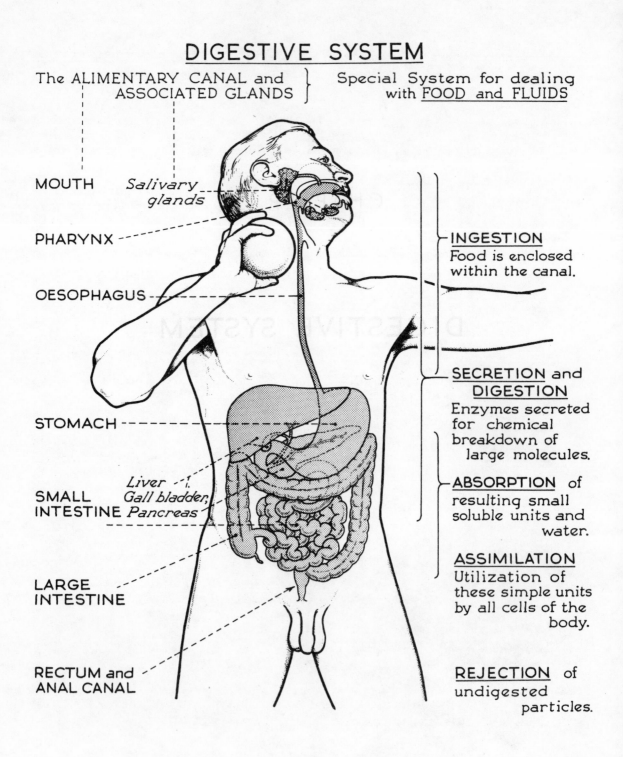

MOUTH

Salivary glands

PHARYNX

OESOPHAGUS

STOMACH

SMALL INTESTINE

Liver
Gall bladder,
Pancreas

LARGE INTESTINE

RECTUM and ANAL CANAL

INGESTION
Food is enclosed within the canal.

SECRETION and DIGESTION
Enzymes secreted for chemical breakdown of large molecules.

ABSORPTION of resulting small soluble units and water.

ASSIMILATION
Utilization of these simple units by all cells of the body.

REJECTION of undigested particles.

PROGRESS of FOOD along ALIMENTARY CANAL

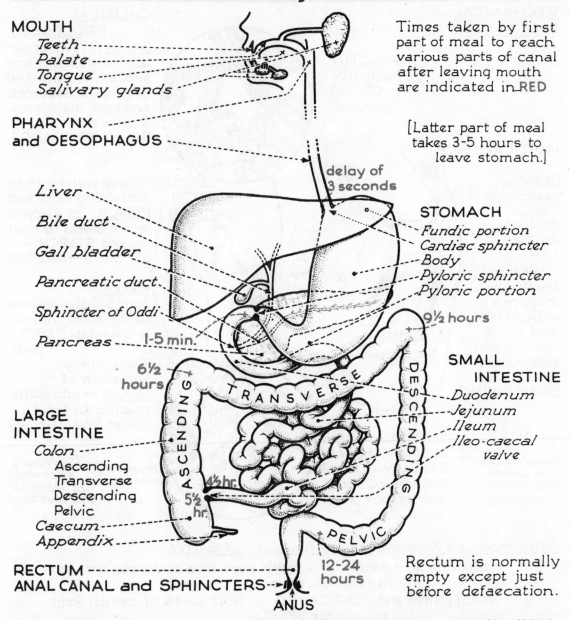

MOUTH
Teeth --------
Palate --------
Tongue --------
Salivary glands --------

Times taken by first
part of meal to reach
various parts of canal
after leaving mouth
are indicated in RED

**PHARYNX
and OESOPHAGUS** --------

[Latter part of meal
takes 3-5 hours to
leave stomach.]

delay of
3 seconds

Liver --------
Bile duct --------
Gall bladder --------
Pancreatic duct --------
Sphincter of Oddi --------
Pancreas -------- 1-5 min.

STOMACH
Fundic portion
Cardiac sphincter
Body
Pyloric sphincter
Pyloric portion

9½ hours

6½
hours

**SMALL
INTESTINE**
Duodenum
Jejunum
Ileum
Ileo-caecal
valve

TRANSVERSE

ASCENDING

DESCENDING

**LARGE
INTESTINE**
Colon --------
  Ascending
  Transverse
  Descending
  Pelvic
Caecum --------
Appendix --------

4½hr.
5½
hr.

PELVIC

RECTUM --------
ANAL CANAL and SPHINCTERS --------

12-24
hours

Rectum is normally
empty except just
before defaecation.

ANUS

*During its progress along the canal FOOD is subjected to MECHANICAL
as well as CHEMICAL changes to render it suitable for absorption and assimilation.*

DIGESTION in the MOUTH

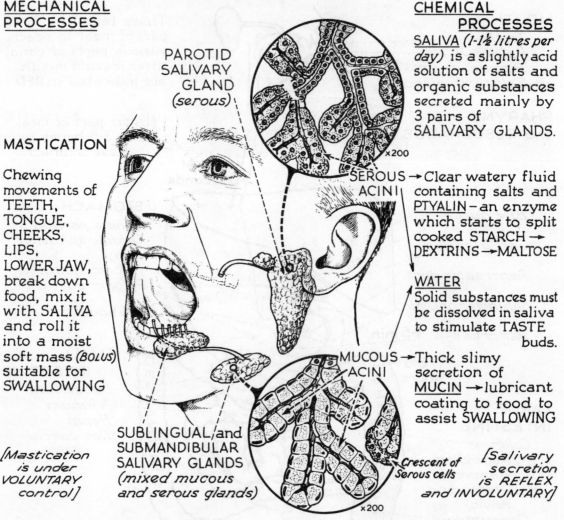

MECHANICAL PROCESSES

PAROTID SALIVARY GLAND (*serous*)

x200

MASTICATION

Chewing movements of TEETH, TONGUE, CHEEKS, LIPS, LOWER JAW, break down food, mix it with SALIVA and roll it into a moist soft mass (*BOLUS*) suitable for SWALLOWING

SUBLINGUAL and SUBMANDIBULAR SALIVARY GLANDS (*mixed mucous and serous glands*)

[Mastication is under VOLUNTARY control]

SEROUS ACINI

MUCOUS ACINI

Crescent of Serous cells

x200

CHEMICAL PROCESSES

SALIVA (*1-1½ litres per day*) is a slightly acid solution of salts and organic substances secreted mainly by 3 pairs of SALIVARY GLANDS.

→Clear watery fluid containing salts and PTYALIN — an enzyme which starts to split cooked STARCH → DEXTRINS →MALTOSE

WATER
Solid substances must be dissolved in saliva to stimulate TASTE buds.

→Thick slimy secretion of MUCIN → lubricant coating to food to assist SWALLOWING

[Salivary secretion is REFLEX and INVOLUNTARY]

Other important (*non-digestive*) functions of SALIVA :-

CLEANSING — Mouth and teeth kept free of debris, etc., to inhibit bacteria.

MOISTENING and LUBRICATING — Soft parts of mouth kept pliable for SPEECH.

EXCRETORY— Many organic substances (*e.g. urea, sugar*) and inorganic substances (*e.g. mercury, lead*) can be excreted in saliva.

CONTROL of SALIVARY SECRETION

Increased secretion at mealtimes is REFLEX *(involuntary)*.
Salivary Reflexes are of two types:—

(a) UNCONDITIONED *(inborn)*

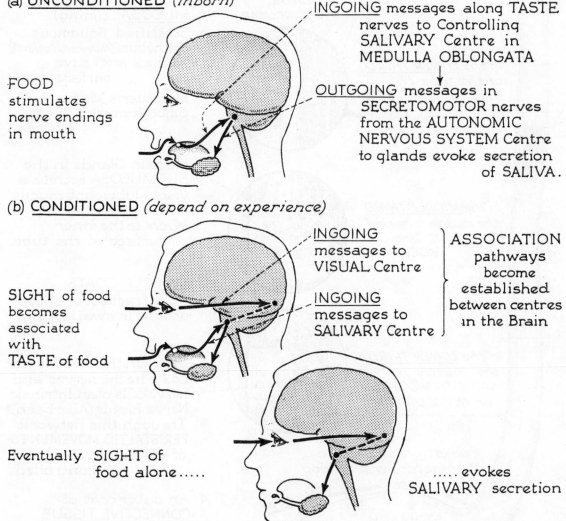

INGOING messages along TASTE nerves to Controlling SALIVARY Centre in MEDULLA OBLONGATA

OUTGOING messages in SECRETOMOTOR nerves from the AUTONOMIC NERVOUS SYSTEM Centre to glands evoke secretion of SALIVA.

FOOD stimulates nerve endings in mouth

(b) CONDITIONED *(depend on experience)*

SIGHT of food becomes associated with TASTE of food

INGOING messages to VISUAL Centre

INGOING messages to SALIVARY Centre

ASSOCIATION pathways become established between centres in the Brain

Eventually SIGHT of food alone.....

..... evokes SALIVARY secretion

Similar conditioned reflexes are established by smell, by thought of food, and even by the sounds of its preparation.

OESOPHAGUS

The Oesophagus is a muscular tube about 25 cm. long which conveys ingested food and fluid from the Mouth to the Stomach.

PHARYNX

The Muscle is arranged in 2 layers:-
Outer LONGITUDINAL
Inner CIRCULAR

In the UPPER THIRD of the tube the muscle is SKELETAL

In the MIDDLE THIRD the muscle is mixed SKELETAL and VISCERAL

In the LOWER THIRD the muscle is SMOOTH, VISCERAL or INVOLUNTARY

×15

×15

×15

Relaxation of Cardiac Sphincter permits food to enter

DIAPHRAGM

STOMACH

The tube has 4 Coats:-

1. MUCOSA (lining)
Stratified Squamous Epithelium (non-keratinized) -a thick protective surface layer.

Muscularis Mucosae — Smooth muscle of the Mucosa.

Mucous Glands in the
2. SUBMUCOSA secrete a viscid lubricant mucus which passes along ducts to the inner surface of the tube.

3. MUSCLE COATS
Contraction of these occurs in swallowing.

Between the two Muscle Coats lie the nerves and nerve cells of an Intrinsic Nerve Plexus (Auerbach). Through this network PERISTALTIC MOVEMENTS of the Muscle Coats are controlled.

4. An outer coat of CONNECTIVE TISSUE blends with that of trachea, etc.

Except during passage of food the oesophagus is flattened and closed; its mucosa thrown into several longitudinal folds.

SWALLOWING

Swallowing is a complex act initiated *voluntarily* and completed
involuntarily (or *reflexly*)

STIMULUS	INGOING PATHWAY	CENTRE IN MEDULLA OBLONGATA	OUTGOING PATHWAY	EFFECT

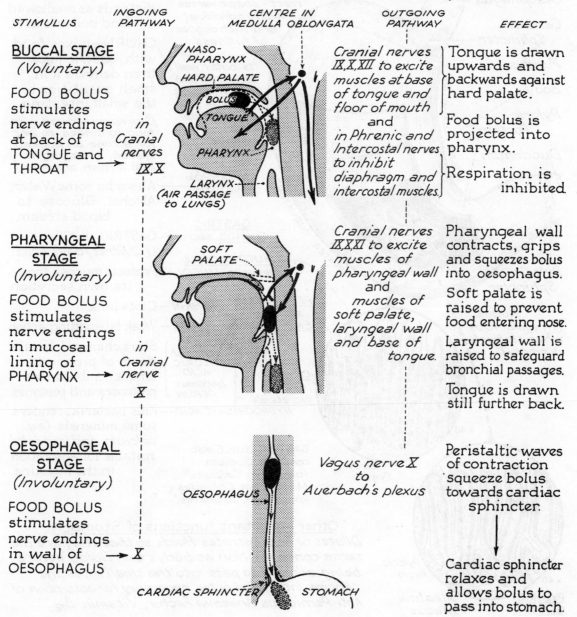

BUCCAL STAGE (*Voluntary*)

FOOD BOLUS stimulates nerve endings at back of TONGUE and THROAT → *in Cranial nerves IX, X*

NASO-PHARYNX · HARD PALATE · BOLUS · TONGUE · PHARYNX · LARYNX—(AIR PASSAGE TO LUNGS)

Cranial nerves IX, X, XII to excite muscles at base of tongue and floor of mouth and in Phrenic and Intercostal nerves to inhibit diaphragm and intercostal muscles.

Tongue is drawn upwards and backwards against hard palate.

Food bolus is projected into pharynx.

Respiration is inhibited.

PHARYNGEAL STAGE (*Involuntary*)

FOOD BOLUS stimulates nerve endings in mucosal lining of PHARYNX → *in Cranial nerve X*

SOFT PALATE

Cranial nerves IX, XI to excite muscles of pharyngeal wall and muscles of soft palate, laryngeal wall and base of tongue.

Pharyngeal wall contracts, grips and squeezes bolus into oesophagus.

Soft palate is raised to prevent food entering nose.

Laryngeal wall is raised to safeguard bronchial passages.

Tongue is drawn still further back.

OESOPHAGEAL STAGE (*Involuntary*)

FOOD BOLUS stimulates nerve endings in wall of OESOPHAGUS → X

OESOPHAGUS · CARDIAC SPHINCTER · STOMACH

Vagus nerve X to Auerbach's plexus

Peristaltic waves of contraction squeeze bolus towards cardiac sphincter.

Cardiac sphincter relaxes and allows bolus to pass into stomach.

57

STOMACH

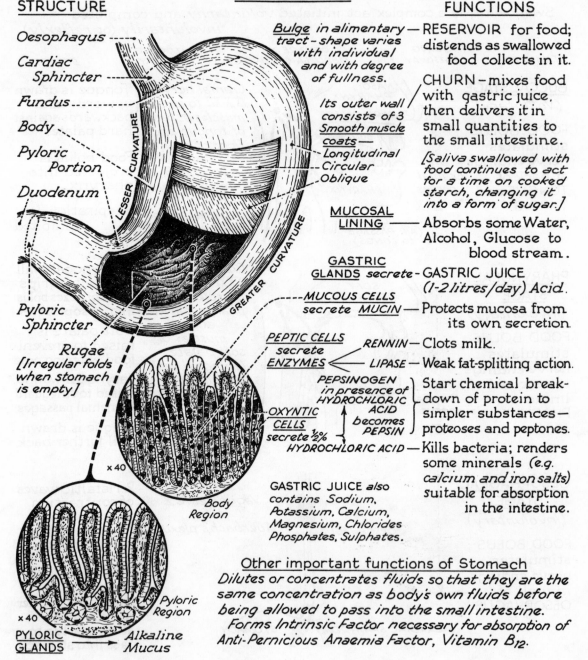

STRUCTURE

- Oesophagus
- Cardiac Sphincter
- Fundus
- Body
- Pyloric Portion
- Duodenum
- Pyloric Sphincter
- Rugae [Irregular folds when stomach is empty]

LESSER CURVATURE

GREATER CURVATURE

Bulge in alimentary tract - shape varies with individual and with degree of fullness.

Its outer wall consists of 3 Smooth muscle coats —
- Longitudinal
- Circular
- Oblique

MUCOSAL LINING

GASTRIC GLANDS secrete

MUCOUS CELLS secrete MUCIN

PEPTIC CELLS secrete ENZYMES — RENNIN — LIPASE

PEPSINOGEN in presence of HYDROCHLORIC ACID becomes PEPSIN

OXYNTIC CELLS secrete ½% HYDROCHLORIC ACID

× 40

Body Region

× 40

PYLORIC GLANDS → Alkaline Mucus

Pyloric Region

GASTRIC JUICE also contains Sodium, Potassium, Calcium, Magnesium, Chlorides Phosphates, Sulphates.

FUNCTIONS

RESERVOIR for food; distends as swallowed food collects in it.

CHURN – mixes food with gastric juice, then delivers it in small quantities to the small intestine.

[Saliva swallowed with food continues to act for a time on cooked starch, changing it into a form of sugar.]

Absorbs some Water, Alcohol, Glucose to blood stream.

GASTRIC JUICE (1-2 litres/day) Acid.

Protects mucosa from its own secretion.

Clots milk.

Weak fat-splitting action.

Start chemical breakdown of protein to simpler substances — proteoses and peptones.

Kills bacteria; renders some minerals (e.g. calcium and iron salts) suitable for absorption in the intestine.

Other important functions of Stomach

Dilutes or concentrates fluids so that they are the same concentration as body's own fluids before being allowed to pass into the small intestine.

Forms Intrinsic Factor necessary for absorption of Anti-Pernicious Anaemia Factor, Vitamin B_{12}.

GASTRIC JUICE

SECRETION of Gastric Juice is under 2 types of CONTROL:-

(a) NERVOUS - Messages are conveyed rapidly from Brain Centre by Nerve Fibres of the AUTONOMIC NERVOUS SYSTEM for *immediate* effect. e.g. stimulation of Parasympathetic (Vagus) nerves to GASTRIC GLANDS → secretion of HIGHLY ACID JUICE containing ENZYMES.

(b) HUMORAL - Message is CHEMICAL and carried in BLOOD STREAM for *slower* and *longer-lasting* control. [Note:- Chemical messengers (Hormones) travel all over the body even though the message stimulates only one part — e.g. in this case the Gastric Glands.]

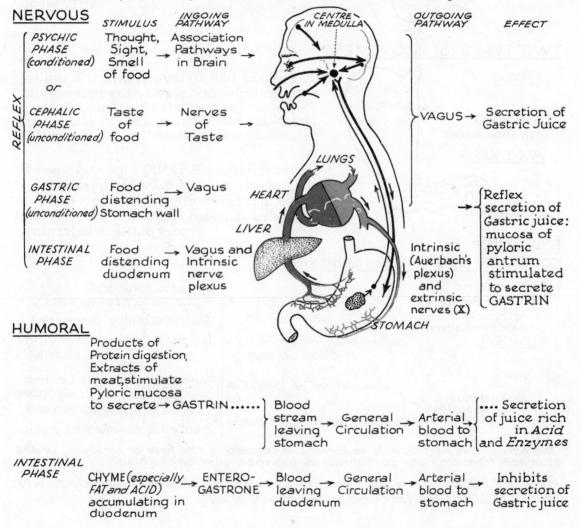

NERVOUS

	STIMULUS	INGOING PATHWAY		CENTRE IN MEDULLA	OUTGOING PATHWAY	EFFECT
PSYCHIC PHASE (conditioned)	Thought, Sight, Smell of food	Association Pathways in Brain →			VAGUS →	Secretion of Gastric Juice
or						
CEPHALIC PHASE (unconditioned)	Taste of food	Nerves of Taste →				
GASTRIC PHASE (unconditioned)	Food distending Stomach wall	Vagus				Reflex secretion of Gastric juice: mucosa of pyloric antrum stimulated to secrete GASTRIN
INTESTINAL PHASE	Food distending duodenum	Vagus and Intrinsic nerve plexus			Intrinsic (Auerbach's plexus) and extrinsic nerves (X)	

REFLEX

HUMORAL

Products of Protein digestion, Extracts of meat, stimulate Pyloric mucosa to secrete → GASTRIN ...... | Blood stream leaving stomach → General Circulation → Arterial blood to stomach | .... Secretion of juice rich in *Acid* and *Enzymes*

INTESTINAL PHASE | CHYME (especially FAT and ACID) accumulating in duodenum → ENTERO-GASTRONE → Blood leaving duodenum → General Circulation → Arterial blood to stomach → Inhibits secretion of Gastric juice

MOVEMENTS of the STOMACH

Very little movement is seen in empty stomach until onset of HUNGER.

FILLING

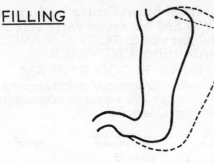

FUNDUS and
GREATER CURVATURE bulge and lengthen
as stomach fills with food. This is
"Receptive Relaxation" – the smooth muscle
cells relax so that the pressure in the
stomach does not rise until the organ is
fairly large.

TWO TYPES of MOVEMENT occur while food is in stomach.

TONUS

HOPPER

'TONE RING'
with tone waves

PERISTALSIS

PYLORIC
SPHINCTER

MILL

BODY – Food lies in layers and the walls
exert a slight but steady pressure on it.
This squeezes food steadily towards the
PYLORUS even when stomach is becoming
relatively empty near the end of a meal.

From INCISURA ANGULARIS vigorous waves
of contraction mix food with digestive
juices and carry chyme through the
normally relaxed PYLORUS into the first
part of the duodenum.

EMPTYING

CONTROL of MOTILITY and EMPTYING

1. Enterogastric
   NERVOUS Reflex
   As Chyme enters and
   distends duodenum
   HUMORAL

   inhibition of VAGUS →

   release of ENTEROGASTRONE →
   to Blood Stream

   Inhibits tone and
   peristalsis temporarily.
   Gastric motility depressed.
   Temporarily slows emptying
   of stomach.

2.
   As duodenum
   empties

   stimulation of VAGUS →

   withdrawal of ENTEROGASTRONE →

   Waves of peristalsis become
   stronger.
   Gastric motility increased.
   Emptying speeded up again.

*The hormone is particularly important in regulating the rate of emptying of the
stomach from moment to moment during the digestion of a meal.*

*[STARCHY FOODS leave the stomach quickly: MEAT leaves relatively slowly:
FATTY FOODS pass through most slowly of all.]*

VOMITING

Vomiting is a <u>REFLEX</u> act.

Nerve pathways involved:-

STIMULUS → INGOING NERVE PATHWAY → VOMITING CENTRE IN MEDULLA OBLONGATA → OUTGOING NERVE PATHWAY → EFFECT

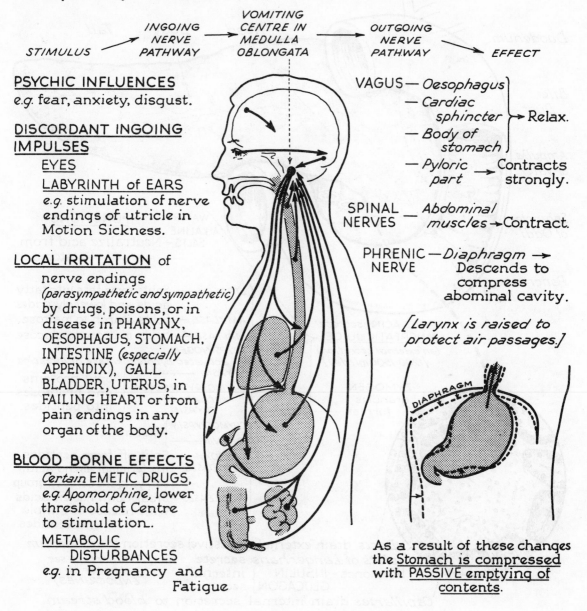

PSYCHIC INFLUENCES
e.g. fear, anxiety, disgust.

DISCORDANT INGOING IMPULSES
EYES
LABYRINTH of EARS
e.g. stimulation of nerve endings of utricle in Motion Sickness.

LOCAL IRRITATION of
nerve endings *(parasympathetic and sympathetic)* by drugs, poisons, or in disease in PHARYNX, OESOPHAGUS, STOMACH, INTESTINE *(especially* APPENDIX), GALL BLADDER, UTERUS, in FAILING HEART or from pain endings in any organ of the body.

BLOOD BORNE EFFECTS
Certain EMETIC DRUGS, *e.g. Apomorphine*, lower threshold of Centre to stimulation.
METABOLIC DISTURBANCES
e.g. in Pregnancy and Fatigue

VAGUS
— *Oesophagus*
— *Cardiac sphincter*
— *Body of stomach*
} → Relax.
— *Pyloric part* → Contracts strongly.

SPINAL NERVES — *Abdominal muscles* → Contract.

PHRENIC NERVE — *Diaphragm* → Descends to compress abominal cavity.

[Larynx is raised to protect air passages.]

DIAPHRAGM

As a result of these changes the <u>Stomach</u> is compressed with <u>PASSIVE</u> emptying of contents.

PANCREAS

The Pancreas is a large gland lying across the Posterior Abdominal Wall.
It has 2 secretions – a *digestive* secretion poured into the duodenum, and
a *hormonal* secretion passed into the blood stream.

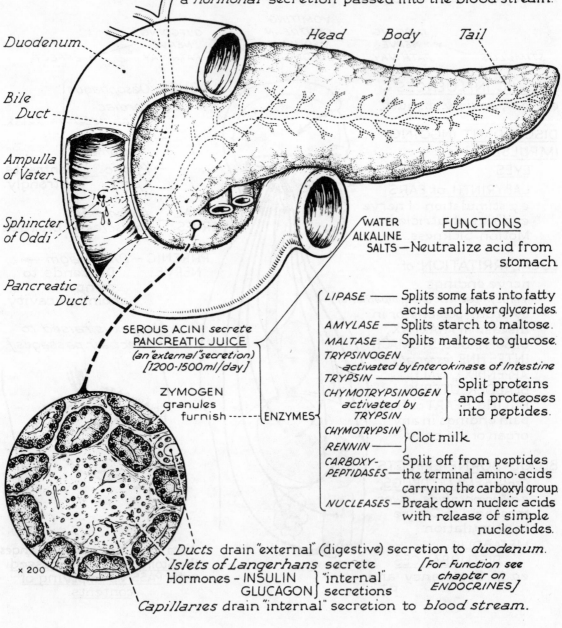

Duodenum

Bile
Duct

Ampulla
of Vater

Sphincter
of Oddi

Pancreatic
Duct

Head   Body   Tail

WATER
ALKALINE
SALTS – Neutralize acid from
stomach.

FUNCTIONS

SEROUS ACINI *secrete*
PANCREATIC JUICE
(an "external" secretion)
[1200·1500 ml/day]

ZYMOGEN
granules
furnish ------ ENZYMES

LIPASE —— Splits some fats into fatty
acids and lower glycerides.

AMYLASE — Splits starch to maltose.

MALTASE — Splits maltose to glucose.

TRYPSINOGEN
  activated by Enterokinase of Intestine
TRYPSIN ——
CHYMOTRYPSINOGEN
  activated by
  TRYPSIN
} Split proteins
and proteoses
into peptides.

CHYMOTRYPSIN
RENNIN —— } Clot milk

*CARBOXY-
PEPTIDASES* — Split off from peptides
the terminal amino-acids
carrying the carboxyl group.

NUCLEASES — Break down nucleic acids
with release of simple
nucleotides.

× 200

Ducts drain "external" (digestive) secretion to *duodenum.*
Islets of Langerhans secrete
Hormones – INSULIN } "internal"
          GLUCAGON } secretions

[For Function see
chapter on
ENDOCRINES]

Capillaries drain "internal" secretion to *blood stream.*

PANCREATIC JUICE

SECRETION of Pancreatic Juice is under 2 types of CONTROL :- NERVOUS and HUMORAL. The humoral mechanism is the more important.

NERVOUS   Stimulation of the Parasympathetic Nerves ⟶

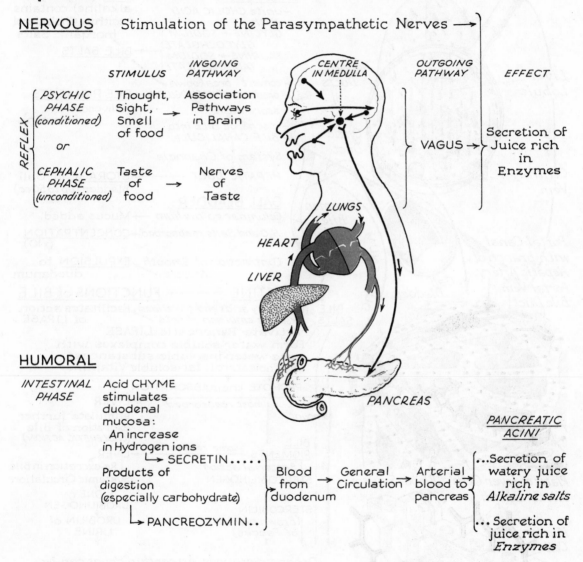

	STIMULUS	INGOING PATHWAY	CENTRE IN MEDULLA	OUTGOING PATHWAY	EFFECT
PSYCHIC PHASE (conditioned)	Thought, Sight, Smell of food	Association Pathways in Brain			
or				VAGUS ⟶	Secretion of Juice rich in Enzymes
CEPHALIC PHASE (unconditioned)	Taste of food	Nerves of Taste			

REFLEX

LUNGS

HEART

LIVER

PANCREAS

HUMORAL

INTESTINAL PHASE   Acid CHYME stimulates duodenal mucosa:
  An increase in Hydrogen ions
    ⟶ SECRETIN .....
Products of digestion (especially carbohydrate)
    ⟶ PANCREOZYMIN ..

Blood from duodenum ⟶ General Circulation ⟶ Arterial blood to pancreas

PANCREATIC ACINI

... Secretion of watery juice rich in *Alkaline salts*

... Secretion of juice rich in *Enzymes*

LIVER and GALL BLADDER

The Liver is a large highly complex organ with many functions. One of these is the production of BILE (500-1000 ml/day)

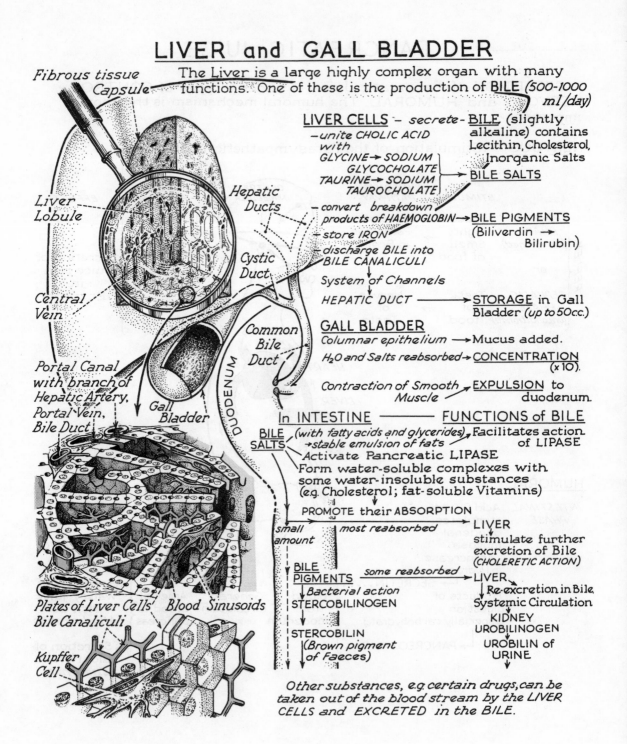

Fibrous tissue Capsule

Liver Lobule

Central Vein

Portal Canal with branch of Hepatic Artery, Portal Vein, Bile Duct

Plates of Liver Cells
Bile Canaliculi

Kupffer Cell

Blood Sinusoids

Hepatic Ducts

Cystic Duct

Common Bile Duct

Gall Bladder

DUODENUM

LIVER CELLS – secrete – BILE (slightly alkaline) contains Lecithin, Cholesterol, Inorganic Salts
— unite CHOLIC ACID with
GLYCINE → SODIUM GLYCOCHOLATE
TAURINE → SODIUM TAUROCHOLATE
→ BILE SALTS

— convert breakdown products of HAEMOGLOBIN → BILE PIGMENTS (Biliverdin → Bilirubin)
— store IRON
— discharge BILE into BILE CANALICULI
System of Channels
HEPATIC DUCT ———→ STORAGE in Gall Bladder (up to 50cc.)

GALL BLADDER
Columnar epithelium → Mucus added.
H_2O and Salts reabsorbed → CONCENTRATION (×10).
Contraction of Smooth Muscle → EXPULSION to duodenum.

In INTESTINE ——— FUNCTIONS of BILE
BILE SALTS (with fatty acids and glycerides) → stable emulsion of fats → Facilitates action of LIPASE
Activate Pancreatic LIPASE
Form water-soluble complexes with some water-insoluble substances (e.g. Cholesterol; fat-soluble Vitamins)
PROMOTE their ABSORPTION

small amount   most reabsorbed ———→ LIVER stimulate further excretion of Bile (CHOLERETIC ACTION)

BILE PIGMENTS   some reabsorbed ——→ LIVER
↓ Bacterial action   Re-excretion in Bile
STERCOBILINOGEN   Systemic Circulation
↓   KIDNEY
STERCOBILIN   UROBILINOGEN
(Brown pigment of Faeces)   UROBILIN of URINE

Other substances, e.g. certain drugs, can be taken out of the blood stream by the LIVER CELLS and EXCRETED in the BILE.

EXPULSION of BILE *(from the Gall Bladder)*

BILE is SECRETED continuously by the LIVER.
It is STORED and CONCENTRATED in the GALL BLADDER.
Periodically *(e.g. during a meal)* it is DISCHARGED into the DUODENUM.
NERVOUS and HUMORAL factors influence this expulsion:-

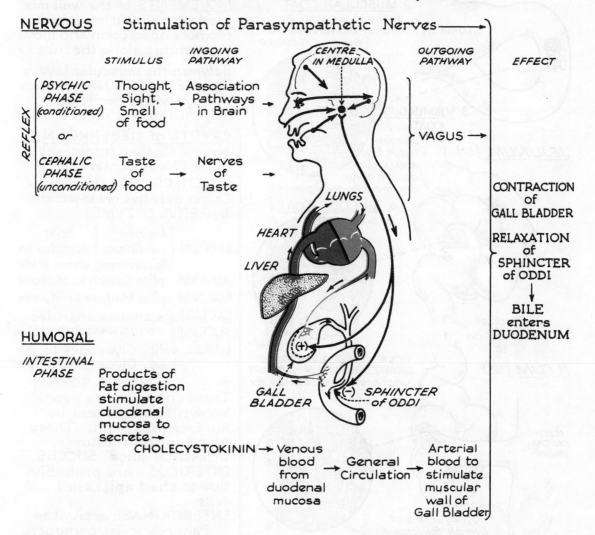

NERVOUS   Stimulation of Parasympathetic Nerves ⟶

	STIMULUS	INGOING PATHWAY	CENTRE IN MEDULLA	OUTGOING PATHWAY	EFFECT

REFLEX

PSYCHIC PHASE *(conditioned)*   Thought, Sight, Smell of food   →   Association Pathways in Brain   →

or

CEPHALIC PHASE *(unconditioned)*   Taste of food   →   Nerves of Taste   →

VAGUS ⟶

LUNGS

HEART

LIVER

CONTRACTION of GALL BLADDER

RELAXATION of SPHINCTER of ODDI

↓

BILE enters DUODENUM

HUMORAL

INTESTINAL PHASE   Products of Fat digestion stimulate duodenal mucosa to secrete→
CHOLECYSTOKININ → Venous blood from duodenal mucosa → General Circulation → Arterial blood to stimulate muscular wall of Gall Bladder

GALL BLADDER

(-) SPHINCTER of ODDI

[Cholecystokinin is probably the main factor controlling Gall Bladder
contraction in man]

SMALL INTESTINE

The Small Intestine is a long muscular tube-over 6 metres in length.
It receives CHYME in small quantities from the STOMACH; PANCREATIC
JUICE from the PANCREAS; BILE from the GALL BLADDER.

DUODENUM (10"-12")

STOMACH

It has 4 Coats:-
1. Outer **SEROUS COAT** of Peritoneum *(with vessels & nerves)*.

2. **MUSCULAR COAT**
Smooth muscle – 2 layers:-
Outer – LONGITUDINAL
Inner – CIRCULAR

×10

JEJUNUM (8')

3. **SUBMUCOUS COAT** *with*
fibrous
tissue, b.v's,
and (in duo-
denum only)
Brunner's (mucous) glands

4. **MUCOUS COAT**
(or lining)

×10

Gradual
transition
to

ILEUM (12')

VILLI-
finger-like projections with
STRIATED BORDER EPITHELIUM
-increase surface area for
ABSORPTION

ILEO-
CAECAL
VALVE

×10

20 to 30
Aggregations of
Lymph follicles in
PEYER'S PATCH *form part of*
the ileum's defence mechanism
against bacteria.

MOVEMENTS of the wall mix food with digestive juices, promote absorption and move the residue along the tube.

Between the muscular layers lies AUERBACH'S Nerve Plexus through which peristaltic movements are controlled.

CRYPTS of LIEBERKÜHN secrete alkaline Intestinal Juice (Succus Entericus).

PANETH CELLS at the base of Crypts were thought to secrete DIGESTIVE ENZYMES

EREPSIN {*Amino-peptidases* / *Dipeptidases*} Split peptides to amino acids

AMYLASE splits Starch to Maltose

MALTASE splits Maltose to Glucose

LACTASE } split disaccharides
SUCRASE } to monosaccharides

LIPASE splits Glycerides to lower Glycerides, fatty acids and glycerol

These enzymes are now known to be present in surface microvilli. Those found in the alkaline intestinal juice- SUCCUS ENTERICUS - are probably due to shed epithelial cells.

ENTEROKINASE activates Pancreatic Trypsinogen.

The BASIC PATTERN of the GUT WALL

The wall of the digestive tube shows a basic structural pattern.
FOUR coats are seen in transverse section.

STRUCTURE

1. MUCOUS COAT *(or MUCOSA)*

SURFACE EPITHELIUM *[type varies with site and function]*
with its GLANDS .

LOOSE FIBROUS TISSUE *(lamina propria)*
with capillaries and lymphatic vessels.

MUSCULARIS MUCOSAE
(thin layers of smooth muscle)

LYMPHATIC TISSUE

2. SUBMUCOUS COAT
    (or SUBMUCOSA)

DENSE FIBROUS TISSUE
in which lie bloodvessels,
lymphatic vessels and
MEISSNER'S NERVE PLEXUS
[Glands in oesophagus
and 1st part of duodenum]
LYMPHATIC TISSUE

3. MUSCULAR COAT
(or MUSCULARIS EXTERNA)

SMOOTH MUSCLE LAYERS
Inner-*Circular*
        arrangement
Outer - *Longitudinal*
        arrangement
Between them, blood- and
lymphatic vessels and
AUERBACH'S NERVE PLEXUS
[Some striated muscle
in oesophagus.]

(After GARVEN)

4. SEROUS COAT *(or SEROSA)*

FIBROUS TISSUE
with fat, blood- and lymphatic vessels
MESOTHELIUM
Where tube is suspended by a MESENTERY
the serosa is formed by visceral layer of Peritoneum

[Where there is no mesentery – replaced by
fibrous tissue which merges with
surrounding fibrous tissue]

FUNCTIONS

——— Layer in close contact with gut
contents: specialized for –
SECRETION
ABSORPTION - nutrients and
    hormones to blood stream.

MOBILITY (can continually change
    degree of folding and the surface
    of contact with gut contents).

DEFENCE against bacteria.

STRONG LAYER of SUPPLY
    to specialized mucosa
(Bloodvessels supply
    needs and remove
    absorbed materials).

CO-ORDINATION of motor
and secretory activities
        of mucosa.

MOVEMENT

Controls diameter of tube.
Mixes contents.
Propagates contents
        along tube.

Nervous elements
co-ordinate secretory and
muscular activities.

Carries nerves, blood- and
lymphatic vessels to and
from mesentery.

Forms smooth, moist
membrane which reduces
friction between contacting
surfaces in the peritoneal
            cavity.

INTESTINAL JUICE

Two types of secretion are formed in the intestine.

1. ALKALINE secretion (with Mucus) secreted by Submucosal Brünner's Glands in first part of Duodenum to protect the mucosa against ACID of Stomach.   ACID CHYME → DUODENAL MUCOSA →? Hormone - DUOCRININ --→ BLOOD STREAM → stimulates Brünner's Glands to secrete.

2. SUCCUS ENTERICUS secreted by Crypts of Lieberkühn. Its enzymes are probably due to the presence of shed epithelial cells with the enzymes contained in their surface microvilli.

NERVOUS   Stimulation of Parasympathetic Nerves ⟶

A Local Reflex

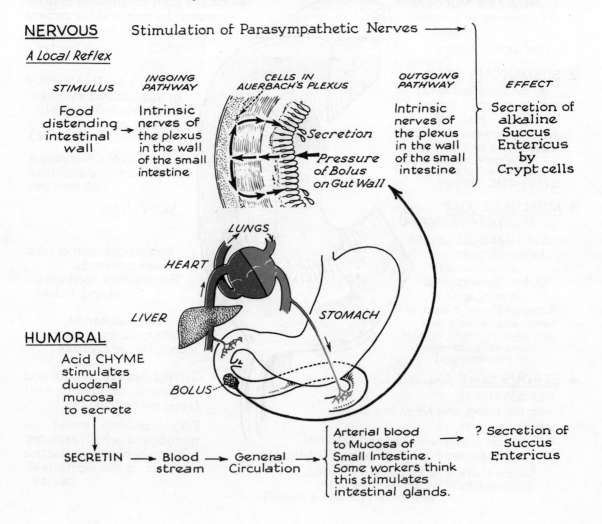

STIMULUS	INGOING PATHWAY	CELLS IN AUERBACH'S PLEXUS	OUTGOING PATHWAY	EFFECT
Food distending intestinal wall	Intrinsic nerves of the plexus in the wall of the small intestine	*Secretion* / *Pressure of Bolus on Gut Wall*	Intrinsic nerves of the plexus in the wall of the small intestine	Secretion of alkaline Succus Entericus by Crypt cells

LUNGS
HEART
LIVER
STOMACH
BOLUS

HUMORAL

Acid CHYME stimulates duodenal mucosa to secrete

SECRETIN ⟶ Blood stream ⟶ General Circulation ⟶ { Arterial blood to Mucosa of Small Intestine. Some workers think this stimulates intestinal glands. } ⟶ ? Secretion of Succus Entericus

MOVEMENTS of the SMALL INTESTINE

The Duodenum receives food in small quantities from the Stomach. The mixture of food and digestive juices — Chyme - is passed along the length of the Small Intestine.

TWO TYPES of MOVEMENT

SEGMENTATION —
Rhythmical alternating contractions and relaxations

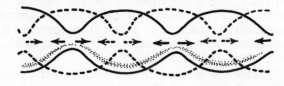

— These "shuttling" movements serve to mix CHYME and to bring it into contact with the absorptive mucosa, i.e. DIGESTION and ABSORPTION are promoted.

This type of movement is MYOGENIC, i.e. it is the property of the smooth muscle cells. It does not depend on a nervous mechanism.

PERISTALSIS —
Food probably acts as stimulus to stretch receptors in muscle and perhaps in mucous membrane

Circular muscle behind bolus *CONTRACTS*   Muscle in front of bolus *RELAXES*

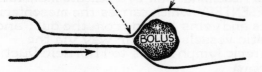

— Waves of this contraction move the food along the canal.

The contraction behind the bolus sweeps it into the relaxed portion of the tube ahead.

This type of movement is NEUROGENIC, i.e. it is carried out through a "Local" Reflex mediated through INTRINSIC nerve plexuses within the wall of the tube.

EMPTYING

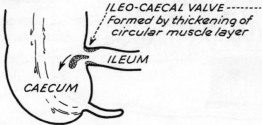

ILEO-CAECAL VALVE ------
Formed by thickening of circular muscle layer

ILEUM

CAECUM

opens and closes during digestion to allow spurts of fluid material from the Ileum to enter the Large Intestine.

This is a "Central" Reflex initiated when food enters the stomach and carried out through EXTRINSIC nerves to wall of tube.

Meals of different composition travel along the intestine at different rates. DIGESTION and ABSORPTION of food are usually complete by the time the residue reaches the Ileo-caecal valve.

ABSORPTION in SMALL INTESTINE

Absorption of most digested foodstuffs occurs in the Small Intestine through the Striated Border Epithelium covering the Villi.

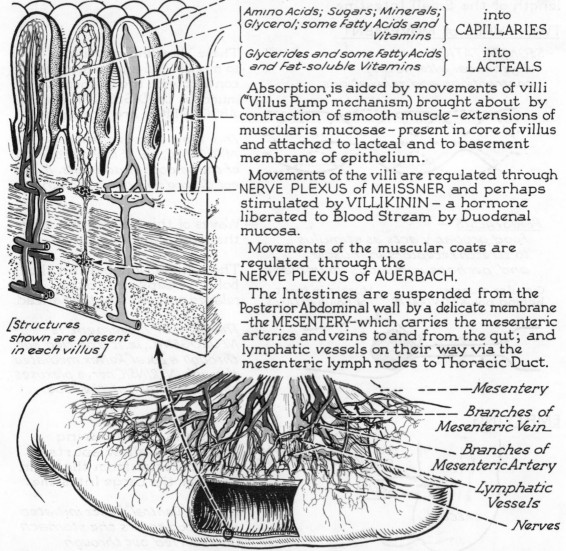

{ Amino Acids; Sugars; Minerals; Glycerol; some Fatty Acids and Vitamins } into CAPILLARIES

{ Glycerides and some Fatty Acids and Fat-soluble Vitamins } into LACTEALS

Absorption is aided by movements of villi ("Villus Pump" mechanism) brought about by contraction of smooth muscle—extensions of muscularis mucosae—present in core of villus and attached to lacteal and to basement membrane of epithelium.

Movements of the villi are regulated through NERVE PLEXUS of MEISSNER and perhaps stimulated by VILLIKININ—a hormone liberated to Blood Stream by Duodenal mucosa.

Movements of the muscular coats are regulated through the NERVE PLEXUS of AUERBACH.

The Intestines are suspended from the Posterior Abdominal wall by a delicate membrane—the MESENTERY—which carries the mesenteric arteries and veins to and from the gut; and lymphatic vessels on their way via the mesenteric lymph nodes to Thoracic Duct.

[Structures shown are present in each villus]

——————Mesentery

———— Branches of Mesenteric Vein

———— Branches of Mesenteric Artery

———— Lymphatic Vessels

——— Nerves

Absorption is not just a process of simple diffusion of substances from areas of high to areas of low concentration. Movement of ions can take place against a concentration gradient. In other words— Absorption is often an active process involving the use of energy by the epithelial cells.

TRANSPORT of ABSORBED FOODSTUFFS

After absorption the NUTRIENTS are transported in:—

(a) **BLOOD** through
Mesenteric Veins
to
Portal Vein
to
LIVER— *Many substances undergo further changes; some are stored; some are passed on via*
Hepatic Veins → Inferior Vena Cava → Heart and Systemic Circulation

(b) **LYMPH** in
Lymphatic Vessels
to
THORACIC DUCT
and via a large vein in the neck to
Systemic Circulation
↓
which then distributes Food to ALL TISSUES of the BODY

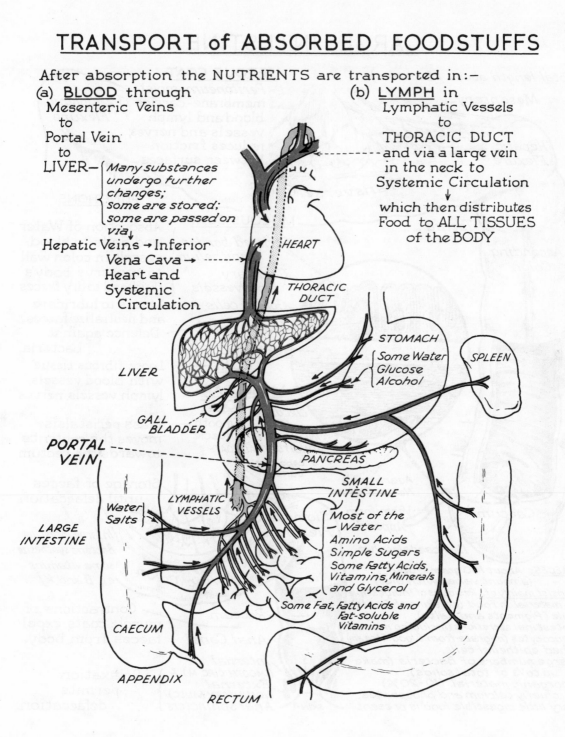

HEART

THORACIC DUCT

LIVER

STOMACH
Some Water Glucose Alcohol

SPLEEN

GALL BLADDER

PORTAL VEIN

PANCREAS

LARGE INTESTINE

Water Salts

LYMPHATIC VESSELS

SMALL INTESTINE
Most of the Water Amino Acids Simple Sugars Some Fatty Acids, Vitamins, Minerals and Glycerol

Some Fat, Fatty Acids and Fat-soluble Vitamins

CAECUM

APPENDIX

RECTUM

LARGE INTESTINE

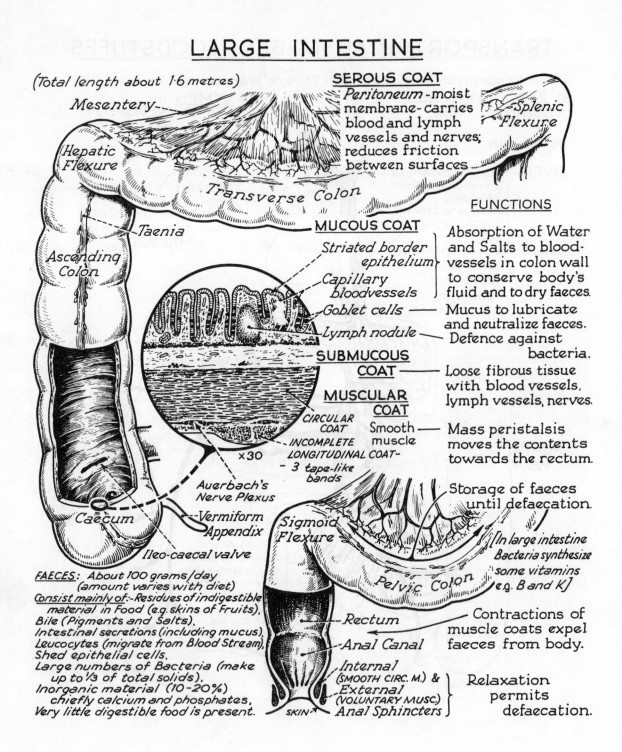

(Total length about 1·6 metres)

Mesentery---

SEROUS COAT
Peritoneum -moist membrane- carries blood and lymph vessels and nerves; reduces friction between surfaces

---Splenic Flexure

Hepatic Flexure

Transverse Colon

---Taenia

Ascending Colon

MUCOUS COAT

Striated border epithelium

Capillary bloodvessels

Goblet cells ---

Lymph nodule ---

SUBMUCOUS COAT ---

MUSCULAR COAT

CIRCULAR COAT

---INCOMPLETE LONGITUDINAL COAT- - 3 tape-like bands

×30

Smooth muscle

Auerbach's Nerve Plexus

--Vermiform Appendix

Caecum

---Ileo-caecal valve

Sigmoid Flexure

Pelvic Colon

--Rectum

--Anal Canal

--Internal (SMOOTH CIRC. M.) & External (VOLUNTARY MUSC.) Anal Sphincters

SKIN

FUNCTIONS

Absorption of Water and Salts to blood-vessels in colon wall to conserve body's fluid and to dry faeces.

Mucus to lubricate and neutralize faeces. Defence against bacteria.

Loose fibrous tissue with blood vessels, lymph vessels, nerves.

Mass peristalsis moves the contents towards the rectum.

Storage of faeces until defaecation.

[In large intestine Bacteria synthesize some vitamins e.g. B and K]

Contractions of muscle coats expel faeces from body.

Relaxation permits defaecation.

FAECES: About 100 grams/day (amount varies with diet)
Consist mainly of:-Residues of indigestible material in Food (e.g. skins of Fruits), Bile (Pigments and Salts), Intestinal secretions (including mucus), Leucocytes (migrate from Blood Stream), Shed epithelial cells, Large numbers of Bacteria (make up to 1/3 of total solids), Inorganic material (10-20%) chiefly calcium and phosphates, Very little digestible food is present.

MOVEMENTS of the LARGE INTESTINE

The Ileo-caecal Valve opens and closes during digestion. Peristaltic waves sweep semi-fluid contents of Ileum through the relaxed valve. During its stay in the Large Intestine faecal matter is subjected to –

TWO TYPES OF MOVEMENT

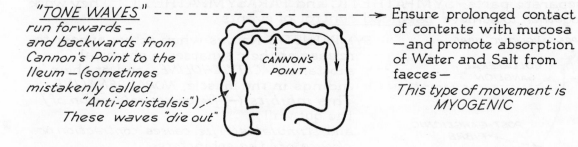

"TONE WAVES" run forwards – and backwards from Cannon's Point to the Ileum – (sometimes mistakenly called *"Anti-peristalsis"*) These waves *"die out"*

CANNON'S POINT

→ Ensure prolonged contact of contents with mucosa —and promote absorption of Water and Salt from faeces — *This type of movement is* MYOGENIC

MASS PERISTALSIS Strong waves at infrequent intervals start at upper end of Ascending Colon

→ Empty Transverse Colon and sweep faeces into the Descending and Pelvic Colons and into Rectum — NEUROGENIC

[Reflex often initiated by passage of food into Stomach (Gastro-colic Reflex)]

EMPTYING
(DEFAECATION)

Complex REFLEX act

Stimulus Passage of Faeces into the Rectum distends wall

[+ passage through anal canal]

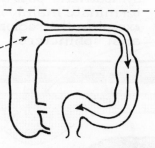

(+)

FAECES

(−)

PARASYMP.

SYMPATHETIC

PELVIC NERVES

PUDENDAL NERVES

(−)

SPINAL CORD

Sensations rise to level of consciousness → Voluntary decision → Impulses to inhibit or permit reflex evacuation.

Outgoing nerve messages → Powerful peristaltic contractions of Descending Colon, Pelvic Colon and Rectum.

[Preceded by inspiratory descent of diaphragm and voluntary contraction of abdominal muscles to raise intra-abdominal pressure]

Contraction of Pelvic Floor muscles with Relaxation of Anal Sphincter
↓
Evacuation of Faeces

INNERVATION of the GUT WALL

The movements of the Alimentary Canal are carried out automatically and, for the most part, beneath the level of consciousness.

Like other "automatic" systems in the body the gut wall is supplied by nerves of the AUTONOMIC NERVOUS SYSTEM. This has TWO separate parts – SYMPATHETIC and PARASYMPATHETIC. They work together to achieve <u>balanced control</u> of events.

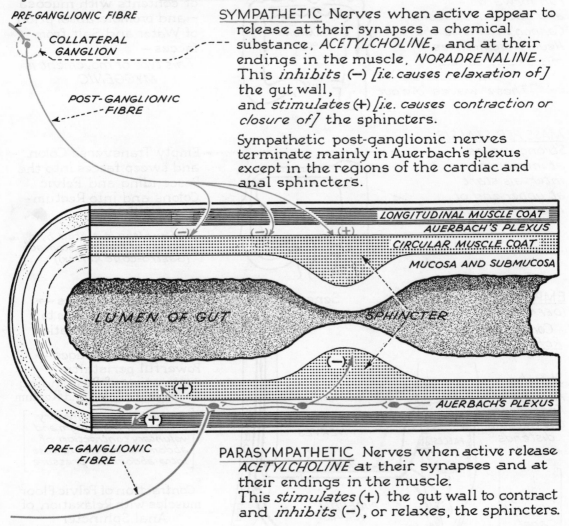

PRE-GANGLIONIC FIBRE

COLLATERAL GANGLION

POST-GANGLIONIC FIBRE

SYMPATHETIC Nerves when active appear to release at their synapses a chemical substance, *ACETYLCHOLINE*, and at their endings in the muscle, *NORADRENALINE*. This *inhibits* (−) *[i.e. causes relaxation of]* the gut wall, and *stimulates* (+) *[i.e. causes contraction or closure of]* the sphincters.

Sympathetic post-ganglionic nerves terminate mainly in Auerbach's plexus except in the regions of the cardiac and anal sphincters.

LONGITUDINAL MUSCLE COAT
AUERBACH'S PLEXUS
CIRCULAR MUSCLE COAT
MUCOSA AND SUBMUCOSA
LUMEN OF GUT
SPHINCTER
AUERBACH'S PLEXUS

PRE-GANGLIONIC FIBRE

PARASYMPATHETIC Nerves when active release *ACETYLCHOLINE* at their synapses and at their endings in the muscle. This *stimulates* (+) the gut wall to contract and *inhibits* (−), or relaxes, the sphincters.

74

NERVOUS CONTROL of GUT MOVEMENTS

Movements in the wall of the Alimentary Canal are
either
- (a) *MYOGENIC* — *a property of the smooth muscle.*
- (b) *NEUROGENIC* — *dependent on the Intrinsic Nerve Plexuses.*

They can occur even after *Extrinsic* nerves to the tract have been cut.   Normally, however, impulses travelling in these nerves of the SYMPATHETIC and PARASYMPATHETIC Systems, from the CONTROLLING CENTRES of the AUTONOMIC NERVOUS SYSTEM in the BRAIN, *influence* and *co-ordinate* events in the whole tract.

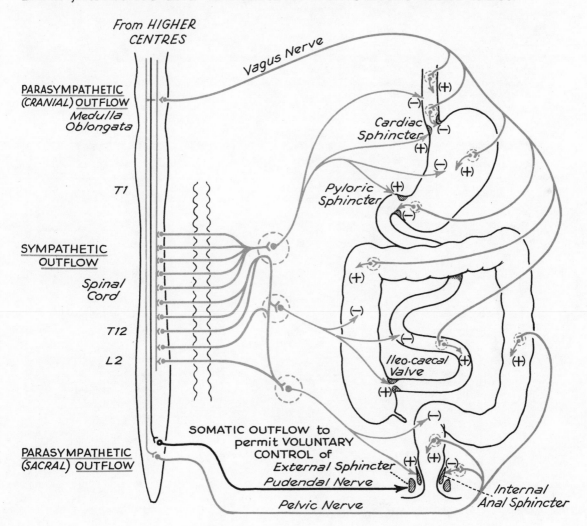

CHAPTER 4

TRANSPORT SYSTEM

The HEART, BLOOD VESSELS
and BODY FLUIDS:
HAEMOPOIETIC SYSTEM

CARDIOVASCULAR SYSTEM

The CIRCULATORY System

Chief TRANSPORT System of the body

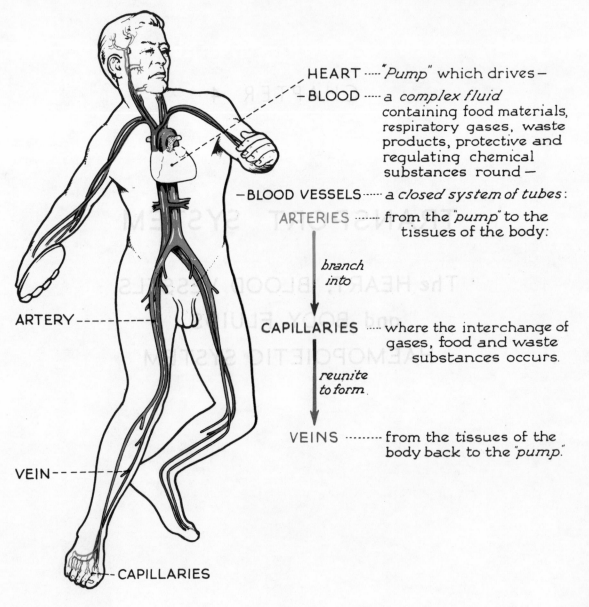

HEART ····· *"Pump"* which drives —

—BLOOD ····· a *complex fluid* containing food materials, respiratory gases, waste products, protective and regulating chemical substances round —

—BLOOD VESSELS ····· a *closed system of tubes*:

ARTERIES ······· from the *"pump"* to the tissues of the body:

branch into

CAPILLARIES ···· where the interchange of gases, food and waste substances occurs.

reunite to form

VEINS ········· from the tissues of the body back to the *"pump."*

ARTERY------

VEIN----

--CAPILLARIES

GENERAL COURSE of the CIRCULATION

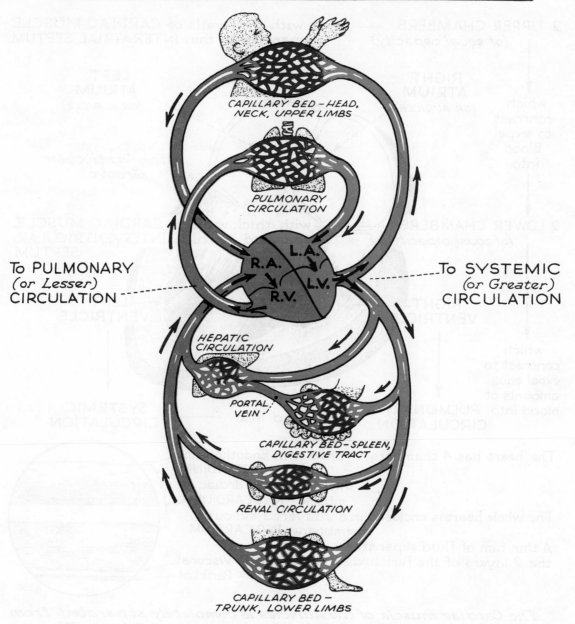

CAPILLARY BED – HEAD, NECK, UPPER LIMBS

PULMONARY CIRCULATION

R.A. L.A.

R.V. L.V.

To PULMONARY (or Lesser) CIRCULATION

To SYSTEMIC (or Greater) CIRCULATION

HEPATIC CIRCULATION

PORTAL VEIN

CAPILLARY BED – SPLEEN, DIGESTIVE TRACT

RENAL CIRCULATION

CAPILLARY BED – TRUNK, LOWER LIMBS

HEART

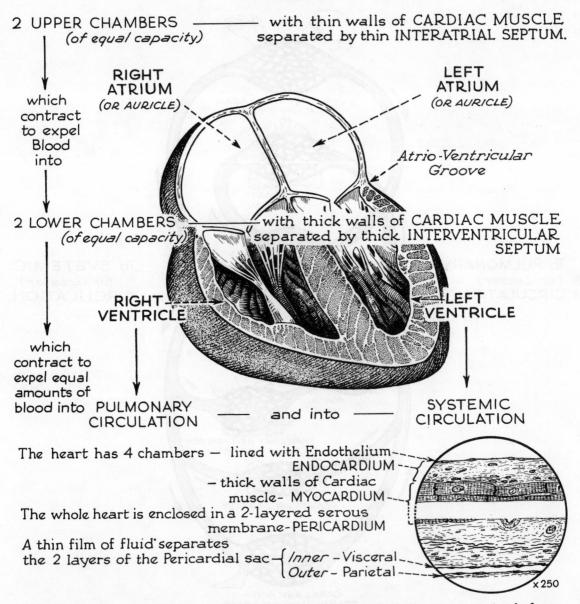

2 UPPER CHAMBERS ——— with thin walls of CARDIAC MUSCLE
(of equal capacity) separated by thin INTERATRIAL SEPTUM.

↓

which
contract
to expel
Blood
into

RIGHT ATRIUM *(OR AURICLE)*

LEFT ATRIUM *(OR AURICLE)*

Atrio-Ventricular Groove

2 LOWER CHAMBERS ——— with thick walls of CARDIAC MUSCLE
(of equal capacity) separated by thick INTERVENTRICULAR SEPTUM

RIGHT VENTRICLE

LEFT VENTRICLE

↓

which
contract to
expel equal
amounts of
blood into

PULMONARY CIRCULATION —— and into —— **SYSTEMIC CIRCULATION**

The heart has 4 chambers — lined with Endothelium—
ENDOCARDIUM
— thick walls of Cardiac muscle- MYOCARDIUM
The whole heart is enclosed in a 2-layered serous membrane-PERICARDIUM

A thin film of fluid separates the 2 layers of the Pericardial sac { *Inner* - Visceral
{ *Outer* - Parietal

x 250

The Cardiac muscle of the Auricles is completely separated from the Cardiac muscle of the Ventricles by a ring of Fibrous Tissue — at the AURICULO-VENTRICULAR (ATRIO-VENTRICULAR) Groove.

HEART

This is a *diagrammatic* section through the heart.

HEART VALVES have a core of Fibrous Tissue - - - - - - - -
covered on both sides with *Endothelium* - - - - -

Extensions from
ATRIO-VENTRICULAR (A-V)
FIBROUS RING

Designed to
allow blood
to flow in
one direction
only —
from ATRIUM
to VENTRICLE—
and on into
ARTERIES

The A-V valves
are attached by
thin CHORDAE
TENDINEAE
to
extensions of
CARDIAC MUSCLE–
PAPILLARY MUSCLES - - - - -

These contract
when ventricles contract
and pull on Chordae
Tendineae so that valve
flaps cannot be everted.

P.A.

AORTA

TO
BODY
TISSUES

PULMONARY ART.

RIGHT
LUNG

LEFT
LUNG

PULM.
VEINS

PULM.
VEINS

S.V.C.

I.V.C.

SEMILUNAR
VALVES
(each with three flaps) prevent
BACKFLOW
from
PULMONARY
ARTERY and
AORTA.

TRICUSPID
VALVE

MITRAL
VALVE

The Great Veins do not have valves guarding their entrance to the heart.
Thickening and contraction of the muscle around their mouths
prevent BACKFLOW of blood from heart.

HEART

The human heart is really a *DOUBLE PUMP* — each quite separate from the other

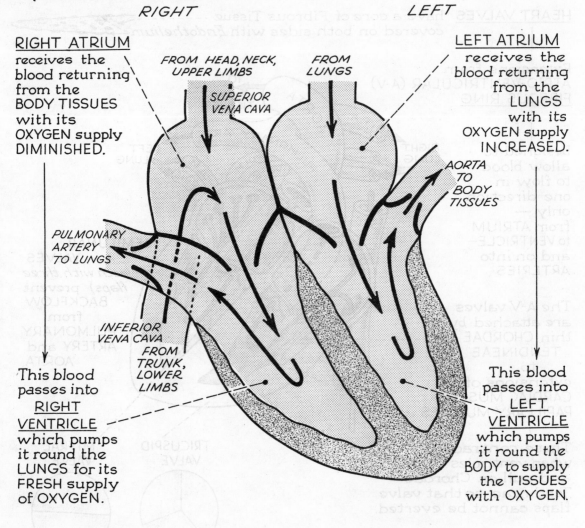

RIGHT

LEFT

RIGHT ATRIUM receives the blood returning from the BODY TISSUES with its OXYGEN supply DIMINISHED.

FROM HEAD, NECK, UPPER LIMBS

SUPERIOR VENA CAVA

FROM LUNGS

LEFT ATRIUM receives the blood returning from the LUNGS with its OXYGEN supply INCREASED.

AORTA TO BODY TISSUES

PULMONARY ARTERY TO LUNGS

INFERIOR VENA CAVA

FROM TRUNK, LOWER LIMBS

This blood passes into RIGHT VENTRICLE which pumps it round the LUNGS for its FRESH supply of OXYGEN.

This blood passes into LEFT VENTRICLE which pumps it round the BODY to supply the TISSUES with OXYGEN.

This diagram simplifies the structure of the heart to make it easier to understand the function of its various parts.

CARDIAC CYCLE

Diagrammatic representation of the sequence of events in the heart during ONE heart beat.

DIASTOLE *[Period of Relaxation - i.e. when heart is resting]*

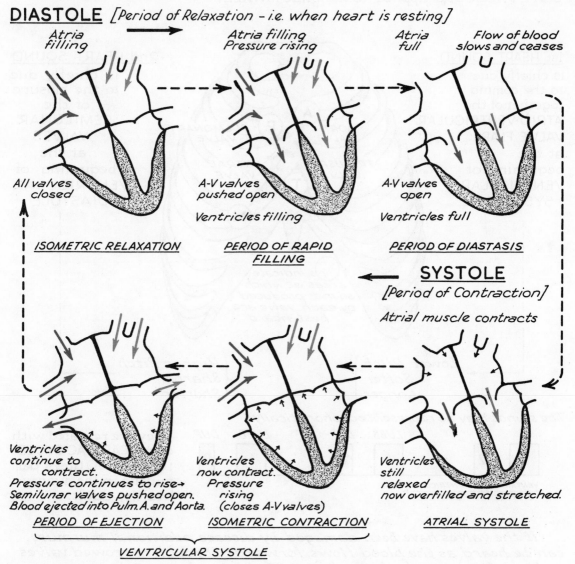

Atria filling

Atria filling
Pressure rising

Atria full

Flow of blood slows and ceases

All valves closed

ISOMETRIC RELAXATION

A-V valves pushed open

Ventricles filling

PERIOD OF RAPID FILLING

A-V valves open

Ventricles full

PERIOD OF DIASTASIS

SYSTOLE
[Period of Contraction]

Atrial muscle contracts

Ventricles continue to contract.
Pressure continues to rise→
Semilunar valves pushed open.
Blood ejected into Pulm. A. and Aorta.

PERIOD OF EJECTION

Ventricles now contract.
Pressure rising
(closes A-V valves)

ISOMETRIC CONTRACTION

Ventricles still relaxed now overfilled and stretched.

ATRIAL SYSTOLE

VENTRICULAR SYSTOLE

The total cycle of events takes about 0.8 second when heart is beating 75 times per minute.

HEART SOUNDS

During each CARDIAC CYCLE 2 HEART SOUNDS can be heard through a STETHOSCOPE applied to the CHEST WALL.

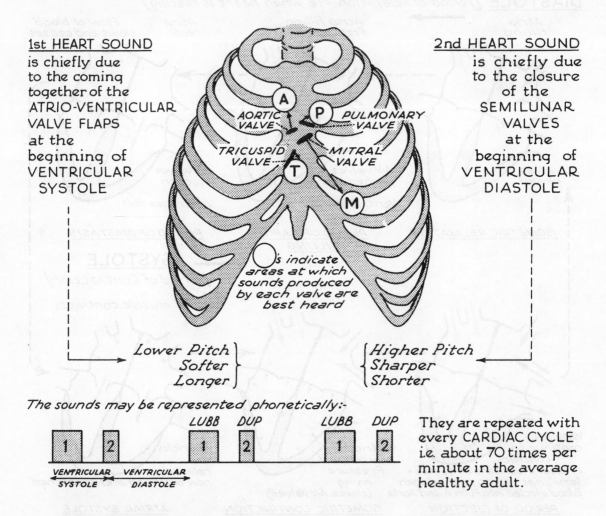

1st HEART SOUND
is chiefly due
to the coming
together of the
ATRIO-VENTRICULAR
VALVE FLAPS
at the
beginning of
VENTRICULAR
SYSTOLE

2nd HEART SOUND
is chiefly due
to the closure
of the
SEMILUNAR
VALVES
at the
beginning of
VENTRICULAR
DIASTOLE

AORTIC VALVE

PULMONARY VALVE

TRICUSPID VALVE

MITRAL VALVE

○'s indicate areas at which sounds produced by each valve are best heard

Lower Pitch
Softer
Longer

Higher Pitch
Sharper
Shorter

The sounds may be represented phonetically:-

LUBB DUP   LUBB DUP

1   2   1   2   1   2

VENTRICULAR   VENTRICULAR
SYSTOLE   DIASTOLE

They are repeated with every CARDIAC CYCLE i.e. about 70 times per minute in the average healthy adult.

If the valves have been damaged by disease additional murmurs can be heard as the blood flows forwards through narrowed valves or leaks backwards through incompetent valves.

Occasionally additional sounds are heard over normal healthy hearts.

ORIGIN and CONDUCTION of the HEART BEAT

The rhythmic contraction of the heart is called the HEART BEAT. The impulse to contract is generated in specialized NODAL TISSUE in the wall of the RIGHT ATRIUM.

Impulses are discharged rhythmically from this SINO-ATRIAL NODE
(The "Pacemaker")

The wave of excitation spreads throughout the muscle of both ATRIA
which are excited to contract

The impulse is picked up by another mass of NODAL TISSUE–the ATRIO-VENTRICULAR NODE and relayed by PURKINJE TISSUE–*(in BUNDLE of HIS and its branches)* lying beneath endocardium on the interventricular septum.
    This relays the impulse to contract to the muscle of both VENTRICLES.

The Right Atrium starts contracting before Left Atrium.

A ring of fibrous tissue separates Atria from Ventricles. The heart beat is not transmitted from Atria to Ventricles directly by ordinary CARDIAC muscle.

Both Ventricles contract together.

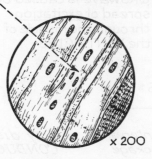

x 200

The wave of excitation is accompanied by an electrical change which is followed within 0·02 seconds by contraction of the cardiac muscle.

ELECTROCARDIOGRAM

The wave of excitation spreading through the heart wall is accompanied by electrical changes. *(Like Nerve and Skeletal Muscle, active Cardiac Muscle is electrically negative relative to resting Cardiac Muscle ahead of the zone of excitation.)* The electrical currents produced are conducted to the surface of the body and can be picked up, amplified and recorded by a special instrument – the <u>ELECTROCARDIOGRAPH</u>.

The record obtained is the *ELECTROCARDIOGRAM: (E.C.G.)*

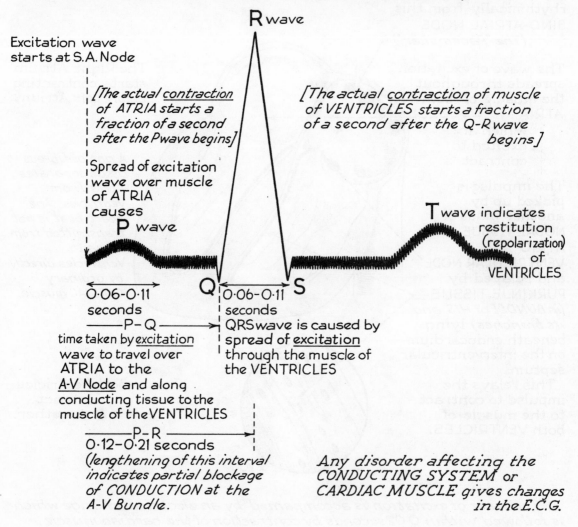

Excitation wave starts at S.A. Node

R wave

[The actual <u>contraction</u> of ATRIA starts a fraction of a second after the P wave begins]

[The actual <u>contraction</u> of muscle of VENTRICLES starts a fraction of a second after the Q-R wave begins]

Spread of excitation wave over muscle of ATRIA causes P wave

T wave indicates restitution (repolarization) of VENTRICLES

Q   S

0.06-0.11 seconds

——P- Q——
time taken by <u>excitation</u> wave to travel over ATRIA to the <u>A-V Node</u> and along conducting tissue to the muscle of the VENTRICLES

0.06-0.11 seconds
QRS wave is caused by spread of <u>excitation</u> through the muscle of the VENTRICLES

——P- R——
0.12–0.21 seconds
(lengthening of this interval indicates partial blockage of CONDUCTION at the A-V Bundle.

Any disorder affecting the CONDUCTING SYSTEM or CARDIAC MUSCLE gives changes in the E.C.G.

86

NERVOUS REGULATION of ACTION of HEART

Although the heart initiates its own impulse to contraction its activity is finely adjusted to meet the body's constantly changing needs by nervous impulses discharged from CONTROLLING CENTRES in the BRAIN and SPINAL CORD along PARASYMPATHETIC and SYMPATHETIC OUTFLOWS.

ACTION of PARASYMPATHETIC —

Continuous stream of impulses →
tends to restrain (—) Heart's Action:-
RATE of CONTRACTION DECREASED,
CONDUCTIVITY DEPRESSED,
FORCE of CONTRACTION DECREASED,
EXCITABILITY DECREASED,
Refractory Period shortened.

PARASYMPATHETIC OUTFLOW

[Variation in this
'VAGAL TONE'
is chief factor
in giving
alteration
of Heart
Rate]

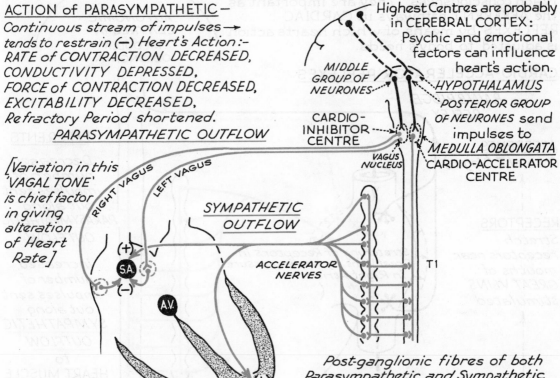

Highest Centres are probably in CEREBRAL CORTEX: Psychic and emotional factors can influence heart's action.

MIDDLE GROUP OF NEURONES

HYPOTHALAMUS
POSTERIOR GROUP OF NEURONES send impulses to MEDULLA OBLONGATA

CARDIO-INHIBITOR CENTRE

VAGUS NUCLEUS

CARDIO-ACCELERATOR CENTRE

RIGHT VAGUS   LEFT VAGUS

SYMPATHETIC OUTFLOW

ACCELERATOR NERVES

T1

S.A.   A.V.

(+)   (—)   (+)

Post-ganglionic fibres of both Parasympathetic and Sympathetic form a CARDIAC PLEXUS of nerves before final distribution to the heart muscle.

ACTION of SYMPATHETIC —

Constantly transmits
streams of impulses →
tend to accelerate (+) Heart's Action:-
RATE of CONTRACTION INCREASED,
CONDUCTIVITY INCREASED,
FORCE of CONTRACTION INCREASED,
EXCITABILITY INCREASED,
Refractory Period lengthened.

The Parasympathetic and Sympathetic have opposite effects on heart's action. Both are constantly in action but finely balanced to meet the body's needs.

Stimulation of Symp.
Inhibition of Para. } →Acceleration } of
Stimulation of Para.   Heart's
Inhibition of Symp. } →Inhibition } Action

CARDIAC REFLEXES

There are INGOING (*SENSORY*) fibres probably in both PARASYMPATHETIC and SYMPATHETIC nerves which convey information to the CENTRES about events taking place in the heart.

These afferent messages do not normally reach consciousness. They are important as the AFFERENT pathways in CARDIAC REFLEXES by means of which heart's action is adjusted to body's needs.

CARDIO-ACCELERATOR REFLEXES

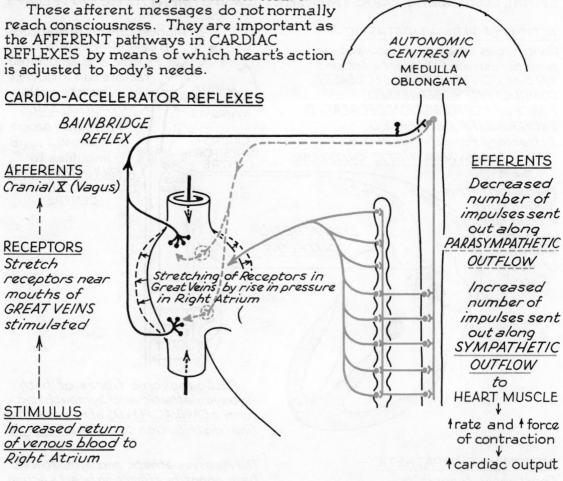

AUTONOMIC CENTRES IN MEDULLA OBLONGATA

BAINBRIDGE REFLEX

AFFERENTS
Cranial X (Vagus)

RECEPTORS
Stretch receptors near mouths of GREAT VEINS stimulated

Stretching of Receptors in Great Veins by rise in pressure in Right Atrium

STIMULUS
Increased return of venous blood to Right Atrium

EFFERENTS
Decreased number of impulses sent out along PARASYMPATHETIC OUTFLOW

Increased number of impulses sent out along SYMPATHETIC OUTFLOW *to* HEART MUSCLE
↑rate and ↑force of contraction
↑cardiac output

This has been regarded for almost 50 years as an important adaptive mechanism whereby heart rate and force of contraction are reflexly adjusted to match the quantity of venous blood returning to the heart.

CARDIAC REFLEXES

CARDIO-INHIBITORY REFLEXES

AUTONOMIC
CENTRES IN
MEDULLA
OBLONGATA

AFFERENTS

["Buffer Nerves"]

IX (GLOSSOPHARYNGEAL)

INTERNAL
CAROTID
ARTERY

X (VAGUS)

RECEPTORS
Stretch (or Baro-)
receptors in

CAROTID
SINUS

ARCH of
AORTA

and probably
in MYOCARDIUM

Stretching of Aorta
by increased blood
pressure

EFFERENTS

Stimulation
of
PARASYMPATHETIC
OUTFLOW

and

Inhibition
of
SYMPATHETIC
OUTFLOW

Pressure
high in
Left
Ventricle

STIMULUS —→ Increased arterial
blood pressure

Stimulation of other afferents e.g. from SPLANCHNIC
region of abdominal cavity and overstimulation of
any PAIN-sensitive nerves may also give reflex —→

EFFECT
Slowing of heart's
action and
↓ force of
contraction

↓ cardiac output

Fall in Blood Pressure

This reflex is constantly in action —
Important Adaptive Mechanism — As Arterial Blood Pressure is raised
by ejection of blood from Ventricle in Systole this reflex is brought
into play to adjust force and frequency with which heart is ejecting
blood into the arterial system —→ prevent undue rise in blood pressure.

CARDIAC OUTPUT

The Volume of Blood expelled from the heart can be measured as follows :-

A semi-rigid tube is inserted into a VEIN in the arm and passed along into the RIGHT ATRIUM of the heart — to obtain a sample of MIXED VENOUS BLOOD — which has given up some of its oxygen to the tissues.
The OXYGEN content is analysed.

100 ml VENOUS Blood hold 14 ml OXYGEN.

The AMOUNT of OXYGEN taken up by the lungs per minute is measured by a SPIROMETER.

·250 ml OXYGEN are removed from the lungs by blood each minute.

A needle is inserted into an ARTERY in the leg and a sample of ARTERIAL BLOOD — which has received its fresh oxygen supply in the lungs — is obtained.
The OXYGEN content is analysed.

100 ml ARTERIAL Blood hold 19 ml OXYGEN.

Each 100 ml blood gains 5 ml OXYGEN as it passes through the lungs. The blood in the lungs takes up 250 ml OXYGEN from the atmosphere per minute.

Therefore there must be $\left(\dfrac{250}{5} \times 100\right)$ *ml [i.e. 5,000 ml] of blood leaving the Right Ventricle and passing through the Lungs to the Left Atrium per minute to take up this* 250 ml OXYGEN.

The same volume of blood must leave the Left Ventricle and enter the Aorta in the same time otherwise blood would soon be dammed back in the lungs. i.e. If heart contracts 72 times per minute STROKE VOLUME = $\dfrac{5000}{72}$ ≑ *70 ml per beat for each Ventricle.*

(After WISHART)

Cardiac Output can be increased many times *(up to 30 litres per min)* in Exercise —
— partly by increase in Heart Rate; partly by increase in Stroke Volume.

BLOOD VESSELS

The system of tubes through which the heart pumps blood.

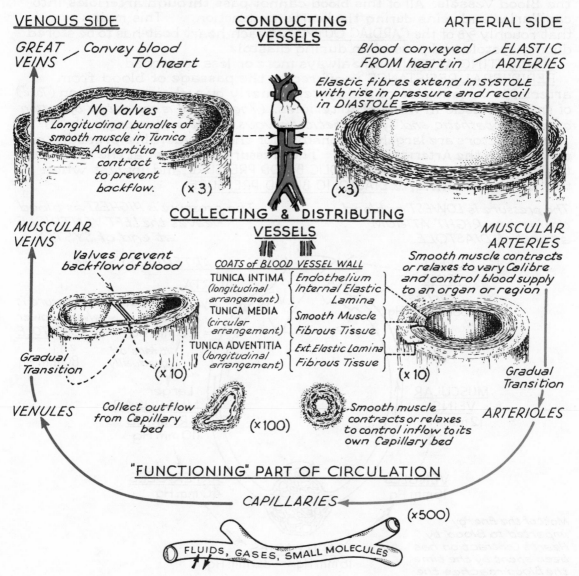

VENOUS SIDE

GREAT VEINS — Convey blood TO heart

No Valves
Longitudinal bundles of smooth muscle in Tunica Adventitia contract to prevent backflow. (× 3)

MUSCULAR VEINS

Valves prevent backflow of blood

Gradual Transition

(× 10)

VENULES

Collect outflow from Capillary bed (× 100)

CONDUCTING VESSELS

COLLECTING & DISTRIBUTING VESSELS

COATS of BLOOD VESSEL WALL

TUNICA INTIMA (longitudinal arrangement)	Endothelium / Internal Elastic Lamina
TUNICA MEDIA (circular arrangement)	Smooth Muscle / Fibrous Tissue
TUNICA ADVENTITIA (longitudinal arrangement)	Ext. Elastic Lamina / Fibrous Tissue

ARTERIAL SIDE

Blood conveyed FROM heart in — ELASTIC ARTERIES

Elastic fibres extend in SYSTOLE with rise in pressure and recoil in DIASTOLE (×3)

MUSCULAR ARTERIES
Smooth muscle contracts or relaxes to vary Calibre and control blood supply to an organ or region

(× 10)

Gradual Transition

ARTERIOLES
Smooth muscle contracts or relaxes to control inflow to its own Capillary bed

"FUNCTIONING" PART OF CIRCULATION

CAPILLARIES (×500)

FLUIDS, GASES, SMALL MOLECULES

Only from CAPILLARIES can Blood give up food and oxygen to tissues; and receive waste products and carbon dioxide from tissues.

BLOOD PRESSURE

Each Ventricle at each heart beat ejects forcibly about 70 ml. of Blood into the Blood Vessels. All of this blood cannot pass through arterioles into capillaries and veins during the heart's contraction. This means that roughly 5/8 of the <u>CARDIAC OUTPUT</u> at each heart beat has to be stored during systole and passed on during diastole.

<u>CONDUCTING ARTERIES</u> are always more or less stretched.

<u>PERIPHERAL RESISTANCE</u> is offered to the passage of blood from arterial to venous side of the system chiefly by partial constriction ("Tone") of smooth muscle in walls of Arterioles. (*The calibre is regulated by action of Parasympathetic and Sympathetic Nervous System — [see pages 95-97]*

These factors are largely responsible for the considerable pressure of the blood in the Arterial System. The pressure is highest at the height of the heart's contraction, i.e. <u>SYSTOLIC BLOOD PRESSURE</u>, and lowest when the heart is relaxing, i.e. <u>DIASTOLIC BLOOD PRESSURE</u>.

The pressure is LOWEST as blood drains into RIGHT ATRIUM at end of DIASTOLE.

The pressure is HIGHEST as blood leaves the LEFT VENTRICLE at end of SYSTOLE.

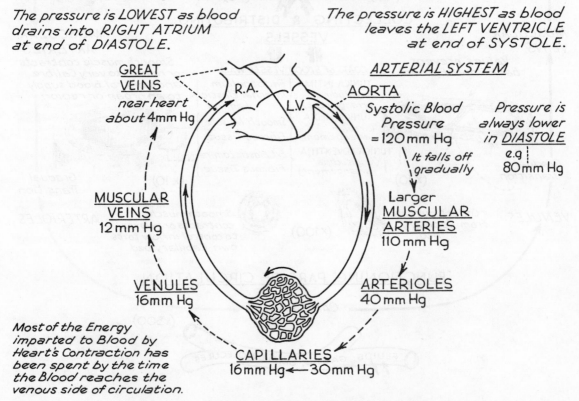

GREAT VEINS
near heart about 4mm Hg

R.A. L.V.

ARTERIAL SYSTEM

AORTA
Systolic Blood Pressure = 120 mm Hg

It falls off gradually

Pressure is *always lower in DIASTOLE* e.g. 80 mm Hg

MUSCULAR VEINS
12 mm Hg

Larger MUSCULAR ARTERIES
110 mm Hg

VENULES
16 mm Hg

ARTERIOLES
40 mm Hg

Most of the Energy imparted to Blood by Heart's Contraction has been spent by the time the Blood reaches the venous side of circulation.

CAPILLARIES
16 mm Hg ← 30 mm Hg

NOTE :- Any alteration in the <u>TOTAL AMOUNT</u> or <u>VISCOSITY</u> of Blood will also affect BLOOD PRESSURE.

MEASUREMENT of ARTERIAL BLOOD PRESSURE

The Arterial Blood Pressure is measured in man by means of a
SPHYGMOMANOMETER.

This consists of a
RUBBER BAG _(covered
with a cloth envelope)_
which is wrapped
round the UPPER ARM
over the BRACHIAL
ARTERY.

One tube connects the
inside of the bag with
a MANOMETER
containing MERCURY.

Another tube connects
the inside of the bag to
a hand operated PUMP
with a release VALVE.

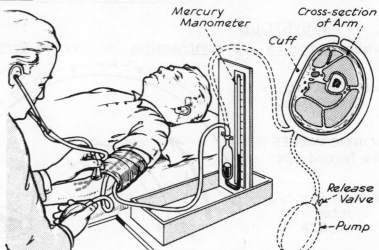

Mercury
Manometer

Cross-section
of Arm

Cuff

Release
Valve

Pump

METHOD

_Air is pumped into the rubber bag till
pressure in cuff is greater than pressure
in artery even during heart's systole —
     — Artery is then closed down during_ SYSTOLE and DIASTOLE
_[At same time air is pushing up mercury
        column in manometer.]_

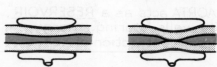

_By releasing valve on pump the pressure
in cuff is gradually reduced till maximum
pressure in artery just overcomes pressure in
cuff—Some blood begins to spurt through during_ SYSTOLE — _Artery still closed
                            during DIASTOLE_

_At this point FAINT rhythmical TAPPING SOUNDS
begin to be heard through STETHOSCOPE. The
height of mercury in millimetres is taken as the
SYSTOLIC Blood Pressure (e.g. 120 mm Hg)_

_Pressure in cuff is reduced still further
till it is just less than the lowest pressure
in artery towards the end of diastole
(i.e. just before next heart beat) —_

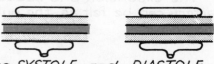

_   — Blood flow is unimpeded during SYSTOLE and DIASTOLE_
_The sounds stop. The height of mercury in the manometer at this point
is taken as the DIASTOLIC Blood Pressure (e.g. about 80mm Hg)_

These values differ with SEX, AGE, EXERCISE, SLEEP, etc.

ELASTIC ARTERIES

The large <u>CONDUCTING ARTERIES</u> near the <u>HEART</u> are
ELASTIC ARTERIES

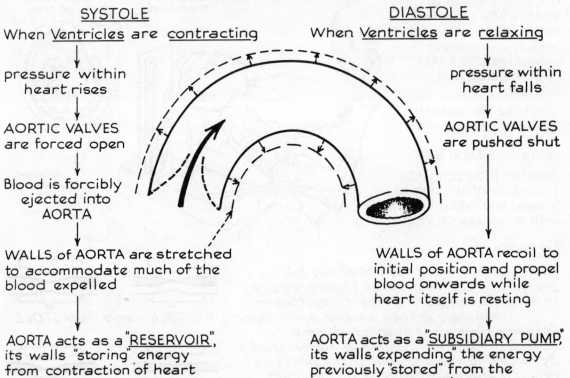

SYSTOLE
When <u>Ventricles</u> are <u>contracting</u>

pressure within
heart rises

↓

AORTIC VALVES
are forced open

↓

Blood is forcibly
ejected into
AORTA

↓

WALLS of AORTA are stretched
to accommodate much of the
blood expelled

↓

AORTA acts as a "<u>RESERVOIR</u>",
its walls "storing" energy
from contraction of heart

DIASTOLE
When <u>Ventricles</u> are <u>relaxing</u>

pressure within
heart falls

↓

AORTIC VALVES
are pushed shut

↓

WALLS of AORTA recoil to
initial position and propel
blood onwards while
heart itself is resting

↓

AORTA acts as a "<u>SUBSIDIARY PUMP</u>",
its walls "expending" the energy
previously "stored" from the
heart's contraction.

As blood is pumped from the heart during systole, this distension
and increase in pressure which starts in the aorta passes along the
whole arterial system as a wave — the *pulse wave*.

The expansion and subsequent relaxation of the wall of the radial
artery can be felt as 'the pulse' at the wrist.

*A great increase in Blood Pressure can result if
these walls lose some of their elasticity with age or disease
and can no longer stretch readily to accommodate so much of
the heart's output during systole.*

NERVOUS REGULATION of ARTERIOLES

The Smooth Muscle in *Arterioles* can contract or relax to alter calibre of the vessels, and thus adjust distribution of blood to meet the constantly changing needs of different parts of the body and also to maintain the normal blood pressure.

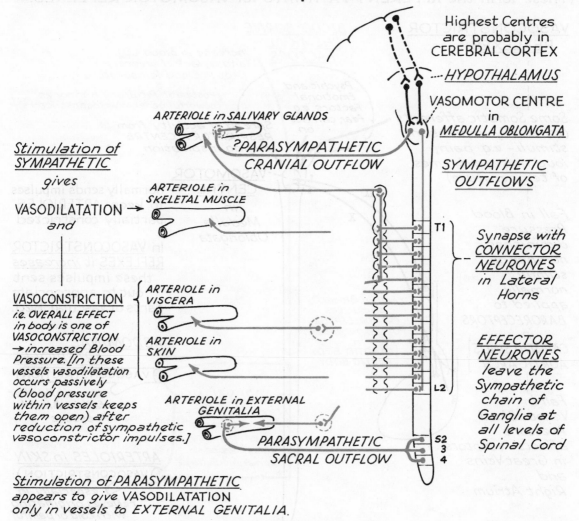

Highest Centres
are probably in
CEREBRAL CORTEX

····*HYPOTHALAMUS*

VASOMOTOR CENTRE
in
MEDULLA OBLONGATA

SYMPATHETIC
OUTFLOWS

ARTERIOLE in SALIVARY GLANDS

?PARASYMPATHETIC
CRANIAL OUTFLOW

Stimulation of
SYMPATHETIC
gives

VASODILATATION →
and

ARTERIOLE in
SKELETAL MUSCLE

T1

Synapse with
CONNECTOR
NEURONES
in Lateral
Horns

VASOCONSTRICTION
i.e. OVERALL EFFECT
in body is one of
VASOCONSTRICTION
→ increased Blood
Pressure. [In these
vessels vasodilatation
occurs passively
(blood pressure
within vessels keeps
them open) after
reduction of sympathetic
vasoconstrictor impulses.]

ARTERIOLE in
VISCERA

ARTERIOLE in
SKIN

EFFECTOR
NEURONES
leave the
Sympathetic
chain of
Ganglia at
all levels of
Spinal Cord

L2

ARTERIOLE in EXTERNAL
GENITALIA

PARASYMPATHETIC
SACRAL OUTFLOW

S2
3
4

Stimulation of PARASYMPATHETIC
appears to give VASODILATATION
only in vessels to EXTERNAL GENITALIA.

Veins have a similar nerve supply. Active Sympathetic Vasoconstriction
mobilizes the blood held in 'venous reservoirs'.

REFLEX and CHEMICAL REGULATION of ARTERIOLAR TONE – I

Afferent impulses are constantly reaching the VASOMOTOR CENTRE in the Medulla Oblongata in ingoing nerves from all parts of the body. These form the AFFERENT PATHWAYS for VASOMOTOR REFLEXES.

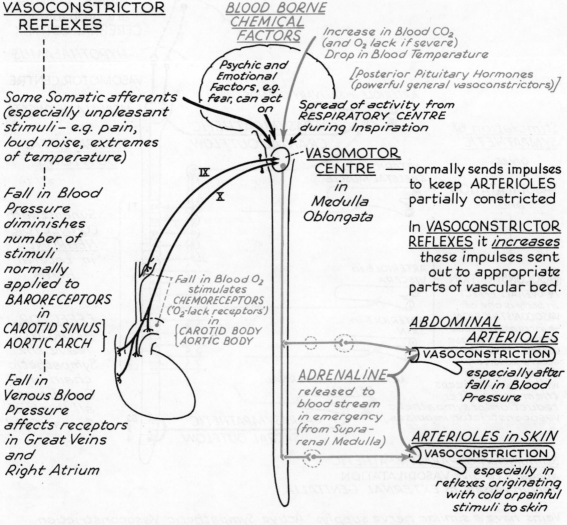

VASOCONSTRICTOR REFLEXES

Some Somatic afferents (especially unpleasant stimuli – e.g. pain, loud noise, extremes of temperature)

Fall in Blood Pressure diminishes number of stimuli normally applied to BARORECEPTORS in CAROTID SINUS] AORTIC ARCH]

Fall in Venous Blood Pressure affects receptors in Great Veins and Right Atrium

BLOOD BORNE CHEMICAL FACTORS

Increase in Blood CO_2 (and O_2 lack if severe) Drop in Blood Temperature

[Posterior Pituitary Hormones (powerful general vasoconstrictors)]

Psychic and Emotional Factors, e.g. fear, can act on

Spread of activity from RESPIRATORY CENTRE during Inspiration

IX

X

Fall in Blood O_2 stimulates CHEMORECEPTORS ('O_2-lack receptors') in [CAROTID BODY] [AORTIC BODY]

VASOMOTOR CENTRE in Medulla Oblongata

— normally sends impulses to keep ARTERIOLES partially constricted

In VASOCONSTRICTOR REFLEXES it *increases* these impulses sent out to appropriate parts of vascular bed.

ABDOMINAL ARTERIOLES

(VASOCONSTRICTION)

especially after fall in Blood Pressure

ADRENALINE released to blood stream in emergency (from Supra-renal Medulla)

ARTERIOLES in SKIN

(VASOCONSTRICTION)

especially in reflexes originating with cold or painful stimuli to skin

An *OVERALL VASOCONSTRICTION*, especially of abdominal vessels, gives rise in Blood Pressure.

REFLEX and CHEMICAL REGULATION of ARTERIOLAR TONE – II

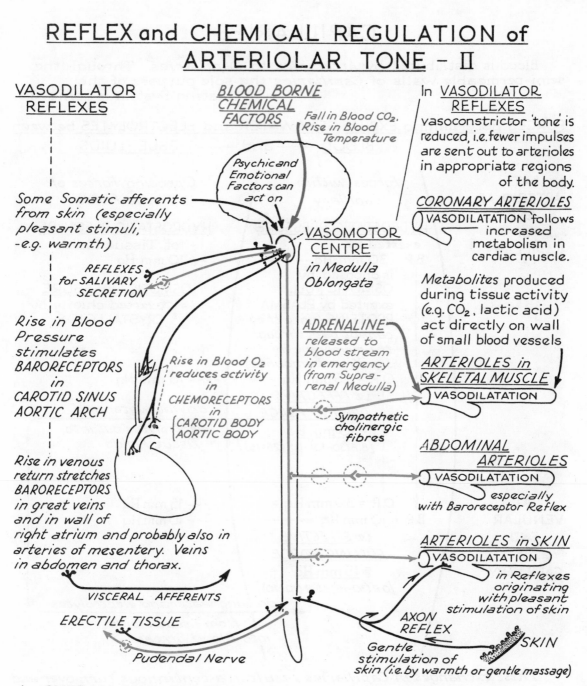

VASODILATOR REFLEXES

BLOOD BORNE CHEMICAL FACTORS

Fall in Blood CO_2. Rise in Blood Temperature

In **VASODILATOR REFLEXES** vasoconstrictor 'tone' is reduced, i.e. fewer impulses are sent out to arterioles in appropriate regions of the body.

Psychic and Emotional Factors can act on

VASOMOTOR CENTRE in Medulla Oblongata

Some Somatic afferents from skin (especially pleasant stimuli, - e.g. warmth)

REFLEXES for SALIVARY SECRETION

Rise in Blood Pressure stimulates BARORECEPTORS in CAROTID SINUS AORTIC ARCH

Rise in Blood O_2 reduces activity in CHEMORECEPTORS in { CAROTID BODY AORTIC BODY

ADRENALINE released to blood stream in emergency (from Supra-renal Medulla)

Sympathetic cholinergic fibres

CORONARY ARTERIOLES
VASODILATATION follows increased metabolism in cardiac muscle.

Metabolites produced during tissue activity (e.g. CO_2, lactic acid) act directly on wall of small blood vessels

ARTERIOLES in SKELETAL MUSCLE
VASODILATATION

Rise in venous return stretches BARORECEPTORS in great veins and in wall of right atrium and probably also in arteries of mesentery. Veins in abdomen and thorax.

VISCERAL AFFERENTS

ABDOMINAL ARTERIOLES
VASODILATATION
especially with Baroreceptor Reflex

ARTERIOLES in SKIN
VASODILATATION
in Reflexes originating with pleasant stimulation of skin

ERECTILE TISSUE

Pudendal Nerve

AXON REFLEX

Gentle stimulation of skin (i.e. by warmth or gentle massage)

SKIN

An *OVERALL VASODILATATION* gives *fall in Blood Pressure.*

CAPILLARIES

Blood is distributed by *Arterioles* to *Capillaries*. Through the semi-permeable walls of *Capillaries* the sole purpose of the Circulating System is achieved.

FORCES determining EXCHANGE of WATER and ELECTROLYTES between
BLOOD —— and —— TISSUE FLUIDS

AT ARTERIOLAR END of CAPILLARY	*Forces within Capillary*	*Opposing Forces of Tissue Fluids*
	HYDROSTATIC PRESSURE exerted by B.P. = 30 mm Hg -----	HYDROSTATIC PRESSURE of Tissue Fluids ----10 mm Hg
	is opposed by OSMOTIC PULL exerted by PLASMA PROTEINS *(since they cannot pass through the membrane in significant amounts)* O.P. = 25 mm Hg ------	OSMOTIC PRESSURE exerted chiefly by CRYSTALLOIDS 15 mm Hg
	i.e. *EFFECTIVE DRIVING FORCE* ≡ 10 mm Hg [B.P.(30-10)-O.P.(25-15)]	*driving water and electrolytes out into Tissue Fluids*
AT VENULAR END of CAPILLARY	O.P. = 30 mm Hg ------ B.P. = 10 mm Hg ------ i.e. *EFFECTIVE PULLING FORCE* ≡ 15 mm Hg [O.P.(30-15)-B.P.(10-10)]	15 mm Hg 10 mm Hg
		drawing water and electrolytes from Tissue Fluids into Blood Stream

These exchanges in Capillaries result in a continuous "turn-over" and renewal of TISSUE FLUIDS.

VEINS: VENOUS RETURN

Capillaries unite to form *Veins* which *convey blood back to the heart.* By the time blood reaches the veins much of the force imparted to it by the heart's contraction has been spent.

VENOUS RETURN to the heart depends on various factors :-

GRAVITY –
from head and neck region.

"RESPIRATORY PUMP"

(a) *Intrathoracic pressure* is always lower than that in atmosphere. This *"negative pressure"* (which fluctuates with Respiratory Movements) exerts a *"suctioning"* pull which tends to draw column of blood upwards.

(b) Descent of *DIAPHRAGM* in Inspiration increases *Intra-abdominal pressure* and forces blood upwards in abdominal veins.

"MUSCULAR PUMP"

Contractions of skeletal muscles help to squeeze veins and move blood upwards.

"CARDIAC PUMP"

Residual force imparted by heart's contraction aids in forcing venous blood towards heart.

RIGHT ATRIUM

DIAPHRAGM

"OVERFILLING" of the system (i.e. adequate BLOOD VOLUME) and ADEQUATE 'TONE' of smooth muscle in venous walls are essential. (e.g. if too many vessels are dilated at any one time adequate venous return to the heart will not be maintained.)

Sympathetic nerves keep veins partially constricted. Increased sympathetic 'tone' (e.g. after haemorrhage) mobilizes blood held in venous reservoirs and helps to maintain venous return to the heart.

ACTION of VALVES
Once the column of blood has moved up *UNIDIRECTIONAL VALVES* in larger veins prevent back flow.

CAPILLARIES   ARTERIAL PRESSURE

VENULES

BLOOD FLOW

The rate of blood flow varies in different parts of the vascular system. It is rapid in large vessels; slower in small vessels. *[The larger the total cross-sectional area represented the slower the rate of flow.]*

GREAT VEINS

Area of cross-section 32 sq. cm

R.A   L.A.

R.V.   L.V.

BLOOD FLOW NOT SO FAST AS IN AORTA

BLOOD FLOW FAST

The AORTA has a wide individual diameter — *area of cross-section about 8 sq. cm.* It divides into arteries. These are smaller than the aorta but their total cross-sectional area is greater. The rate of blood flow in them is slower than that in the aorta itself.

As MUSCULAR ARTERIES branch to give more and more small vessels so the cross-section of all these added together increases.

MUSCULAR VEINS

very numerous

SYSTEMIC CIRCULATION

BLOOD FLOW, SLOW

CAPILLARIES have a very small diameter but there are enormous numbers of them. This represents a total area in cross-section of — *from 800 to 3,200 sq. cm*

PULMONARY CIRCULATION

BLOOD FLOW is faster in <u>PULMONARY CAPILLARIES</u> — Total cross-sectional area is smaller than that of *Systemic Capillaries.* Capillary pressure is less. This means that the osmotic 'pull' of the plasma proteins (25 mm Hg) exceeds the 'driving force' of the B.P. (5-10 mm Hg) along the whole length of the pulmonary capillary and fluid, therefore, cannot pass out into the alveoli.

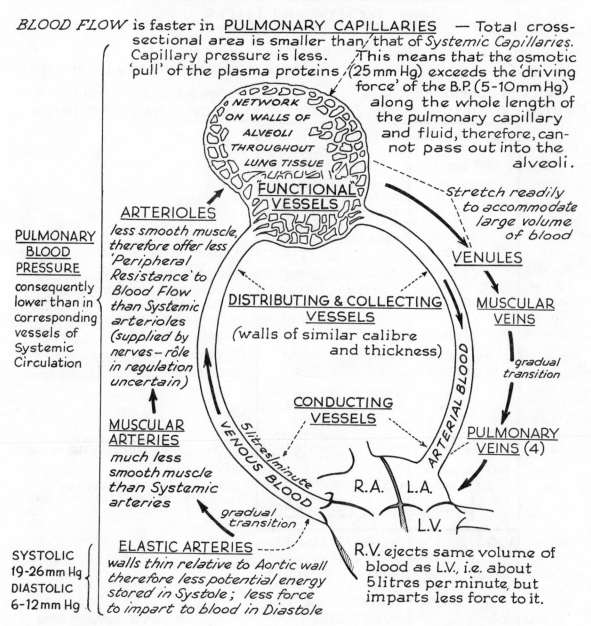

NETWORK ON WALLS OF ALVEOLI THROUGHOUT LUNG TISSUE — "FUNCTIONAL" VESSELS

Stretch readily to accommodate large volume of blood

<u>ARTERIOLES</u>
less smooth muscle, therefore offer less 'Peripheral Resistance' to Blood Flow than Systemic arterioles (supplied by nerves – rôle in regulation uncertain)

<u>PULMONARY BLOOD PRESSURE</u>
consequently lower than in corresponding vessels of Systemic Circulation

DISTRIBUTING & COLLECTING VESSELS
(walls of similar calibre and thickness)

<u>VENULES</u>

<u>MUSCULAR VEINS</u>

gradual transition

ARTERIAL BLOOD

5 litres/minute VENOUS BLOOD

<u>CONDUCTING VESSELS</u>

<u>PULMONARY VEINS</u> (4)

<u>MUSCULAR ARTERIES</u>
much less smooth muscle than Systemic arteries

R.A.   L.A.

L.V.

gradual transition

<u>ELASTIC ARTERIES</u>
walls thin relative to Aortic wall therefore less potential energy stored in Systole; less force to impart to blood in Diastole

R.V. ejects same volume of blood as L.V., i.e. about 5 litres per minute, but imparts less force to it.

SYSTOLIC 19-26 mm Hg
DIASTOLIC 6-12 mm Hg

A great dilatation of Pulmonary vessels occurs in exercise.

DISTRIBUTION of WATER and ELECTROLYTES in BODY FLUIDS

Water makes up about 70% of Adult Human Body —
— i.e about 46 litres in 70 kg man.

Some lies outside the cells.
EXTRACELLULAR — 16 litres

A large part is within the cells.
INTRACELLULAR — 30 litres

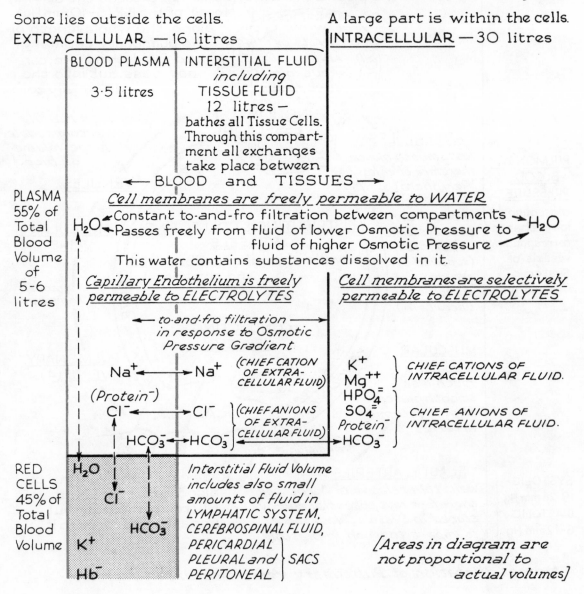

BLOOD PLASMA	INTERSTITIAL FLUID
3·5 litres	*including* TISSUE FLUID 12 litres — bathes all Tissue Cells. Through this compartment all exchanges take place between

PLASMA 55% of Total Blood Volume of 5-6 litres

← BLOOD and TISSUES →

Cell membranes are freely permeable to WATER

H_2O ← Constant to-and-fro filtration between compartments → H_2O
← Passes freely from fluid of lower Osmotic Pressure to fluid of higher Osmotic Pressure →

This water contains substances dissolved in it.

Capillary Endothelium is freely permeable to ELECTROLYTES

Cell membranes are selectively permeable to ELECTROLYTES

← to-and-fro filtration →
in response to Osmotic Pressure Gradient

Na^+ ← → Na^+ (CHIEF CATION OF EXTRA-CELLULAR FLUID)

K^+
Mg^{++} } CHIEF CATIONS OF INTRACELLULAR FLUID.

(Protein⁻)

Cl^- ← → Cl^-
HCO_3^- ← → HCO_3^- } (CHIEF ANIONS OF EXTRA-CELLULAR FLUID)

$HPO_4^=$
$SO_4^=$
Protein⁻
HCO_3^- } CHIEF ANIONS OF INTRACELLULAR FLUID.

RED CELLS 45% of Total Blood Volume

H_2O
Cl^-
HCO_3^-
K^+
Hb^-

Interstitial Fluid Volume includes also small amounts of fluid in LYMPHATIC SYSTEM, CEREBROSPINAL FLUID, PERICARDIAL PLEURAL and PERITONEAL } SACS

[Areas in diagram are not proportional to actual volumes]

WATER BALANCE

In health the *total amount of body water (and salt)* is kept *reasonably constant* in spite of wide fluctuations in daily intake.

A **BALANCE** is struck between

<u>FLUID INTAKE</u> ———— and ———— <u>FLUID OUTPUT</u>

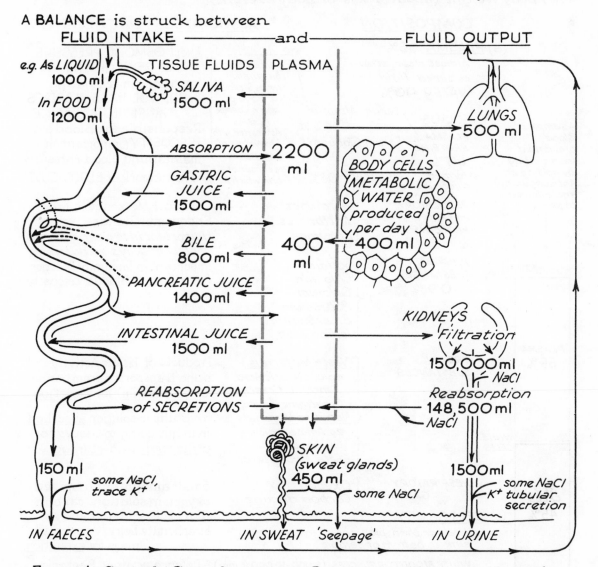

e.g. As LIQUID 1000 ml
In FOOD 1200 ml

TISSUE FLUIDS   PLASMA

SALIVA 1500 ml

ABSORPTION   2200 ml

GASTRIC JUICE 1500 ml

BILE 800 ml

PANCREATIC JUICE 1400 ml

INTESTINAL JUICE 1500 ml

REABSORPTION of SECRETIONS

150 ml   *some NaCl, trace K⁺*

BODY CELLS *METABOLIC WATER produced per day* 400 ml   400 ml

LUNGS 500 ml

KIDNEYS 'Filtration' 150,000 ml   *NaCl* Reabsorption 148,500 ml *NaCl*

SKIN (sweat glands) 450 ml   *some NaCl*

1500 ml   *some NaCl K⁺ tubular secretion*

IN FAECES   *IN SWEAT* 'Seepage'   *IN URINE*

Except in Growth, Convalescence or Pregnancy, when new tissue is being formed, an INCREASE or DECREASE in INTAKE leads to an appropriate INCREASE or DECREASE in OUTPUT to maintain the BALANCE.

BLOOD

Blood is the specialized fluid tissue of the *TRANSPORT SYSTEM*. *[Specific Gravity, 1·055–1·065; pH, 7·3–7·4; Average amount, 5 litres, varies with body weight (about 7·7% of body weight).*

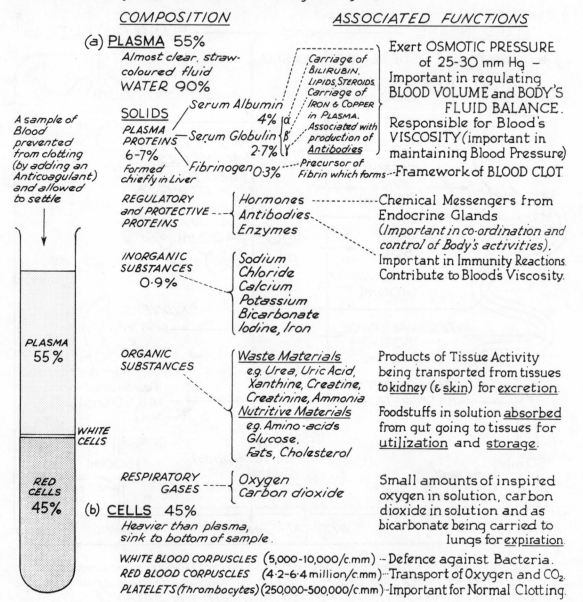

COMPOSITION

(a) PLASMA 55%
Almost clear, straw-coloured fluid
WATER 90%

SOLIDS
PLASMA PROTEINS 6-7%
formed chiefly in Liver
— Serum Albumin 4% α
— Serum Globulin 2·7% β γ
— Fibrinogen 0·3%

Carriage of BILIRUBIN, LIPIDS, STEROIDS.
Carriage of IRON & COPPER in Plasma.
Associated with production of Antibodies
Precursor of Fibrin which forms

REGULATORY and PROTECTIVE PROTEINS
— Hormones
— Antibodies
— Enzymes

INORGANIC SUBSTANCES 0·9%
Sodium
Chloride
Calcium
Potassium
Bicarbonate
Iodine, Iron

ORGANIC SUBSTANCES
Waste Materials
e.g. Urea, Uric Acid, Xanthine, Creatine, Creatinine, Ammonia.
Nutritive Materials
e.g. Amino-acids Glucose, Fats, Cholesterol

RESPIRATORY GASES
Oxygen
Carbon dioxide

(b) CELLS 45%
Heavier than plasma, sink to bottom of sample.

A sample of Blood prevented from clotting (by adding an Anticoagulant) and allowed to settle

PLASMA 55%

WHITE CELLS

RED CELLS 45%

ASSOCIATED FUNCTIONS

Exert OSMOTIC PRESSURE of 25-30 mm Hg –
Important in regulating BLOOD VOLUME and BODY'S FLUID BALANCE.
Responsible for Blood's VISCOSITY (important in maintaining Blood Pressure)
Framework of BLOOD CLOT.

Chemical Messengers from Endocrine Glands
(Important in co-ordination and control of Body's activities).
Important in Immunity Reactions.
Contribute to Blood's Viscosity.

Products of Tissue Activity being transported from tissues to kidney (& skin) for excretion.

Foodstuffs in solution absorbed from gut going to tissues for utilization and storage.

Small amounts of inspired oxygen in solution, carbon dioxide in solution and as bicarbonate being carried to lungs for expiration.

WHITE BLOOD CORPUSCLES (5,000-10,000/c.mm) – Defence against Bacteria.
RED BLOOD CORPUSCLES (4·2-6·4 million/c.mm) – Transport of Oxygen and CO_2.
PLATELETS (Thrombocytes) (250,000-500,000/c.mm) – Important for Normal Clotting.

BLOOD COAGULATION

Blood does not normally clot within healthy blood vessels.
When blood is shed it undergoes a series of changes which end in <u>CLOTTING</u>.

The <u>CLASSICAL THEORY of MORAWITZ</u> recognized the interaction of <u>4 FACTORS</u>

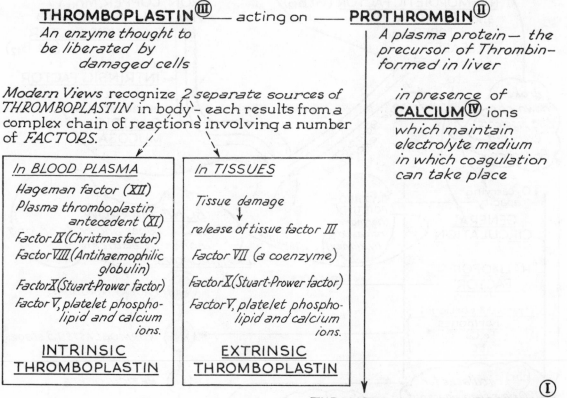

THROMBOPLASTIN (III) —— acting on —— **PROTHROMBIN** (II)

An enzyme thought to be liberated by damaged cells

A plasma protein — the precursor of Thrombin - formed in liver

Modern Views recognize *2 separate sources of THROMBOPLASTIN* in body - each results from a complex chain of reactions involving a number of *FACTORS*.

in presence of **CALCIUM** (IV) *ions which maintain electrolyte medium in which coagulation can take place*

In BLOOD PLASMA	In TISSUES
Hageman factor (XII)	
Plasma thromboplastin antecedent (XI)	*Tissue damage*
	↓
Factor IX (Christmas factor)	*release of tissue factor III*
Factor VIII (Antihaemophilic globulin)	*Factor VII (a coenzyme)*
Factor X (Stuart-Prower factor)	*Factor X (Stuart-Prower factor)*
Factor V, platelet phospho-lipid and calcium ions.	*Factor V, platelet phospho-lipid and calcium ions.*
INTRINSIC THROMBOPLASTIN	**EXTRINSIC THROMBOPLASTIN**

THROMBIN which acts on **FIBRINOGEN** (I)
(*A SOLUBLE PROTEIN FORMED IN LIVER*)

↓

Insoluble **FIBRIN** deposited as fine threads to form <u>FRAMEWORK of BLOOD CLOT</u>

↓

Platelets cling to inter-sections of fibrin threads

Both mechanisms appear to be necessary for <u>PHYSIOLOGICAL HAEMOSTASIS</u> (*i.e. cessation of bleeding when wound occurs*) —— *also requires* ——→ <u>ADHESION and SHRINKAGE of CLOT</u>

<u>EXPRESSION of SERUM</u>

5

FACTORS REQUIRED for NORMAL HAEMOPOIESIS

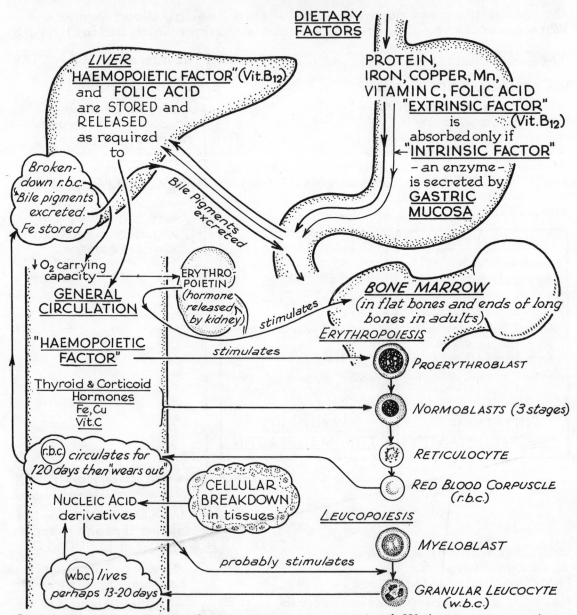

DIETARY FACTORS

LIVER
"HAEMOPOIETIC FACTOR" (Vit. B$_{12}$)
and FOLIC ACID
are STORED and
RELEASED
as required
to

PROTEIN,
IRON, COPPER, Mn,
VITAMIN C, FOLIC ACID
"EXTRINSIC FACTOR"
is   (Vit. B$_{12}$)
absorbed only if
"INTRINSIC FACTOR"
– an enzyme –
is secreted by
GASTRIC MUCOSA

Broken-down r.b.c. Bile pigments excreted. Fe stored

Bile Pigments excreted

↓O$_2$ carrying capacity

ERYTHRO-POIETIN *(hormone released by kidney)*

GENERAL CIRCULATION

stimulates

BONE MARROW
(in flat bones and ends of long bones in adults)

ERYTHROPOIESIS

"HAEMOPOIETIC FACTOR"

stimulates

PROERYTHROBLAST

Thyroid & Corticoid Hormones
Fe, Cu
Vit. C

NORMOBLASTS *(3 stages)*

r.b.c. circulates for 120 days then "wears out"

RETICULOCYTE

NUCLEIC ACID derivatives

CELLULAR BREAKDOWN in tissues

RED BLOOD CORPUSCLE *(r.b.c.)*

LEUCOPOIESIS

MYELOBLAST

w.b.c. lives perhaps 13-20 days

probably stimulates

GRANULAR LEUCOCYTE *(w.b.c.)*

In health the number of r.b.c. and the amount of Hb in them remain fairly constant. Destruction of old red cells is balanced by formation of new.

HAEMOPOIESIS

In the adult the formed elements of the Blood Stream develop from primitive RETICULAR CELLS chiefly in RED BONE MARROW.

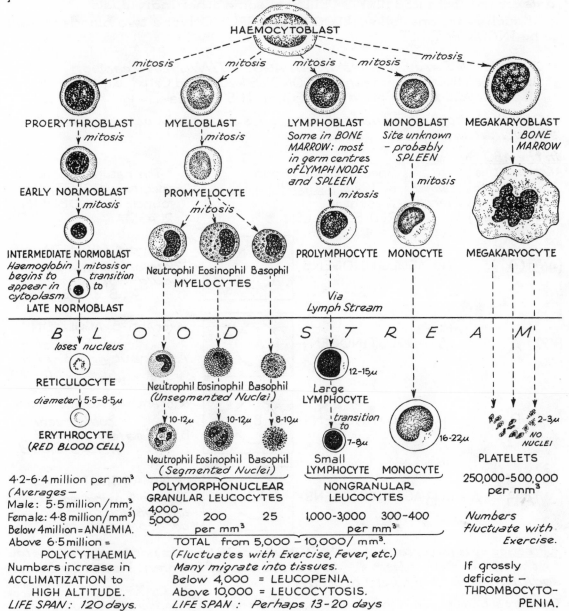

HAEMOCYTOBLAST

mitosis   *mitosis*   *mitosis*   *mitosis*   *mitosis*

PROERYTHROBLAST   MYELOBLAST   LYMPHOBLAST   MONOBLAST   MEGAKARYOBLAST

mitosis   *mitosis*   *Some in BONE MARROW: most in germ centres of LYMPH NODES and SPLEEN*   *Site unknown – probably SPLEEN*   BONE MARROW

EARLY NORMOBLAST   PROMYELOCYTE   *mitosis*   *mitosis*

mitosis   *mitosis*

INTERMEDIATE NORMOBLAST   Neutrophil Eosinophil Basophil   PROLYMPHOCYTE   MONOCYTE   MEGAKARYOCYTE

MYELOCYTES

Haemoglobin begins to appear in cytoplasm   *mitosis or transition to*

LATE NORMOBLAST   *Via Lymph Stream*

B L O O D   S T R E A M

loses nucleus

RETICULOCYTE   Neutrophil Eosinophil Basophil   12–15μ

diameter 5·5–8·5μ   (Unsegmented Nuclei)   Large LYMPHOCYTE

  10–12μ   10–12μ   8–10μ   *transition to*

ERYTHROCYTE (RED BLOOD CELL)   Neutrophil Eosinophil Basophil   7–8μ   16–22μ   2–3μ NO NUCLEI

  (Segmented Nuclei)   Small LYMPHOCYTE MONOCYTE   PLATELETS

4·2–6·4 million per mm³ (Averages — Male: 5·5 million/mm³, Female: 4·8 million/mm³) Below 4 million = ANAEMIA. Above 6·5 million = POLYCYTHAEMIA. Numbers increase in ACCLIMATIZATION to HIGH ALTITUDE. LIFE SPAN: 120 days.

POLYMORPHONUCLEAR GRANULAR LEUCOCYTES

4,000–5,000	200	25
per mm³		

NONGRANULAR LEUCOCYTES

1,000–3,000	300–400
per mm³	

TOTAL from 5,000 – 10,000/ mm³. (Fluctuates with Exercise, Fever, etc.) Many migrate into tissues. Below 4,000 = LEUCOPENIA. Above 10,000 = LEUCOCYTOSIS. LIFE SPAN: Perhaps 13–20 days

250,000–500,000 per mm³

Numbers fluctuate with Exercise.

If grossly deficient — THROMBOCYTO-PENIA.

BLOOD GROUPS

There are present in the PLASMA of some individuals substances which can cause the AGGLUTINATION *(clumping together)* and subsequent HAEMOLYSIS *(breakdown)* of the RED BLOOD CELLS of some other individuals.

If such reactions follow BLOOD TRANSFUSION the two bloods are said to be INCOMPATIBLE.

Two FACTORS are involved in an AGGLUTINATION reaction :-
An AGGLUTINOGEN present in DONOR'S Red Blood Cell } e.g. A } or B}
A specific AGGLUTININ present in RECIPIENT'S Plasma } α } or β}

Obviously no such combination occurs naturally otherwise auto-agglutination would result.

<u>In the ABO System</u>

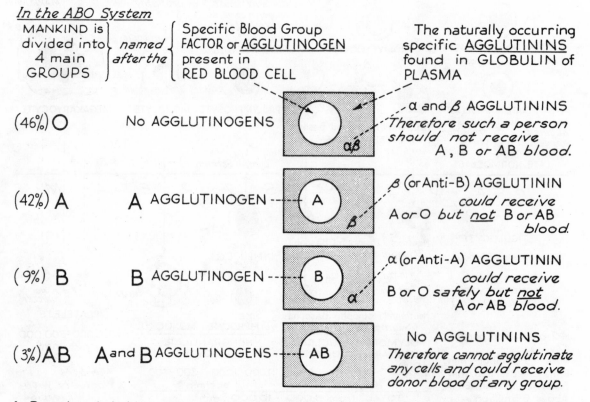

MANKIND is divided into 4 main GROUPS }	*named after the* }	Specific Blood Group FACTOR or <u>AGGLUTINOGEN</u> present in RED BLOOD CELL	The naturally occurring specific <u>AGGLUTININS</u> found in GLOBULIN of PLASMA
(46%) O		No AGGLUTINOGENS	α and β AGGLUTININS *Therefore such a person should not receive A , B or AB blood.*
(42%) A		A AGGLUTINOGEN ---- A	β (or Anti-B) AGGLUTININ *could receive A or O but not B or AB blood.*
(9%) B		B AGGLUTINOGEN ---- B	α (or Anti-A) AGGLUTININ *could receive B or O safely but not A or AB blood.*
(3%) AB		A and B AGGLUTINOGENS ---- AB	No AGGLUTININS *Therefore cannot agglutinate any cells and could receive donor blood of any group.*

<u>In Practice</u> *it is important that the DONOR's cells should not be agglutinated by the RECIPIENT's plasma. Agglutination of Recipient's cells by Donor agglutinins is less likely to occur.*

[work.

There are sub-groups within these main groups. These have to be considered in blood transfusion

BLOOD GROUPS

To determine the Blood Group to which an individual belongs *TWO TEST SERA* only are required and the *RED BLOOD CELLS* to be grouped.

DROP of GROUP A SERUM & GROUP B SERUM
(on glass slide)   [N.B. These droplets are of PLASMA – *NOT* Red Blood Cells. *i.e.* only AGGLUTININS are present.]

To each test serum a saline suspension of RED BLOOD CELLS is added. *i.e. only AGGLUTINOGENS are added.*

GROUP A SERUM | GROUP B SERUM

GROUP O Blood Cells give NO agglutination since NO AGGLUTINOGENS are present in these cells to be clumped by test sera AGGLUTININS.

GROUP B Blood Cells (B AGGLUTINOGEN present) give agglutination with GROUP A serum–since the specific Anti-B (β) AGGLUTININ is present in the first test serum.

GROUP A Blood Cells give agglutination with GROUP B serum — since the specific Anti-A (α) AGGLUTININ is present in this serum.

GROUP AB Blood Cells give agglutination with both test sera.

As well as determining Blood Group in this way, the Blood of Donor is always matched directly with Blood of Patient to avoid sub-group incompatibility.

AGGLUTINATION is usually visible under the microscope within a few minutes. The clumped cells look like grains of cayenne pepper in a clear liquid. If no agglutination occurs the fluid remains uniformly pink.

If the wrong blood is given to a patient, clumps of red blood cells may block small blood vessels in vital organs, e.g. lung or brain. The subsequent haemolysis (breakdown) of agglutinated cells may lead to Haemoglobin in the Urine and eventually to Kidney Failure and Death.

RHESUS FACTOR

Over 50 different BLOOD GROUP FACTORS have been demonstrated. Not all are important in blood transfusion work.

The <u>RHESUS FACTOR</u> is important

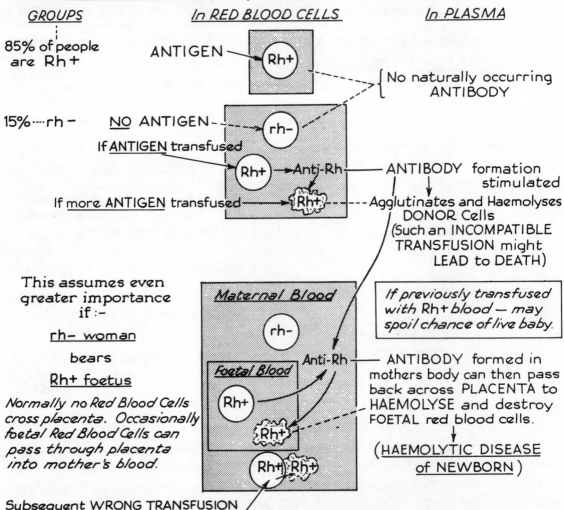

GROUPS   _In RED BLOOD CELLS_   _In PLASMA_

85% of people are Rh+

ANTIGEN → Rh+

No naturally occurring ANTIBODY

15%····rh−   <u>NO</u> ANTIGEN → rh−

If <u>ANTIGEN</u> transfused

Rh+ → Anti-Rh → ANTIBODY formation stimulated

If <u>more</u> ANTIGEN transfused → Rh+ ····· Agglutinates and Haemolyses DONOR Cells (Such an INCOMPATIBLE TRANSFUSION might LEAD to DEATH)

This assumes even greater importance if :−

<u>rh− woman</u>

bears

<u>Rh+ foetus</u>

Normally no Red Blood Cells cross placenta. Occasionally foetal Red Blood Cells can pass through placenta into mother's blood.

Maternal Blood

rh−

Anti-Rh

Foetal Blood

Rh+

Rh+

Rh+ Rh+

If previously transfused with Rh+ blood — may spoil chance of live baby.

ANTIBODY formed in mothers body can then pass back across PLACENTA to HAEMOLYSE and destroy FOETAL red blood cells.

(<u>HAEMOLYTIC DISEASE of NEWBORN</u>)

Subsequent WRONG TRANSFUSION of Rh+ blood to mother could lead to dangerous AGGLUTINATION and HAEMOLYSIS of DONOR CELLS within mother's own body.

There are many sub-groups in this system.

INHERITANCE of RHESUS FACTOR

This BLOOD GROUP FACTOR is inherited on Mendelian principles.

HOMOZYGOUS
Rh Positive
<u>FATHER</u>

LACK of the FACTOR is a
Mendelian recessive
rh negative
<u>MOTHER</u>

All Spermatozoa carry the FACTOR: *All Ova lack the FACTOR*

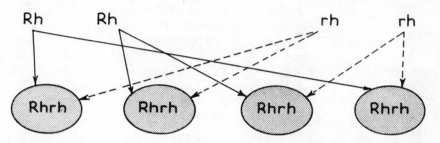

All children will be Rhrh, i.e. all carry the FACTOR and are
Rh Positive. *i.e.* During pregnancy the rh negative woman will have a
Rh Positive foetus.

HETEROZYGOUS
Rh Positive
<u>FATHER</u>

rh negative
<u>MOTHER</u>

Half the Spermatozoa carry the FACTOR: *All Ova lack the FACTOR*

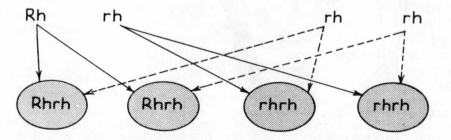

Possibility that some children will be Rhrh Positive, like the
father; others rhrh negative like the mother.

LYMPHATIC SYSTEM

ALL CELLS --------
are bathed by TISSUE FLUID.---------
This diffuses from CAPILLARIES. -------
Some returns to CAPILLARIES.
Some drains into blind-ending,
thin-walled LYMPHATICS.------------
It is then known as LYMPH
(similar to plasma but less protein).

A network of LYMPHATIC VESSELS drains tissue spaces throughout the body (except in Central Nervous System). They unite to form LARGER and LARGER vessels ⟶ RIGHT LYMPHATIC DUCT and LEFT THORACIC DUCT ⟶ SUBCLAVIAN VEINS (i.e. lymph is returned to the blood stream here)

In the course of LARGER vessels, LYMPH is filtered through __LYMPH NODES__

AFFERENT LYMPHATICS– pour their LYMPH into RETICULAR FRAMEWORK of loose Sinus tissue.

MACROPHAGE cells ingest foreign material (e.g. carbon in lungs) or harmful bacteria.

LYMPH NODULES produce LYMPHOCYTES and PLASMA CELLS

important in antibody formation and immuno-logical reactions

Capsule of Fibrous Tissue

EFFERENT LYMPHATIC — receives lymph after its slow passage through node.

MOVEMENT of LYMPH towards HEART depends partly on compression of lymphatic vessels by muscles of limbs and partly on 'suction' created by movements of respiration. Valves within the vessels prevent backflow. The lymphatic tissue of the body forms an important part of the Body's Defence against invading agents such as PROTOZOA, BACTERIA, VIRUSES, or their poisonous TOXINS. These act as ANTIGENS stimulating ANTIBODY FORMATION — which can subsequently destroy or neutralize the antigen.

SPLEEN

The Spleen is a vascular organ, weighing about 200 grams. It is situated in the left side of the abdomen behind the stomach and above the kidney.

FIBROUS TISSUE CAPSULE
with extensions into substance
of organ
TRABECULAE

SPLENIC ARTERY
and VEIN
and branches
TRABECULAR
ARTERY and VEIN

Near their termination
ARTERIOLES
are surrounded
by collections of
LYMPHATIC TISSUE
("WHITE PULP")–contains
the plasma cells which
manufacture many antibodies and produces
LYMPHOCYTES → added to blood in its
passage through spleen.

"RED PULP"– Framework of Reticular Tissue — acts
as RESERVOIR for Blood.
(Some smooth muscle in Capsule and Trabeculae may aid in
expelling blood from reservoir to general circulation.)

PHAGOCYTIC CELLS of | Destroy worn out Red Blood
ELLIPSOID and other | Corpuscles, platelets and other
Reticulo-Endothelial Cells | (foreign) particles, etc.

It is difficult to trace further pathways of blood through spleen and there is much disagreement about the exact route taken. Probably Arterioles link up with VENOUS SINUSES (in the way shown).

Blood appears to percolate through the narrow spaces in the wall of the SINUS into the Red Pulp— and can get back into Sinus in same way.

During foetal life the spleen forms red and white blood cells.
It is not apparently essential to life in the adult.

CEREBROSPINAL FLUID

COMPOSITION:- Cerebrospinal Fluid (C.S.F.) is similar to blood plasma but does not clot. It is a clear watery alkaline fluid. It contains salts, glucose, some urea and creatinine, very little protein and very few lymphocytes.
VOLUME:- 120-150 ml. in man.   _SPECIFIC GRAVITY_:- 1·005-1·008.

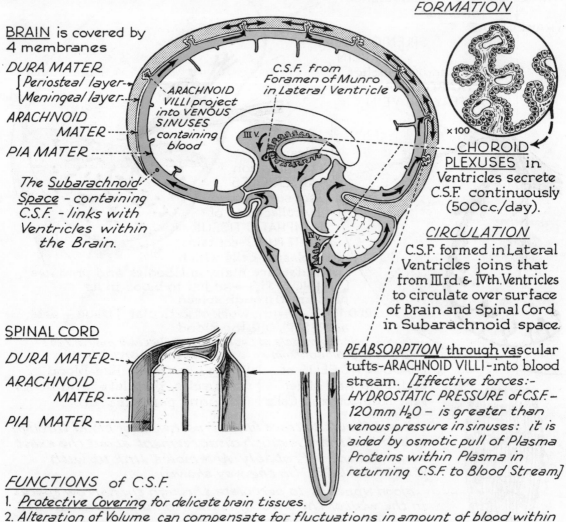

FORMATION

BRAIN is covered by 4 membranes

DURA MATER
[Periosteal layer →
[Meningeal layer →

ARACHNOID MATER ----

PIA MATER --------

ARACHNOID VILLI project into VENOUS SINUSES containing blood

C.S.F. from Foramen of Munro in Lateral Ventricle

III V.

IV. V.

×100

CHOROID PLEXUSES in Ventricles secrete C.S.F. continuously (500 c.c./day).

The _Subarachnoid Space_ - containing C.S.F. - links with Ventricles within the Brain.

CIRCULATION
C.S.F. formed in Lateral Ventricles joins that from IIIrd & IVth. Ventricles to circulate over surface of Brain and Spinal Cord in Subarachnoid space.

SPINAL CORD

DURA MATER ----

ARACHNOID MATER ----

PIA MATER ----

REABSORPTION through vascular tufts-ARACHNOID VILLI-into blood stream. [_Effective forces_:- HYDROSTATIC PRESSURE of C.S.F.- 120 mm H_2O - is greater than venous pressure in sinuses: it is aided by osmotic pull of Plasma Proteins within Plasma in returning C.S.F. to Blood Stream]

FUNCTIONS of C.S.F.
1. _Protective Covering_ for delicate brain tissues.
2. _Alteration of Volume_ can compensate for fluctuations in amount of blood within skull and thus keep total volume of cranial contents constant.
3. _Exchange of Metabolic substances_ between Nerve Cells and C.S.F.
  (i.e. it receives some waste products.)

CHAPTER 5

RESPIRATORY SYSTEM

RESPIRATORY SYSTEM

All living cells require to get OXYGEN from the fluid around them and to get rid of CARBON DIOXIDE to it.

INTERNAL RESPIRATION is the exchange of these gases between tissue cells and their fluid environment.

EXTERNAL RESPIRATION is the exchange of these gases *(oxygen and carbon dioxide)* between the body and the external environment.

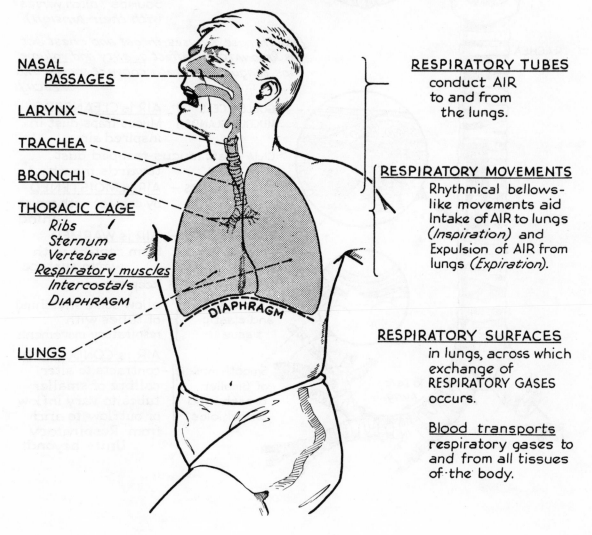

NASAL
   PASSAGES

LARYNX

TRACHEA

BRONCHI

THORACIC CAGE
  Ribs
  Sternum
  Vertebrae
  Respiratory muscles
    Intercostals
    DIAPHRAGM

LUNGS

DIAPHRAGM

RESPIRATORY TUBES
conduct AIR
to and from
the lungs.

RESPIRATORY MOVEMENTS
Rhythmical bellows-
like movements aid
Intake of AIR to lungs
(*Inspiration*) and
Expulsion of AIR from
lungs (*Expiration*).

RESPIRATORY SURFACES
in lungs, across which
exchange of
RESPIRATORY GASES
occurs.

Blood transports
respiratory gases to
and from all tissues
of the body.

AIR CONDUCTING PASSAGES

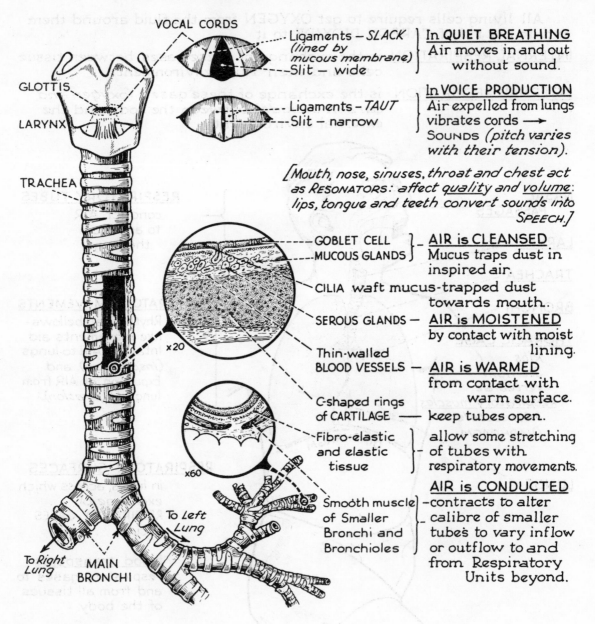

VOCAL CORDS

Ligaments — *SLACK*
(lined by
mucous membrane)
Slit — wide

GLOTTIS

LARYNX

Ligaments — *TAUT*
Slit — narrow

In QUIET BREATHING
Air moves in and out
without sound.

In VOICE PRODUCTION
Air expelled from lungs
vibrates cords ⟶
SOUNDS (*pitch varies
with their tension*).

*[Mouth, nose, sinuses, throat and chest act
as RESONATORS: affect quality and volume:
lips, tongue and teeth convert sounds into
SPEECH.]*

TRACHEA

GOBLET CELL
MUCOUS GLANDS

AIR is CLEANSED
Mucus traps dust in
inspired air.

CILIA waft mucus-trapped dust
towards mouth.

SEROUS GLANDS —

AIR is MOISTENED
by contact with moist
lining.

Thin-walled
BLOOD VESSELS —

AIR is WARMED
from contact with
warm surface.

x20

C-shaped rings
of CARTILAGE —— keep tubes open.

Fibro-elastic
and elastic
tissue

allow some stretching
of tubes with
respiratory movements.

x60

AIR is CONDUCTED

Smooth muscle
of Smaller
Bronchi and
Bronchioles

-contracts to alter
calibre of smaller
tubes to vary inflow
or outflow to and
from Respiratory
Units beyond.

To Left
Lung

To Right
Lung   MAIN
BRONCHI

LUNGS: RESPIRATORY SURFACES

The Trachea and the Bronchial 'Tree' conduct Air down to the
RESPIRATORY SURFACES.

*There is no exchange of gases in
these tubes.*

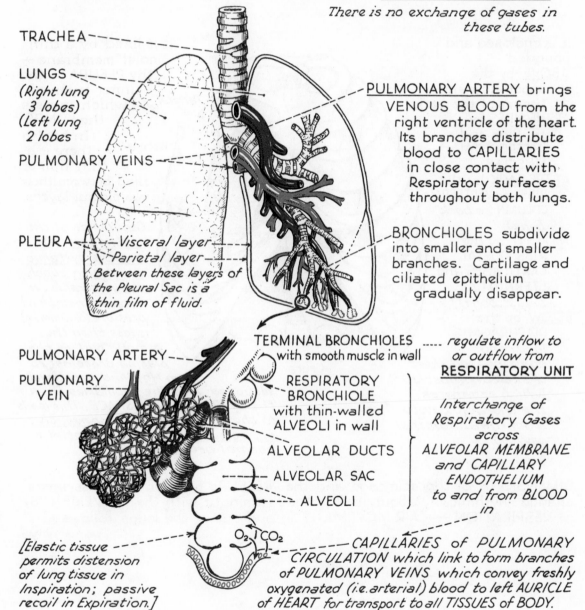

TRACHEA

LUNGS
(Right lung
3 lobes)
(Left lung
2 lobes

PULMONARY VEINS

PLEURA — *Visceral layer*
Parietal layer
*Between these layers of
the Pleural Sac is a
thin film of fluid.*

PULMONARY ARTERY brings
VENOUS BLOOD from the
right ventricle of the heart.
Its branches distribute
blood to CAPILLARIES
in close contact with
Respiratory surfaces
throughout both lungs.

BRONCHIOLES subdivide
into smaller and smaller
branches. Cartilage and
ciliated epithelium
gradually disappear.

PULMONARY ARTERY

PULMONARY
VEIN

TERMINAL BRONCHIOLES ..... *regulate inflow to
with smooth muscle in wall* *or outflow from*
RESPIRATORY UNIT

RESPIRATORY
BRONCHIOLE
with thin-walled
ALVEOLI in wall

ALVEOLAR DUCTS

ALVEOLAR SAC

ALVEOLI

*Interchange of
Respiratory Gases
across
ALVEOLAR MEMBRANE
and CAPILLARY
ENDOTHELIUM
to and from BLOOD
in*

O_2 | CO_2

*[Elastic tissue
permits distension
of lung tissue in
Inspiration; passive
recoil in Expiration.]*

*CAPILLARIES of PULMONARY
CIRCULATION which link to form branches
of PULMONARY VEINS which convey freshly
oxygenated (i.e. arterial) blood to left AURICLE
of HEART for transport to all TISSUES of BODY.*

THORAX

The Thorax (or chest) is the closed cavity which contains the LUNGS, HEART and Great Vessels.

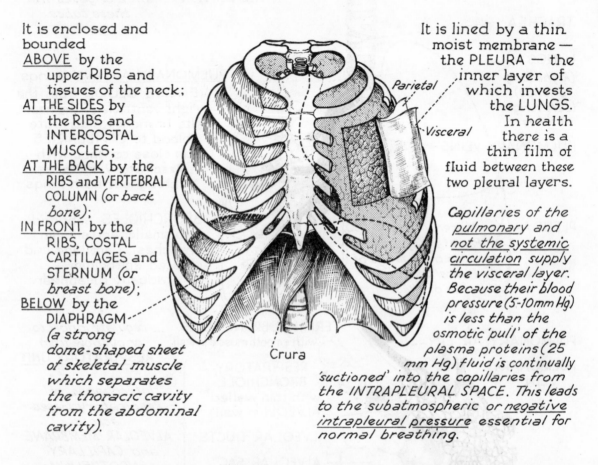

It is enclosed and bounded
ABOVE by the upper RIBS and tissues of the neck;
AT THE SIDES by the RIBS and INTERCOSTAL MUSCLES;
AT THE BACK by the RIBS and VERTEBRAL COLUMN (or *back bone*);
IN FRONT by the RIBS, COSTAL CARTILAGES and STERNUM (or *breast bone*);
BELOW by the DIAPHRAGM (*a strong dome-shaped sheet of skeletal muscle which separates the thoracic cavity from the abdominal cavity*).

Parietal

Visceral

Crura

It is lined by a thin moist membrane — the PLEURA — the inner layer of which invests the LUNGS. In health there is a thin film of fluid between these two pleural layers.

Capillaries of the pulmonary and not the systemic circulation supply the visceral layer. Because their blood pressure (5-10mm Hg) is less than the osmotic 'pull' of the plasma proteins (25 mm Hg) fluid is continually 'suctioned' into the capillaries from the INTRAPLEURAL SPACE. This leads to the subatmospheric or negative intrapleural pressure essential for normal breathing.

DIMENSIONS of thoracic cage and the PRESSURE between pleural surfaces change rhythmically about 18-20 times a minute with the MOVEMENTS of RESPIRATION —— AIR MOVEMENT in and out of the lungs follows passively.

MECHANISM of BREATHING

The rhythmical changes in the CAPACITY of the Thorax are brought about by MUSCULAR ACTION. The changes in LUNG VOLUME with INTAKE or EXPULSION of air follow passively.

In NORMAL QUIET BREATHING

INSPIRATION

EXTERNAL INTERCOSTAL MUSCLES actively contract –
– ribs and sternum move upwards and outwards
– width of chest increases from side to side and from front to back.

DIAPHRAGM contracts –
– descends
– depth of chest increases.

CAPACITY of THORAX is INCREASED
↓
PRESSURE between PLEURAL SURFACES (already negative) is REDUCED from –2 to –6 mm Hg (i.e. an increased "suction pull" is exerted on LUNG TISSUE)
↓
ELASTIC TISSUE of LUNGS is STRETCHED
↓
LUNGS EXPAND to fill THORACIC CAVITY
↓
AIR PRESSURE within ALVEOLI is now less than atmospheric pressure
↓
AIR is sucked into ALVEOLI from ATMOSPHERE

EXPIRATION

EXTERNAL INTERCOSTAL MUSCLES relax –
– ribs and sternum move downwards and inwards
– width of chest diminishes.

DIAPHRAGM relaxes –
– ascends
– depth of chest diminishes.

CAPACITY of THORAX is DECREASED
↓
PRESSURE between PLEURAL SURFACES is INCREASED from –6 to –2 mm Hg (i.e. less pull is exerted on LUNG TISSUE)
↓
ELASTIC TISSUE of LUNGS RECOILS
↓
AIR PRESSURE within ALVEOLI is now greater than atmospheric pressure
↓
AIR is forced out of ALVEOLI to ATMOSPHERE

In FORCED BREATHING

Muscles of Nostrils and round Glottis may contract to aid entrance of air to lungs.
Extensors of Vertebral Column may aid inspiration.
Muscles of Neck contract –
– move 1ST rib upwards
(and sternum upwards and forwards)

Internal Intercostals may contract –
– move ribs downwards more actively.
Abdominal muscles contract –
– actively aid ascent of diaphragm.

ARTIFICIAL RESPIRATION

Many methods aim at alternately DECREASING and INCREASING the CAPACITY of the THORAX 15-18 times per minute in imitation of the normal expiratory and inspiratory movements.

Two well-known methods illustrate the mechanism:-
[In each case mouth and throat must be clear of obstruction and tongue drawn forwards]

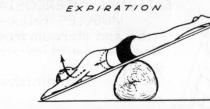
EXPIRATION

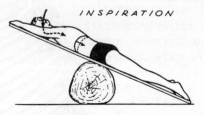
INSPIRATION

1. **EVE'S ROCKING STRETCHER** — uses the abdominal organs like a piston against the diaphragm.

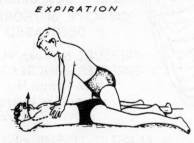
EXPIRATION

INSPIRATION

2. **SCHAFER** —— pressure applied to compress lower part of rib cage and so decrease capacity of thorax — *AIR MOVES OUT*.
On release of pressure thorax returns to previous position and capacity — *AIR SUCKED IN*.

In the MOUTH-to-MOUTH method the applicator pinches shut the patient's nostrils, seals his lips round the patient's mouth and blows in air[C]. When he removes his mouth the patient breathes out passively [D]. *[The residual oxygen in the applicator's own expired air is thus made available to patient]*

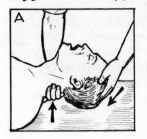
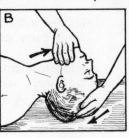
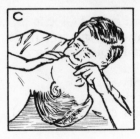
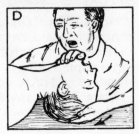

122

CAPACITY of LUNGS

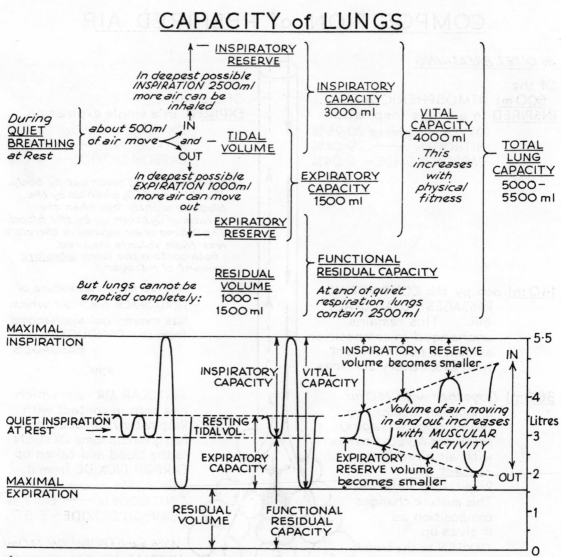

During QUIET BREATHING at Rest — about 500ml of air move IN and OUT

In deepest possible INSPIRATION 2500ml more air can be inhaled — INSPIRATORY RESERVE

In deepest possible EXPIRATION 1000ml more air can move out — EXPIRATORY RESERVE

TIDAL VOLUME

INSPIRATORY CAPACITY 3000 ml

EXPIRATORY CAPACITY 1500 ml

VITAL CAPACITY 4000 ml — *This increases with physical fitness*

TOTAL LUNG CAPACITY 5000 – 5500 ml

But lungs cannot be emptied completely: RESIDUAL VOLUME 1000 – 1500 ml

FUNCTIONAL RESIDUAL CAPACITY — *At end of quiet respiration lungs contain 2500ml*

MAXIMAL INSPIRATION

QUIET INSPIRATION AT REST

RESTING TIDAL VOL.

INSPIRATORY CAPACITY

VITAL CAPACITY

EXPIRATORY CAPACITY

MAXIMAL EXPIRATION

RESIDUAL VOLUME

FUNCTIONAL RESIDUAL CAPACITY

INSPIRATORY RESERVE volume becomes smaller

Volume of air moving in and out increases with MUSCULAR ACTIVITY

EXPIRATORY RESERVE volume becomes smaller

IN — OUT

5.5, 4, 3, 2, 1, 0 Litres

(After PAPPENHEIMER, J.R., et al (1950) Fed. Proc, 9,602)

At rest a normal male adult breathes in and out about 16 times per minute. The amount of air breathed in per minute is therefore 500ml × 16, i.e. 8000ml or 8 litres – This is the RESPIRATORY MINUTE VOLUME or PULMONARY VENTILATION. In exercise it may go up to as much as 200 litres. These values are about 25% lower in women.

In deep breathing the volume of ATMOSPHERIC AIR INSPIRED with each inspiration and the amount which reaches the ALVEOLI increase.

COMPOSITION of RESPIRED AIR

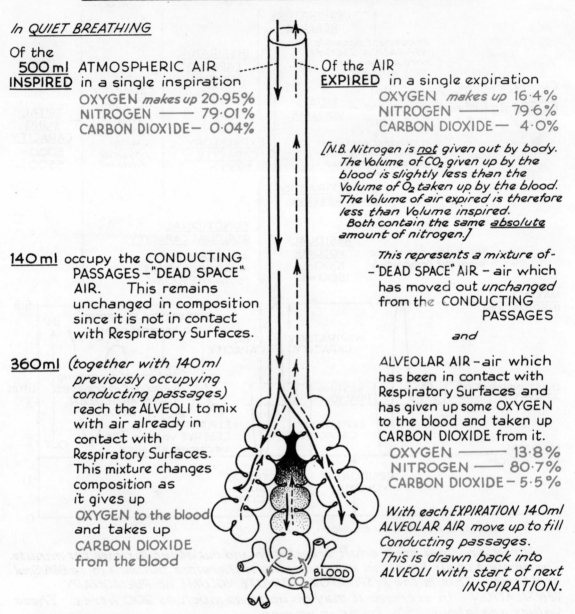

In *QUIET BREATHING*

Of the
500 ml ATMOSPHERIC AIR ----- Of the AIR
INSPIRED in a single inspiration
    OXYGEN *makes up* 20·95%
    NITROGEN ——— 79·01%
    CARBON DIOXIDE— 0·04%

Of the AIR
EXPIRED in a single expiration
    OXYGEN *makes up* 16·4%
    NITROGEN ——— 79·6%
    CARBON DIOXIDE— 4·0%

*[N.B. Nitrogen is <u>not</u> given out by body.
The Volume of CO_2 given up by the
blood is slightly less than the
Volume of O_2 taken up by the blood.
The Volume of air expired is therefore
less than Volume inspired.
  Both contain the same <u>absolute</u>
amount of nitrogen.]*

140 ml occupy the CONDUCTING
PASSAGES—"DEAD SPACE"
AIR.   This remains
unchanged in composition
since it is not in contact
with Respiratory Surfaces.

This represents a mixture of—
—"DEAD SPACE" AIR – air which
has moved out *unchanged*
from the CONDUCTING
          PASSAGES

and

360 ml *(together with 140 ml
previously occupying
conducting passages)*
reach the ALVEOLI to mix
with air already in
contact with
Respiratory Surfaces.
This mixture changes
composition as
it gives up
OXYGEN to the blood
and takes up
CARBON DIOXIDE
from the blood

ALVEOLAR AIR – air which
has been in contact with
Respiratory Surfaces and
has given up some OXYGEN
to the blood and taken up
CARBON DIOXIDE from it.
    OXYGEN ——— 13·8%
    NITROGEN ——— 80·7%
    CARBON DIOXIDE— 5·5%

*With each EXPIRATION 140 ml
ALVEOLAR AIR move up to fill
Conducting passages.
This is drawn back into
ALVEOLI with start of next
INSPIRATION.*

O_2
CO_2 BLOOD

In <u>*DEEP BREATHING*</u> *the composition of Air breathed out with each
expiration comes closer to that of Alveolar Air.*

MOVEMENT of RESPIRATORY GASES

A gas moves from an area where it is present at *higher concentration* to an area where it is present at *lower concentration*. The movement of gas molecules continues till the pressure exerted by them is the same throughout both areas. DRY Atmospheric Air *(at sea level)* has a pressure of 760 mm Hg

EXPIRED AIR
is laden with water vapour
exerting a pressure = 47 mm Hg
O_2, 16·4% of (760-47) = 116·2 mm Hg
CO_2, 4% of 713 = 28·5 mm Hg

INSPIRED AIR
O_2, 20·95% of 760 = 159 mm Hg
CO_2, 0·04% of 760 = 0·3 mm Hg

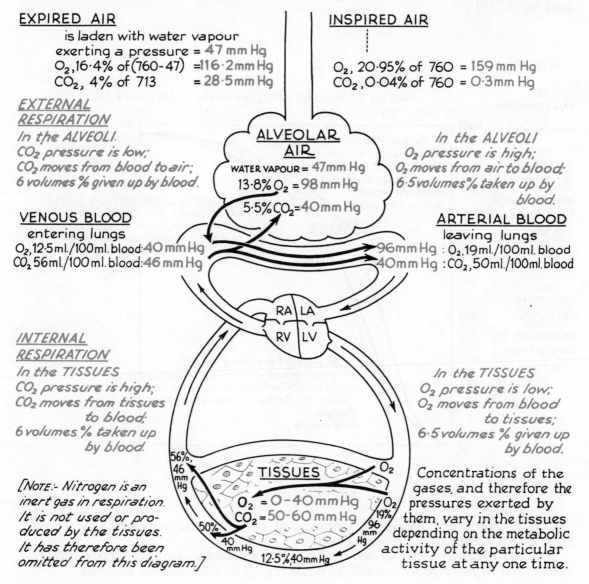

EXTERNAL RESPIRATION

In the ALVEOLI.
CO_2 pressure is low;
CO_2 moves from blood to air;
6 volumes % given up by blood.

ALVEOLAR AIR
WATER VAPOUR = 47mm Hg
13·8% O_2 = 98 mm Hg
5·5% CO_2 = 40mm Hg

In the ALVEOLI
O_2 pressure is high;
O_2 moves from air to blood;
6·5 volumes % taken up by blood.

VENOUS BLOOD
entering lungs
O_2, 12·5 ml./100ml. blood : 40 mm Hg
CO_2 56 ml./100 ml. blood : 46 mm Hg

ARTERIAL BLOOD
leaving lungs
96 mm Hg : O_2, 19 ml./100 ml. blood
40 mm Hg : CO_2, 50 ml./100 ml. blood

RA | LA
RV | LV

INTERNAL RESPIRATION

In the TISSUES
CO_2 pressure is high;
CO_2 moves from tissues to blood;
6 volumes % taken up by blood.

In the TISSUES
O_2 pressure is low;
O_2 moves from blood to tissues;
6·5 volumes % given up by blood.

TISSUES
O_2 = 0-40 mm Hg
CO_2 = 50-60 mm Hg

56%, 46 mm Hg
O_2
O_2 19% 96 mm Hg
50%
40 mm Hg
12·5%, 40 mm Hg

[NOTE:- Nitrogen is an inert gas in respiration. It is not used or produced by the tissues. It has therefore been omitted from this diagram.]

Concentrations of the gases, and therefore the pressures exerted by them, vary in the tissues depending on the metabolic activity of the particular tissue at any one time.

DISSOCIATION of OXYGEN from HAEMOGLOBIN

The amount of O_2 taken up by HAEMOGLOBIN in the LUNGS or given up by OXYHAEMOGLOBIN in the TISSUES depends on the PARTIAL PRESSURE of the O_2 in the immediate environment.

It is also influenced by the PARTIAL PRESSURE of CO_2, by TEMPERATURE and by ACIDITY.

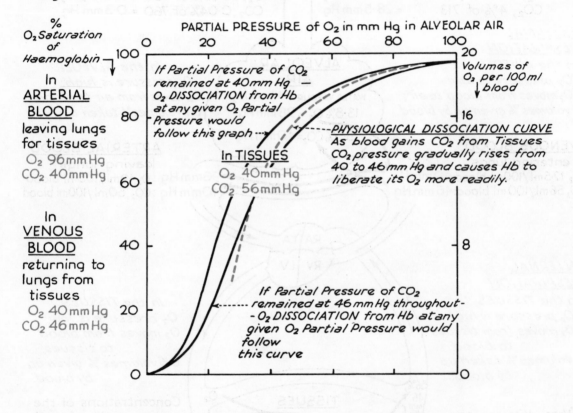

e.g.

%
O_2 Saturation
of
Haemoglobin →

In
ARTERIAL
BLOOD
leaving lungs
for tissues
O_2 96 mm Hg
CO_2 40 mm Hg

In
VENOUS
BLOOD
returning to
lungs from
tissues
O_2 40 mm Hg
CO_2 46 mm Hg

PARTIAL PRESSURE of O_2 in mm Hg in ALVEOLAR AIR

If Partial Pressure of CO_2 remained at 40 mm Hg — O_2 DISSOCIATION from Hb at any given O_2 Partial Pressure would follow this graph ···

In TISSUES
O_2 40 mm Hg
CO_2 56 mm Hg

Volumes of
O_2 per 100 ml
↓ blood

PHYSIOLOGICAL DISSOCIATION CURVE
As blood gains CO_2 from Tissues CO_2 pressure gradually rises from 40 to 46 mm Hg and causes Hb to liberate its O_2 more readily.

If Partial Pressure of CO_2 remained at 46 mm Hg throughout – – O_2 DISSOCIATION from Hb at any given O_2 Partial Pressure would follow this curve

This effect of CO_2 partial pressure on dissociation of O_2 from Hb (the Böhr effect) is advantageous.

e.g. An increase in CO_2 partial pressure locally during tissue activity causes Hb to part more readily with its O_2 to the active tissues.

UPTAKE and RELEASE of CARBON DIOXIDE

CO_2 is carried by both red blood corpuscles and plasma. The bulk of it is carried as BICARBONATE formed chiefly in red cells and carried largely by plasma. The amount of CO_2 taken up by the blood in the tissues or released by the blood in the lungs depends on the PARTIAL PRESSURES of the CO_2 in the immediate environment and also on the state of the HAEMOGLOBIN or on the amounts of OXYGEN linked to Hb at any one time.

In R.B.C.		In Plasma

Hb ——————— and ——— Plasma Proteins act as *weak acids.*
KHb *(reduced Hb)* — and ——— Na Proteinate are salts of these acids.
K^+——————— and ——— Na^+ are easily displaced by stronger acids
(e.g. by H_2CO_3 formed from union of CO_2+H_2O)

$KHbO_2$ *(oxygenated Hb)* is a *stronger* acid and
K^+ is <u>less</u> readily displaced from it by H_2CO_3.

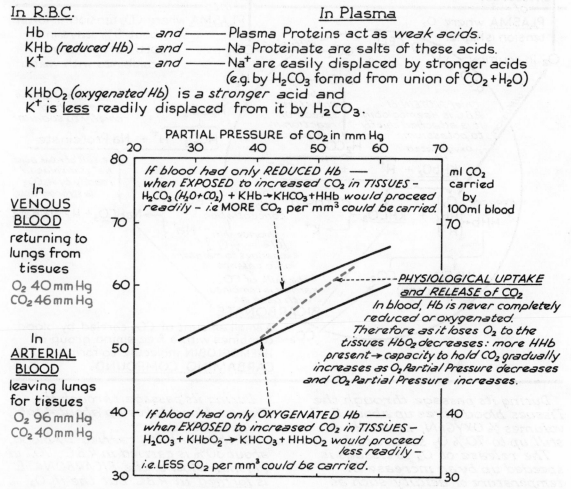

PARTIAL PRESSURE of CO_2 in mm Hg

In
<u>VENOUS</u>
<u>BLOOD</u>
returning to
lungs from
tissues
O_2 40 mm Hg
CO_2 46 mm Hg

In
<u>ARTERIAL</u>
<u>BLOOD</u>
leaving lungs
for tissues
O_2 96 mm Hg
CO_2 40 mm Hg

*If blood had only REDUCED Hb —
when EXPOSED to increased CO_2 in TISSUES -
H_2CO_3 (H_2O+CO_2) + KHb → $KHCO_3$+HHb would proceed
readily - i.e MORE CO_2 per mm³ could be carried.*

ml CO_2
carried
by
100ml blood

<u>PHYSIOLOGICAL UPTAKE
and RELEASE of CO_2</u>
*In blood, Hb is never completely
reduced or oxygenated.
Therefore as it loses O_2 to the
tissues HbO_2 decreases: more HHb
present → capacity to hold CO_2 gradually
increases as O_2 Partial Pressure decreases
and CO_2 Partial Pressure increases.*

*If blood had only OXYGENATED Hb —
when EXPOSED to increased CO_2 in TISSUES -
H_2CO_3 + $KHbO_2$ → $KHCO_3$ + $HHbO_2$ would proceed
less readily -
i.e.LESS CO_2 per mm³ could be carried.*

*i.e. The more OXYGEN the blood holds the less CO_2 it can hold and vice-versa.
This facilitates release of CO_2 in lungs and uptake of CO_2 in tissues.*

CARRIAGE and TRANSFER of OXYGEN and CARBON DIOXIDE

When ARTERIAL BLOOD is delivered by SYSTEMIC CAPILLARIES to the TISSUES it is exposed to:-

OXYGEN at very low tensions (1 - 40 mm Hg)

CARBON DIOXIDE at high tensions (50 - 60 mm Hg)

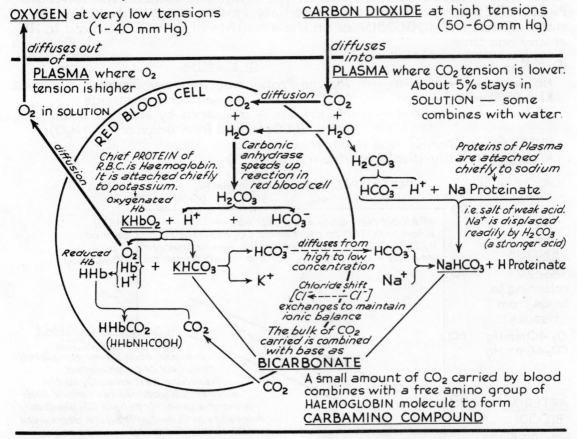

diffuses out of
PLASMA where O_2 tension is higher

O_2 in SOLUTION

RED BLOOD CELL

Chief PROTEIN of R.B.C. is Haemoglobin. It is attached chiefly to potassium.

Oxygenated Hb
$KHbO_2 + H^+ + HCO_3^-$

Reduced Hb
$HHb \begin{Bmatrix} Hb^- \\ H^+ \end{Bmatrix} + KHCO_3$

O_2

$HHbCO_2$ (HHbNHCOOH)

CO_2

CO_2 diffusion CO_2
$+$
H_2O

Carbonic anhydrase speeds up reaction in red blood cell

H_2CO_3

diffuses from high to low concentration
HCO_3^-
K^+

Chloride shift $[Cl^- \leftarrow ---\rightarrow Cl^-]$ exchanges to maintain ionic balance

The bulk of CO_2 carried is combined with base as BICARBONATE

diffuses into
PLASMA where CO_2 tension is lower. About 5% stays in SOLUTION — some combines with water.

CO_2
$+$
H_2O

H_2CO_3

Proteins of Plasma are attached chiefly to sodium

$HCO_3^- \quad H^+ + Na$ Proteinate

i.e. salt of weak acid. Na^+ is displaced readily by H_2CO_3 (a stronger acid)

HCO_3^-
Na^+

$NaHCO_3 + H$ Proteinate

A small amount of CO_2 carried by blood combines with a free amino group of HAEMOGLOBIN molecule to form CARBAMINO COMPOUND

During its passage through the Tissues blood gives up about 5 - 7 volumes % OXYGEN, i.e. its Hb is still up to 70% O_2 saturated.

The release of O_2 to tissues is speeded up by an increase in temperature or acidity such as occurs when tissues are active.

During its passage through the Tissues blood takes up about 4 to 6 volumes % CO_2.

Of the total CO_2 carried by blood, about 30% is carried in R.B.C., 70% in Plasma. Most of the BICARBONATE is formed in R.B.C. but the HCO_3^- diffusion gradient caused by carbonic anhydrase means that most of it is carried in plasma.

CARRIAGE and TRANSFER of OXYGEN and CARBON DIOXIDE

When <u>VENOUS BLOOD</u> flows through the <u>PULMONARY CAPILLARIES</u> it is exposed to:-

<u>OXYGEN</u> at high pressures in *ALVEOLAR AIR* (98 mm Hg)

CARBON DIOXIDE at low pressures (40 mm Hg)

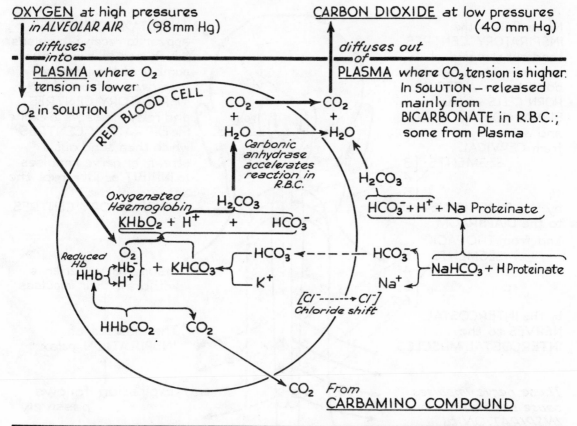

As blood passes through capillaries of lungs it takes up approximately 5 to 7 volumes % of OXYGEN. O_2 combines with Haemoglobin (Hb) molecule. It becomes about 95-97% saturated with Oxygen.

As blood passes through capillaries of lungs it gives up approximately 4 to 6 volumes % of CARBON DIOXIDE. A small amount is released from combination with the free amino group in Haemoglobin molecule - Carbamino compound. Most comes from BICARBONATE in R.B.C. and Plasma by processes indicated in diagram.

NERVOUS CONTROL of RESPIRATORY MOVEMENTS

Normal Respiratory Movements are Involuntary. They are carried out automatically (i.e. without conscious control) through the rhythmical discharge of nerve impulses from <u>CONTROLLING CENTRES in the BRAIN</u>.

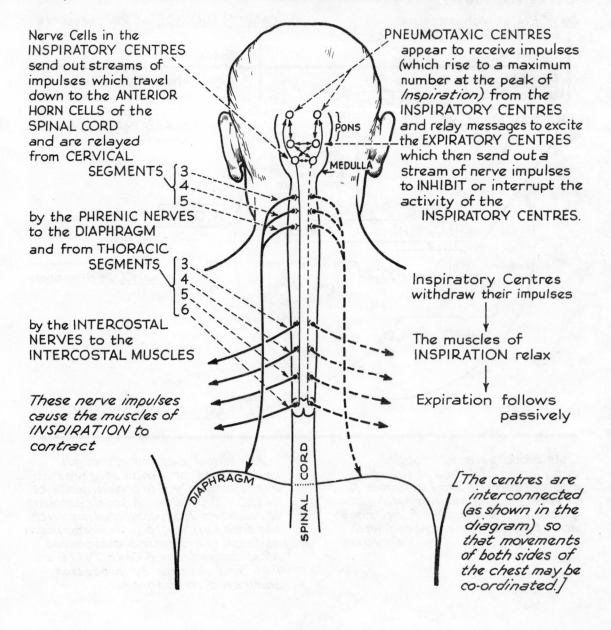

Nerve Cells in the INSPIRATORY CENTRES send out streams of impulses which travel down to the ANTERIOR HORN CELLS of the SPINAL CORD and are relayed from CERVICAL SEGMENTS 3 4 5

by the PHRENIC NERVES to the DIAPHRAGM and from THORACIC SEGMENTS 3 4 5 6

by the INTERCOSTAL NERVES to the INTERCOSTAL MUSCLES

These nerve impulses cause the muscles of INSPIRATION to contract

PONS

MEDULLA

PNEUMOTAXIC CENTRES appear to receive impulses (which rise to a maximum number at the peak of *Inspiration*) from the INSPIRATORY CENTRES and relay messages to excite the EXPIRATORY CENTRES which then send out a stream of nerve impulses to INHIBIT or interrupt the activity of the INSPIRATORY CENTRES.

Inspiratory Centres withdraw their impulses

↓

The muscles of INSPIRATION relax

↓

Expiration follows passively

DIAPHRAGM

SPINAL CORD

[The centres are interconnected (as shown in the diagram) so that movements of both sides of the chest may be co-ordinated.]

130

CHEMICAL REGULATION of RESPIRATION

The activity of the Respiratory Centres is affected by the chemistry and temperature of the blood supplying them.

The most important factor in regulating the activity of the respiratory centres is the level of CARBON DIOXIDE in the blood passing through them. —— This leads to an alteration in outgoing impulses to respiratory muscles.

Increase in CO_2
Increase in H^+ concentration } stimulates RESPIRATORY CENTRES —→ increases *depth* of breathing (and eventually increases *rate* of breathing).

Fall in blood CO_2 depresses – – – – – –→ shallow breathing.

Increase in temperature of blood – – –→ quickens but does not deepen respiration.

Cooling of blood – – – – – – – –→ slows respiration.

Normally respiration is little affected by slight changes in O_2 content of blood:

Severe lack of O_2 depresses respiratory centre

CHEMO-REFLEXES:

In addition to the direct effect of CO_2 and O_2 on centre

IX

X

Lack of O_2 – – – stimulates CHEMORECEPTORS – – reflexly stimulates
e.g. as at low (oxygen-lack receptors) respiration
atmospheric pressure in CAROTID BODY
(high altitude) and AORTIC BODY

NOTE:- These reflexes are usually powerful enough to override the direct depressant action of lack of O_2 on respiratory centres themselves.

The CHEMICAL and NERVOUS means of regulating the activity of RESPIRATORY CENTRES act together to adjust rate and depth of breathing to the needs of the body.

VOLUNTARY and REFLEX FACTORS in the REGULATION of RESPIRATION

Although fundamentally automatic and regulated by chemical factors in the blood, ingoing impulses from many parts of the body also modify the activity of the RESPIRATORY CENTRES and consequently alter the outgoing impulses to the Respiratory muscles to co-ordinate RHYTHM, RATE or DEPTH of breathing with other activities of the body.

Impulses from **HIGHER CENTRES** – PSYCHIC and EMOTIONAL INFLUENCES

- Voluntary alterations in Breathing.
- Interruptions of expiration in SPEECH and SINGING.
- Deep inspiration then short spasmodic expirations in LAUGHTER and WEEPING.
- Prolonged expiration in SIGHING.
- Deep inspiration with mouth open in YAWNING.
- Slow shallow breathing in SUSPENSE and CONCENTRATION.
- Rapid breathing in FEAR and EXCITEMENT.

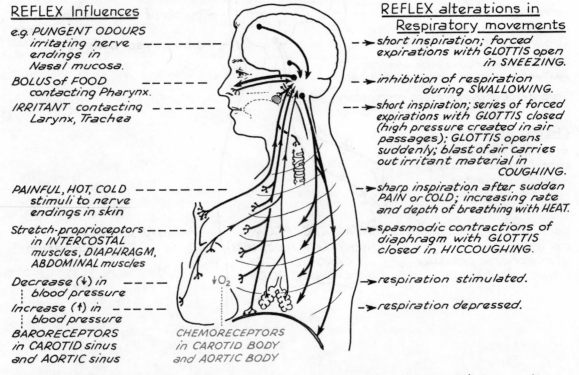

REFLEX Influences

e.g. PUNGENT ODOURS irritating nerve endings in Nasal mucosa.

BOLUS of FOOD contacting Pharynx.

IRRITANT contacting Larynx, Trachea

PAINFUL, HOT, COLD stimuli to nerve endings in skin

Stretch-proprioceptors in INTERCOSTAL muscles, DIAPHRAGM, ABDOMINAL muscles

Decrease (↓) in blood pressure

Increase (↑) in blood pressure

BARORECEPTORS in CAROTID sinus and AORTIC sinus

↓O_2

CHEMORECEPTORS in CAROTID BODY and AORTIC BODY

REFLEX alterations in Respiratory movements

- short inspiration; forced expirations with GLOTTIS open in SNEEZING.
- inhibition of respiration during SWALLOWING.
- short inspiration; series of forced expirations with GLOTTIS closed (high pressure created in air passages); GLOTTIS opens suddenly; blast of air carries out irritant material in COUGHING.
- sharp inspiration after sudden PAIN or COLD; increasing rate and depth of breathing with HEAT.
- spasmodic contractions of diaphragm with GLOTTIS closed in HICCOUGHING.
- respiration stimulated.
- respiration depressed.

Proprioceptors stimulated during muscle movements send impulses to respiratory centre → ↑rate and depth of breathing. (N.B. This occurs with active or passive movements of limbs.)

In normal breathing respiratory rate and rhythm are thought to be influenced rhythmically by the Hering-Breuer Reflex.

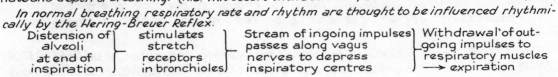

| Distension of alveoli at end of inspiration | stimulates stretch receptors in bronchioles | Stream of ingoing impulses passes along vagus nerves to depress inspiratory centres | Withdrawal of outgoing impulses to respiratory muscles → expiration |

CHAPTER 6

EXCRETORY SYSTEM

EXCRETORY SYSTEM

Together with the *Respiratory System* and the *Skin*, the KIDNEYS are the chief *Excretory organs* of the body.

The Kidneys excrete waste products of metabolism and adjust loss of Water and Electrolytes from the body to keep body fluids relatively constant in amount and composition.

To understand the way in which the kidney carries out these functions, it is essential to understand first the way in which it is supplied with blood. About 25% of Left Ventricle's output of blood in each Cardiac Cycle is distributed to kidneys for Filtration.

KIDNEYS
FORMATION of URINE

URETERS

BLADDER
STORAGE and EXPULSION of URINE

URETHRA

The **RENAL ARTERY** divides into **INTERLOBULAR ARTERIES**

divide into **ARCUATE ARTERIES**

give rise to **STRAIGHT ARTERIES**

from which arise **AFFERENT ARTERIOLES**

Each Afferent Arteriole divides into about **50 CAPILLARIES** which stay close together to form the **GLOMERULUS**.

All the Glomeruli lie within the CORTEX of the Kidney.

KIDNEY

Each Kidney contains approximately one million microscopic units — *NEPHRONS* — which form URINE.

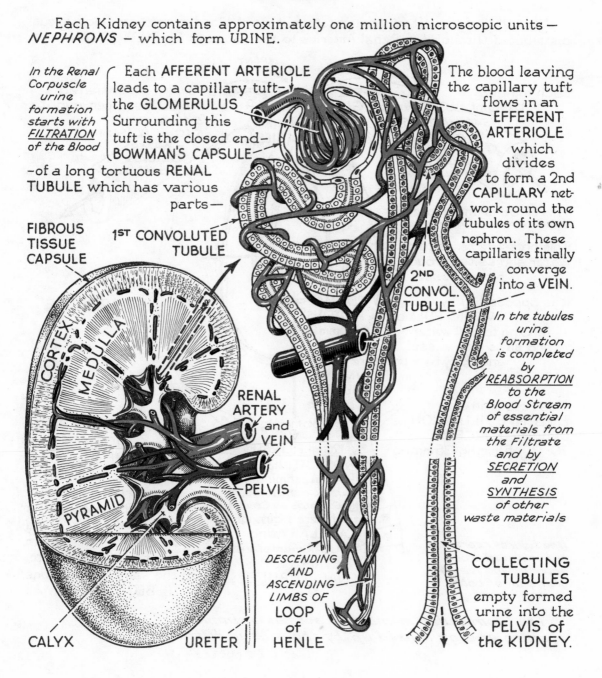

In the Renal Corpuscle urine formation starts with FILTRATION of the Blood

Each **AFFERENT ARTERIOLE** leads to a capillary tuft— the **GLOMERULUS** Surrounding this tuft is the closed end— **BOWMAN'S CAPSULE**

—of a long tortuous **RENAL TUBULE** which has various parts—

The blood leaving the capillary tuft flows in an **EFFERENT ARTERIOLE** which divides to form a 2nd **CAPILLARY** network round the tubules of its own nephron. These capillaries finally converge into a **VEIN**.

FIBROUS TISSUE CAPSULE

1ST CONVOLUTED TUBULE

2ND CONVOL. TUBULE

CORTEX

MEDULLA

In the tubules urine formation is completed by REABSORPTION to the Blood Stream of essential materials from the Filtrate and by SECRETION and SYNTHESIS of other waste materials

RENAL ARTERY and VEIN

PELVIS

PYRAMID

DESCENDING AND ASCENDING LIMBS OF **LOOP of HENLE**

COLLECTING TUBULES empty formed urine into the **PELVIS of the KIDNEY.**

CALYX

URETER

FORMATION of URINE — 1. FILTRATION

About 25% of the left ventricle's total output of blood in each cardiac cycle is distributed through the Renal Arteries to the Kidneys for FILTRATION.

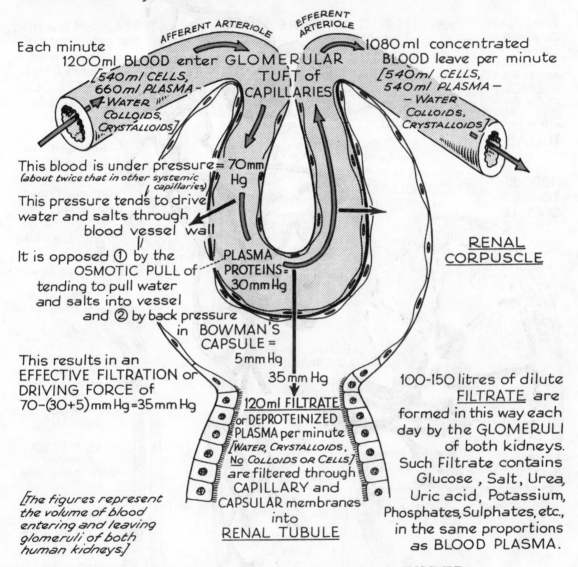

Each minute 1200ml BLOOD enter GLOMERULAR TUFT of CAPILLARIES
[540ml CELLS, 660ml PLASMA — WATER, COLLOIDS, CRYSTALLOIDS]

AFFERENT ARTERIOLE

EFFERENT ARTERIOLE

1080 ml concentrated BLOOD leave per minute
[540ml CELLS, 540ml PLASMA — WATER, COLLOIDS, CRYSTALLOIDS]

This blood is under pressure = 70mm Hg
(about twice that in other systemic capillaries)

This pressure tends to drive water and salts through blood vessel wall

It is opposed ① by the OSMOTIC PULL of PLASMA PROTEINS = 30mm Hg tending to pull water and salts into vessel and ② by back pressure in BOWMAN'S CAPSULE = 5mm Hg

RENAL CORPUSCLE

This results in an EFFECTIVE FILTRATION or DRIVING FORCE of 70−(30+5)mm Hg = 35mm Hg

35 mm Hg

120 ml FILTRATE or DEPROTEINIZED PLASMA per minute [WATER, CRYSTALLOIDS, No COLLOIDS OR CELLS] are filtered through CAPILLARY and CAPSULAR membranes into RENAL TUBULE

[The figures represent the volume of blood entering and leaving glomeruli of both human kidneys.]

100-150 litres of dilute FILTRATE are formed in this way each day by the GLOMERULI of both kidneys. Such Filtrate contains Glucose, Salt, Urea, Uric acid, Potassium, Phosphates, Sulphates, etc., in the same proportions as BLOOD PLASMA.

The Glomerular membrane acts as a simple FILTER —
— i.e. no energy is used up by the cells in filtration.

136

FORMATION of URINE—2. CONCENTRATION

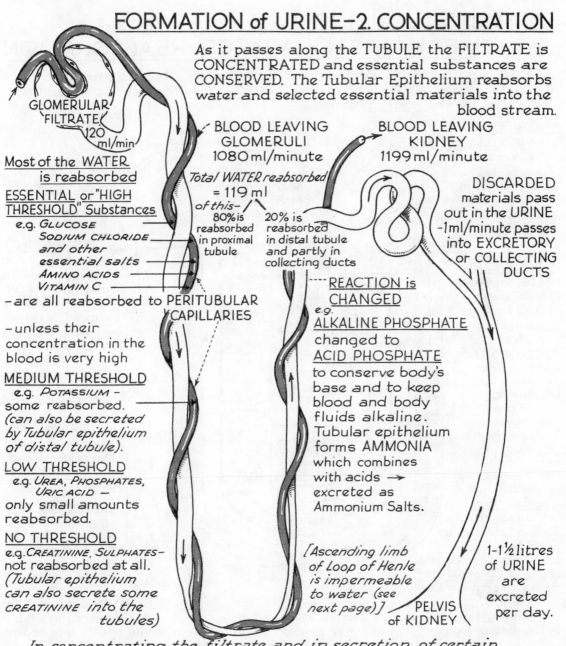

As it passes along the TUBULE the FILTRATE is CONCENTRATED and essential substances are CONSERVED. The Tubular Epithelium reabsorbs water and selected essential materials into the blood stream.

GLOMERULAR FILTRATE 120 ml/min

BLOOD LEAVING GLOMERULI 1080 ml/minute

BLOOD LEAVING KIDNEY 1199 ml/minute

Most of the WATER is reabsorbed

Total WATER reabsorbed = 119 ml

of this—
80% is reabsorbed in proximal tubule

20% is reabsorbed in distal tubule and partly in collecting ducts

ESSENTIAL or "HIGH THRESHOLD" Substances
e.g. GLUCOSE
SODIUM CHLORIDE
and other
essential salts
AMINO ACIDS
VITAMIN C
—are all reabsorbed to PERITUBULAR CAPILLARIES

—unless their concentration in the blood is very high

MEDIUM THRESHOLD
e.g. POTASSIUM —
some reabsorbed.
(can also be secreted by Tubular epithelium of distal tubule).

LOW THRESHOLD
e.g. UREA, PHOSPHATES, URIC ACID —
only small amounts reabsorbed.

NO THRESHOLD
e.g. CREATININE, SULPHATES—
not reabsorbed at all.
(Tubular epithelium can also secrete some CREATININE into the tubules)

DISCARDED materials pass out in the URINE —1ml/minute passes into EXCRETORY or COLLECTING DUCTS

REACTION is CHANGED
e.g.
ALKALINE PHOSPHATE changed to ACID PHOSPHATE to conserve body's base and to keep blood and body fluids alkaline. Tubular epithelium forms AMMONIA which combines with acids → excreted as Ammonium Salts.

[Ascending limb of Loop of Henle is impermeable to water (see next page)]

PELVIS of KIDNEY

1–1½ litres of URINE are excreted per day.

In concentrating the filtrate and in secretion of certain substances, the tubular epithelial cells use energy, i.e. they do work — much of it against an osmotic gradient.

FORMATION of URINE:
MECHANISM of WATER REABSORPTION

Water is <u>not</u> actively reabsorbed by tubular cells. Its movements are determined <u>passively</u> by the <u>osmotic gradient</u> set up by solutes — chiefly by the <u>sodium</u> salts.

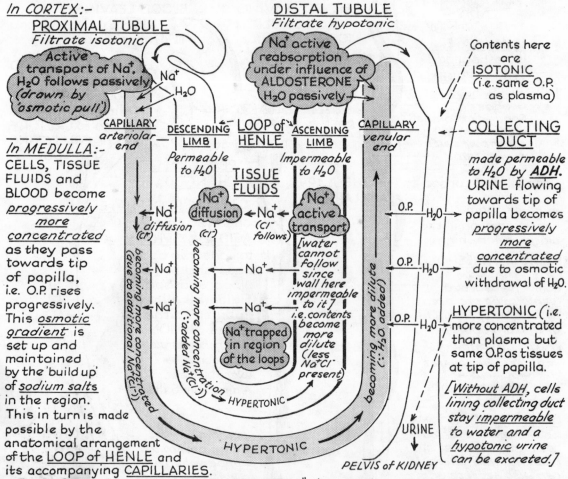

In CORTEX:-
PROXIMAL TUBULE
Filtrate isotonic
Active transport of Na⁺. H₂O follows passively *(drawn by 'osmotic pull')*
Na⁺
H₂O

DISTAL TUBULE
Filtrate hypotonic
Na⁺ active reabsorption under influence of ALDOSTERONE H₂O passively

Contents here are ISOTONIC *(i.e. same O.P. as plasma)*

CAPILLARY *arteriolar end*
DESCENDING LIMB *Permeable to H₂O*
LOOP of HENLE
ASCENDING LIMB *Impermeable to H₂O*
CAPILLARY *venular end*

COLLECTING DUCT *made permeable to H₂O by ADH.* URINE flowing towards tip of papilla becomes *progressively more concentrated* due to osmotic withdrawal of H₂O.

In MEDULLA:- CELLS, TISSUE FLUIDS and BLOOD become *progressively more concentrated* as they pass towards tip of papilla, i.e. O.P. rises progressively. This *osmotic gradient* is set up and maintained by the 'build up' of *sodium salts* in the region. This in turn is made possible by the anatomical arrangement of the LOOP of HENLE and its accompanying CAPILLARIES.

TISSUE FLUIDS
Na⁺ diffusion
Na⁺ active transport
[water cannot follow since wall here impermeable to it] i.e. contents become more dilute (less Na⁺Cl⁻ present)
Na⁺ trapped in region of the loops

Na⁺ *diffusion* (Cl⁻)
Na⁺ (Cl⁻)
Na⁺ (Cl⁻ follows)
Na⁺

becoming more concentrated (due to additional Na⁺(Cl⁻))
becoming more concentration (∴ added Na⁺(Cl⁻))
becoming more dilute (∴ H₂O added)

HYPERTONIC
HYPERTONIC

O.P. H₂O→
O.P. H₂O→
O.P. H₂O→

HYPERTONIC *(i.e. more concentrated than plasma but same O.P. as tissues at tip of papilla.*

[Without ADH, cells lining collecting duct stay impermeable to water and a hypotonic urine can be excreted.]

URINE
PELVIS of KIDNEY

These form a "COUNTER-CURRENT SYSTEM" (i.e <u>outflowing</u> fluid flows counter to (yet near) <u>inflowing</u>). Active transport of sodium ions out of the ascending limb (without accompanying loss of water through its 'water-tight' walls) and its subsequent diffusion into the descending limb is the key to the chain of events shown.

This mechanism permits maximal {reabsorption of water / concentration of urine} with minimal expenditure of energy by tubular cells.

The "CLEARANCE" of INULIN in the NEPHRON

The rate of Glomerular Filtration can be measured by injecting a substance which is known to be *filtered* off by the Renal Corpuscle but *not secreted* or *reabsorbed* by the Tubular epithelium.
INULIN is such a substance. It is *filtered* from the Glomerular Capillaries.

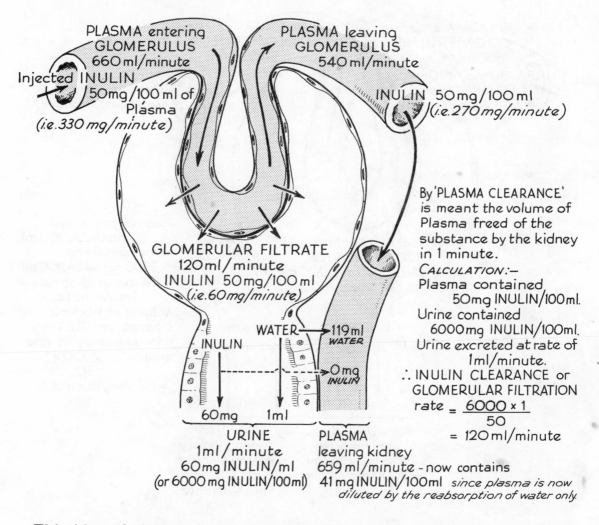

PLASMA entering
GLOMERULUS
660 ml/minute
Injected INULIN
50 mg/100 ml of Plasma
(*i.e.* 330 mg/minute)

PLASMA leaving
GLOMERULUS
540 ml/minute
INULIN 50 mg/100 ml
(*i.e.* 270 mg/minute)

GLOMERULAR FILTRATE
120 ml/minute
INULIN 50 mg/100 ml
(*i.e.* 60 mg/minute)

WATER →119 ml WATER

INULIN

0 mg INULIN

60 mg   1 ml

URINE
1 ml/minute
60 mg INULIN/ml
(or 6000 mg INULIN/100 ml)

PLASMA
leaving kidney
659 ml/minute - now contains
41 mg INULIN/100 ml *since plasma is now diluted by the reabsorption of water only.*

By 'PLASMA CLEARANCE' is meant the volume of Plasma freed of the substance by the kidney in 1 minute.

CALCULATION:−
Plasma contained
50 mg INULIN/100 ml.
Urine contained
6000 mg INULIN/100 ml.
Urine excreted at rate of
1 ml/minute.
∴ INULIN CLEARANCE or GLOMERULAR FILTRATION
$$\text{rate} = \frac{6000 \times 1}{50}$$
$$= 120 \, \text{ml/minute}$$

This idea of clearance can be applied to other substances naturally present such as UREA, or artificially introduced, such as DIODONE.

UREA "CLEARANCE"

UREA, like Inulin, is filtered from the Plasma in the Renal Corpuscle. Unlike Inulin some Urea diffuses back into the Blood Stream from the Tubules.

PLASMA entering
GLOMERULUS
660 ml/minute
UREA 30 mg/100 ml of
Plasma
(i.e. 198 mg/minute)

PLASMA leaving
GLOMERULUS
540 ml/minute
UREA 30 mg/100 ml
(i.e. 162 mg/minute)

GLOMERULAR FILTRATE
120 ml/minute
UREA 30 mg/100 ml
(i.e. 36 mg/minute)

WATER — 119 ml
UREA — 16 mg

1 ml   20 mg

Plasma contained
30 mg UREA/100 ml.
Urine contained
2000 mg UREA/100 ml.
Urine excreted at rate of
1 ml/minute.
∴ Volume of Plasma
cleared of UREA by
normal kidney in one
minute = $\dfrac{2000 \times 1}{30}$
= 66·7 ml/minute.

URINE
1 ml/minute
20 mg UREA/ml
(or 2000 mg/100 ml)

PLASMA
leaving kidney
659 ml/minute
with 178 mg
UREA/minute

UREA clearance is used as a test of normal renal function.

DIODONE "CLEARANCE"

Certain substances are secreted by the tubules during the formation of urine. Creatinine, which is normally found in urine, is one of these.

Some foreign substances injected into the body are excreted in this way too, e.g. DIODONE *(an iodine-containing material)*.

Such substances can form the basis of tests for checking the efficiency of the Tubular epithelium in any one individual.

Some Diodone is normally *filtered*; some is *excreted* by the tubules.

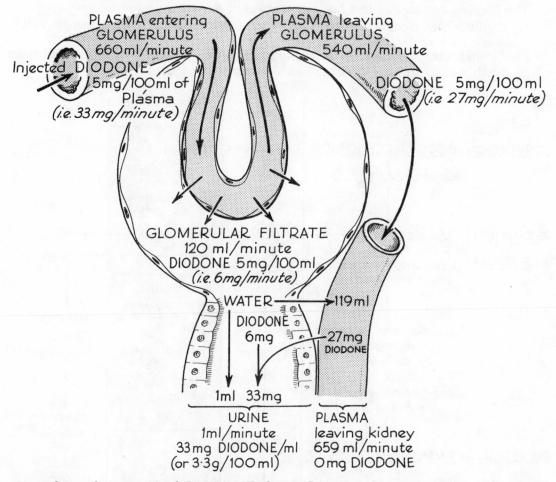

Complete clearance of DIODONE from Plasma in one passage through normal kidney, gauges not only Glomerular Filtrating power but also the efficiency of the Tubular epithelium to secrete.

MAINTENANCE of ACID-BASE BALANCE

Large amounts of ACIDS are produced in cells during METABOLISM. In spite of this the pH of the blood and body fluids is kept relatively constant (7.4)

Together with the Respiratory System, the Kidneys are responsible for maintaining this equilibrium.

The ACIDS are first neutralized by BUFFERING AGENTS in the Blood Stream. They cannot be excreted in this form otherwise the body would soon lose its stocks of these ALKALIS.

The Kidney *conserves the body's reserves of Alkali* in several ways :-
1. By *Tubular Secretion* of *Hydrogen ions* and their *Exchange* with *Base* in the *Filtrate*.
2. By *Tubular Formation and Secretion* of *Ammonia* which combines with *Acids* to be excreted as Ammonium Salts

E.g.

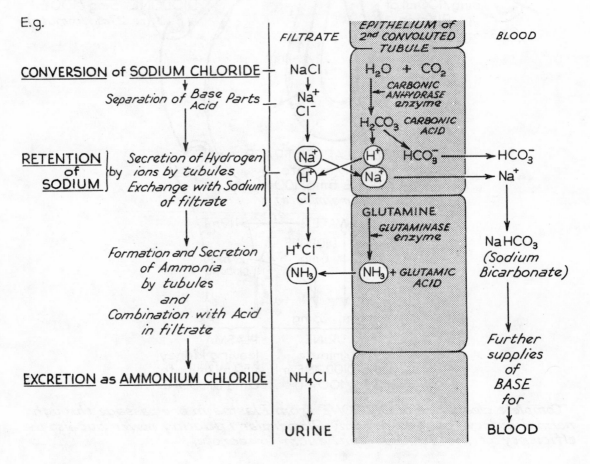

MAINTENANCE of ACID-BASE BALANCE

The chief buffering substances of the blood which appear in the filtrate are *Phosphates and Bicarbonates*. The Kidney is believed to *conserve* these as follows:-

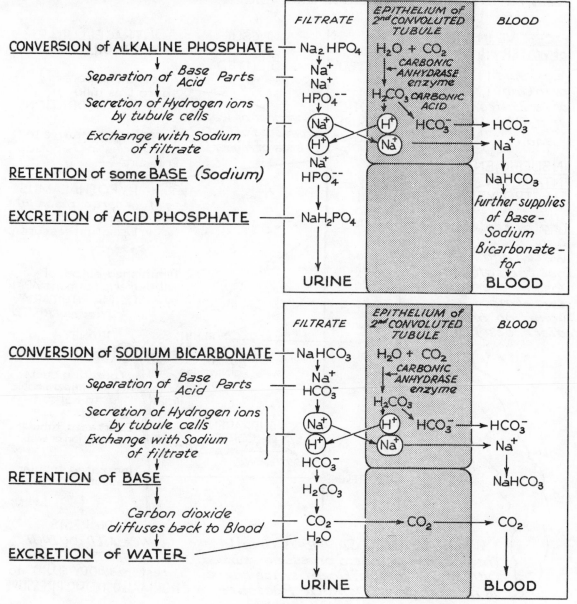

143

REGULATION of WATER BALANCE

As the amounts of water and/or electrolytes in the body fluctuate, excretion of them is adjusted by the *Kidney* so that Body Fluids are restored to normal composition and volume.

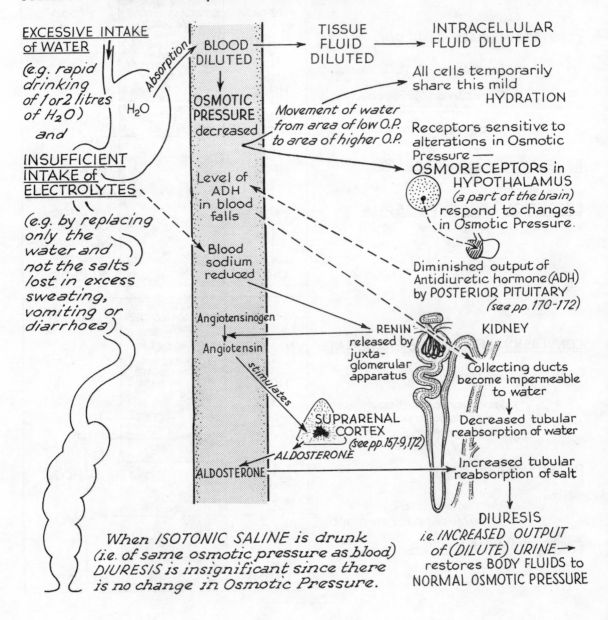

EXCESSIVE INTAKE of WATER

(e.g. rapid drinking of 1 or 2 litres of H_2O) and

INSUFFICIENT INTAKE of ELECTROLYTES

(e.g. by replacing only the water and not the salts lost in excess sweating, vomiting or diarrhoea)

H_2O

Absorption

BLOOD DILUTED

OSMOTIC PRESSURE decreased

Level of ADH in blood falls

Blood sodium reduced

Angiotensinogen

Angiotensin

stimulates

SUPRARENAL CORTEX (see pp.157-9,172)

ALDOSTERONE

ALDOSTERONE

TISSUE FLUID DILUTED

INTRACELLULAR FLUID DILUTED

All cells temporarily share this mild HYDRATION

Movement of water from area of low O.P. to area of higher O.P.

Receptors sensitive to alterations in Osmotic Pressure — OSMORECEPTORS in HYPOTHALAMUS (a part of the brain) respond to changes in Osmotic Pressure.

Diminished output of Antidiuretic hormone (ADH) by POSTERIOR PITUITARY (see pp. 170-172)

RENIN released by juxta-glomerular apparatus

KIDNEY

Collecting ducts become impermeable to water

Decreased tubular reabsorption of water

Increased tubular reabsorption of salt

DIURESIS i.e. INCREASED OUTPUT of (DILUTE) URINE → restores BODY FLUIDS to NORMAL OSMOTIC PRESSURE

When ISOTONIC SALINE is drunk (i.e. of same osmotic pressure as blood) DIURESIS is insignificant since there is no change in Osmotic Pressure.

144

REGULATION of WATER BALANCE

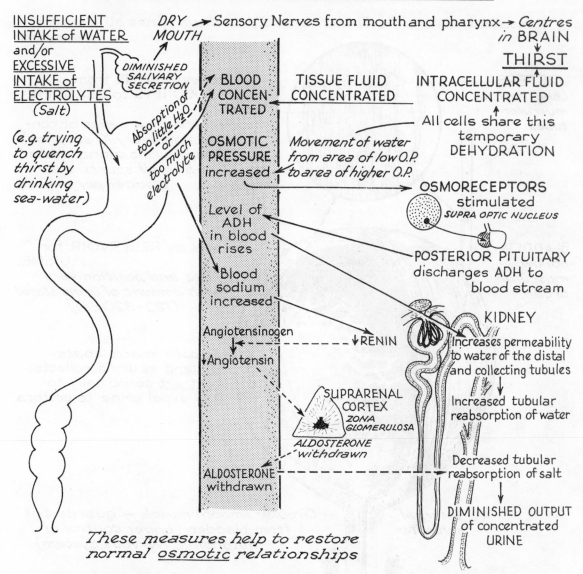

INSUFFICIENT INTAKE of WATER and/or EXCESSIVE INTAKE of ELECTROLYTES (Salt)

(e.g. trying to quench thirst by drinking sea-water)

DRY MOUTH → Sensory Nerves from mouth and pharynx → Centres in BRAIN → THIRST

DIMINISHED SALIVARY SECRETION

Absorption of too little H₂O or too much electrolyte

BLOOD CONCENTRATED

TISSUE FLUID CONCENTRATED

INTRACELLULAR FLUID CONCENTRATED

All cells share this temporary DEHYDRATION

OSMOTIC PRESSURE increased

Movement of water from area of low O.P. to area of higher O.P.

OSMORECEPTORS stimulated
SUPRA OPTIC NUCLEUS

Level of ADH in blood rises

POSTERIOR PITUITARY discharges ADH to blood stream

Blood sodium increased

KIDNEY
Increases permeability to water of the distal and collecting tubules

Angiotensinogen

↓RENIN

↓Angiotensin

Increased tubular reabsorption of water

SUPRARENAL CORTEX
ZONA GLOMERULOSA

ALDOSTERONE withdrawn

Decreased tubular reabsorption of salt

ALDOSTERONE withdrawn

DIMINISHED OUTPUT of concentrated URINE

These measures help to restore normal osmotic relationships

In the maintenance of water balance, loss (or conservation) of water and of electrolytes are closely linked. Another example, shown on page 172, deals with ALDOSTERONE, its rôle in governing reabsorption of sodium by kidney tubules, and its tie-up with ADH in maintaining the normal volume of extracellular fluids including blood.

URINARY BLADDER and URETERS

A resistant, distensible *Transitional Epithelium* lines all *Urinary Passages.*

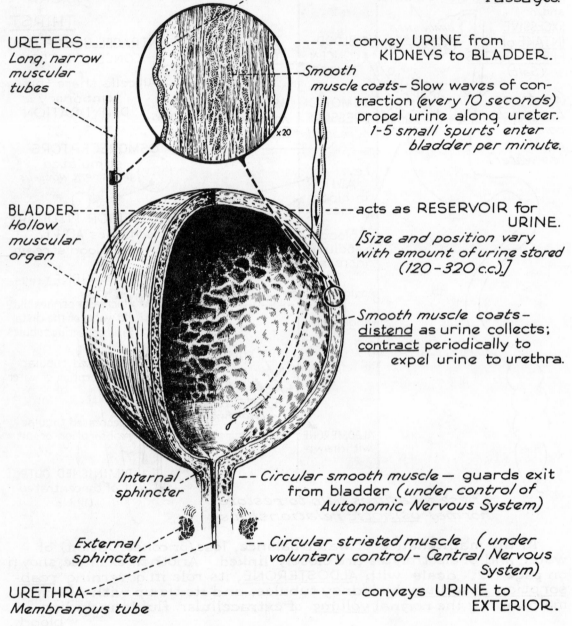

URETERS -------- *Long, narrow muscular tubes*

-------- convey URINE from KIDNEYS to BLADDER.

Smooth muscle coats— Slow waves of contraction (*every 10 seconds*) propel urine along ureter. *1-5 small 'spurts' enter bladder per minute.*

×20

BLADDER ---- *Hollow muscular organ*

---- acts as RESERVOIR for URINE.

[Size and position vary with amount of urine stored (120-320 c.c.).]

Smooth muscle coats— distend as urine collects; contract periodically to expel urine to urethra.

Internal sphincter

-- Circular smooth muscle — guards exit from bladder (*under control of Autonomic Nervous System*)

External sphincter

-- Circular striated muscle (*under voluntary control - Central Nervous System*)

URETHRA ---- *Membranous tube*

---- conveys URINE to EXTERIOR.

STORAGE and EXPULSION of URINE

URINE is formed continuously by the KIDNEYS. It collects, drop by drop, in the URINARY BLADDER which expands to hold about 300 c.c. When the Bladder is full the desire to *void urine* is experienced.

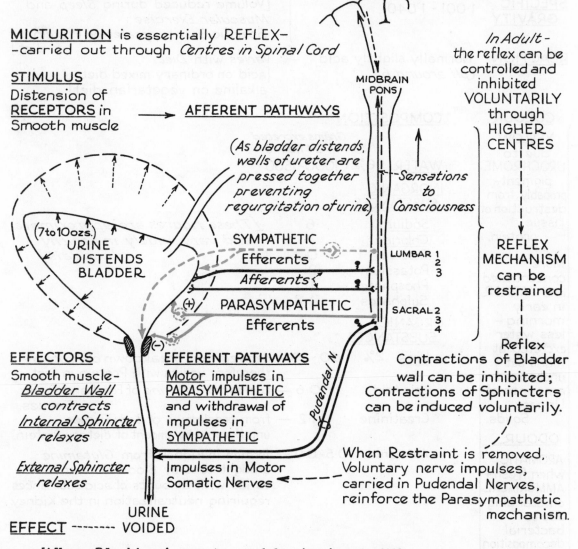

MICTURITION is essentially REFLEX—
—carried out through *Centres in Spinal Cord*

STIMULUS
Distension of **RECEPTORS** in Smooth muscle ⟶ **AFFERENT PATHWAYS**

In Adult –
the reflex can be controlled and inhibited VOLUNTARILY through HIGHER CENTRES

MIDBRAIN PONS

(As bladder distends, walls of ureter are pressed together preventing regurgitation of urine.)

–Sensations to Consciousness

(7 to 10 ozs.) URINE DISTENDS BLADDER

SYMPATHETIC Efferents
Afferents
PARASYMPATHETIC Efferents

LUMBAR 1 2 3

SACRAL 2 3 4

REFLEX MECHANISM can be restrained

Reflex Contractions of Bladder wall can be inhibited; Contractions of Sphincters can be induced voluntarily.

EFFECTORS
Smooth muscle-
Bladder Wall contracts
Internal Sphincter relaxes

External Sphincter relaxes

EFFERENT PATHWAYS
Motor impulses in PARASYMPATHETIC and withdrawal of impulses in SYMPATHETIC

Impulses in Motor Somatic Nerves ⟵

Pudendal N.

When Restraint is removed, Voluntary nerve impulses, carried in Pudendal Nerves, reinforce the Parasympathetic mechanism.

URINE

EFFECT ------- VOIDED

When Bladder is empty and beginning to fill –
Inhibition of Parasympathetic ⎫
Activation of Sympathetic ⎭ → Relaxation of Bladder Wall.
Constriction of Sphincters.

URINE

VOLUME : *In Adult*
1000 - 1500 ml/24 hours

SPECIFIC GRAVITY 1·001 - 1·040

} Vary with *Fluid Intake* and with *Fluid Output* from other routes — *Skin, Lungs, Gut*.
[Volume reduced during *Sleep* and *Muscular Exercise* :
Specific Gravity greater on Protein diet.]

REACTION Normally slightly acid — *Varies* with *Diet*
(*pH around 6*) [acid on ordinary mixed diet: alkaline on vegetarian diet].

COLOUR

YELLOW due to UROCHROME pigment-probably from destruction of tissue protein.

More concentrated and DARKER in early morning — less water excreted at night but unchanged amounts of Urinary Solids.

ODOUR

AROMATIC when fresh→AMMONIACAL on standing due to bacterial decomposition of UREA to AMMONIA.

COMPOSITION

Grams excreted in 24 hours

WATER 96%   1000-1500

INORGANIC SUBSTANCES

Sodium	6
Chloride	7
Calcium	0·2
Potassium	2
Phosphates	1·7
Sulphates	1·8

[These figures are approximate and vary widely in healthy individuals]

ORGANIC SUBSTANCES

Urea 2%   20-30 — derived from breakdown of *Protein* — therefore varies with Protein in Diet.

Uric Acid   0·6 — comes from *Purine* of Food and Body Tissues.

Creatinine   1·2 — from breakdown of Body Tissues; uninfluenced by amount of dietary Protein.

Ammonia   0·5-0·9 — formed in Kidney from *Glutamine* brought to it by Blood Stream; varies with amounts of acid substances requiring neutralization in the Kidney.

[In the Newborn, Volume and Specific Gravity are low and Composition varies.]

CHAPTER 7

ENDOCRINE SYSTEM

ENDOCRINE SYSTEM

The <u>DUCTLESS GLANDS</u> produce HORMONES ("chemical messengers") which they pass into the BLOOD STREAM for GENERAL CIRCULATION to excite or inhibit activity of other organs or tissues.

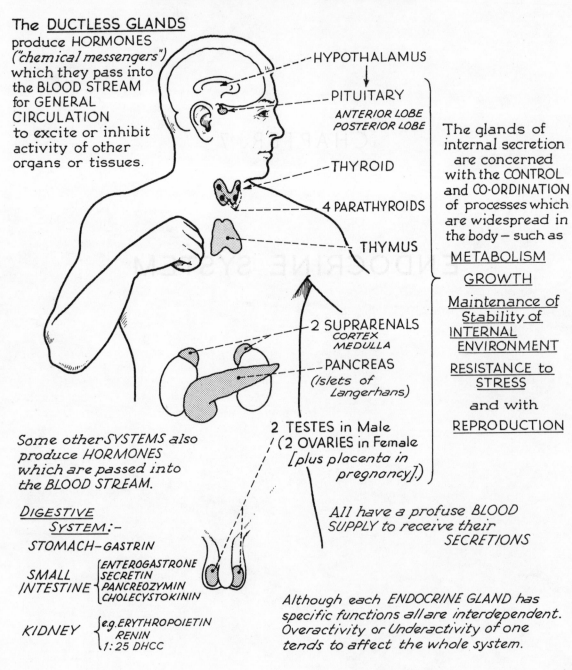

HYPOTHALAMUS

PITUITARY
ANTERIOR LOBE
POSTERIOR LOBE

THYROID

4 PARATHYROIDS

THYMUS

2 SUPRARENALS
CORTEX
MEDULLA

PANCREAS
(Islets of Langerhans)

2 TESTES in Male
(2 OVARIES in Female
[plus placenta in pregnancy].)

The glands of internal secretion are concerned with the CONTROL and CO-ORDINATION of processes which are widespread in the body — such as

<u>METABOLISM</u>

<u>GROWTH</u>

Maintenance of <u>Stability</u> of <u>INTERNAL ENVIRONMENT</u>

<u>RESISTANCE to STRESS</u>

and with

<u>REPRODUCTION</u>

Some other SYSTEMS also produce HORMONES which are passed into the BLOOD STREAM.

DIGESTIVE SYSTEM:-
STOMACH—GASTRIN

SMALL INTESTINE { *ENTEROGASTRONE SECRETIN PANCREOZYMIN CHOLECYSTOKININ*

KIDNEY { *e.g.ERYTHROPOIETIN RENIN 1:25 DHCC*

All have a profuse BLOOD SUPPLY to receive their SECRETIONS

Although each ENDOCRINE GLAND has specific functions all are interdependent. Overactivity or Underactivity of one tends to affect the whole system.

THYROID

STRUCTURE:

2 LOBES *(joined by ISTHMUS)* composed of
FOLLICLES
lined by
CUBICAL
EPITHELIUM

lie in
front of
TRACHEA

(NO DUCTS)

Blood
Vessels
to and
from
THYROID

(×150)

*Gland weighs
about 25g in
adult*

*Richly supplied with
BLOOD CAPILLARIES*

PARAFOLLICULAR
or 'C' CELLS

FUNCTION:

CUBICAL EPITHELIUM extracts from
the Blood stream (and concentrates)
INORGANIC IODIDE

IODINE
links with
TYROSINE

MONOIODOTYROSINE
DIIODO-
TYROSINE

BLOOD

TRIIODO-
THYRONINE
TETRAIODO-
THYRONINE
(Thyroxine)

*Stored in colloid
linked with protein
as 'THYROGLOBULIN*

*when required
a protein-splitting
enzyme releases
'active agents'*

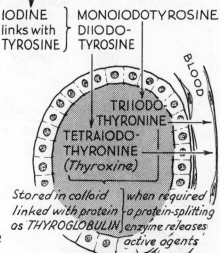

The ACTIVE HORMONES are
released to the GENERAL
CIRCULATION for distribution by the blood stream to
all BODY TISSUES where they
act as CATALYSTS
hastening OXIDATION pro-
cesses in the TISSUE CELLS.
i.e. They probably regulate the
various ENZYMES which
control ENERGY METABOLISM
and thus influence all
METABOLIC PROCESSES
including GROWTH.

[PARAFOLLICULAR Cells
secrete THYROCALCITONIN –
which LOWERS a high BLOOD
CALCIUM by suppressing
calcium mobilization from bone.]

REGULATION of SECRETION:

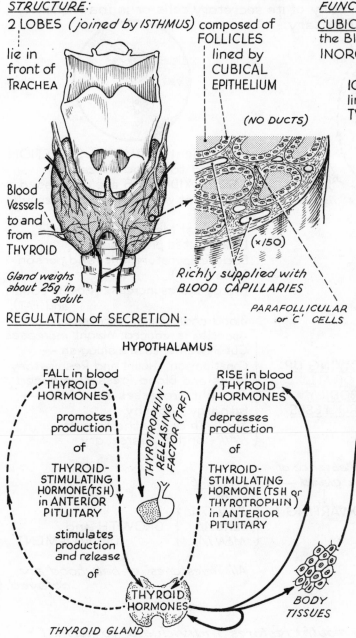

HYPOTHALAMUS

FALL in blood
THYROID
HORMONES

promotes
production
of
THYROID-
STIMULATING
HORMONE (TSH)
in ANTERIOR
PITUITARY

stimulates
production
and release
of

THYROTROPHIN-
RELEASING
FACTOR (TRF)

RISE in blood
THYROID
HORMONES

depresses
production
of
THYROID-
STIMULATING
HORMONE (TSH or
THYROTROPHIN)
in ANTERIOR
PITUITARY

THYROID
HORMONES

THYROID GLAND

BODY
TISSUES

UNDERACTIVITY of THYROID

If the THYROID shows *atrophy* of its secretory cells or is inadequately stimulated by the Anterior Pituitary:-

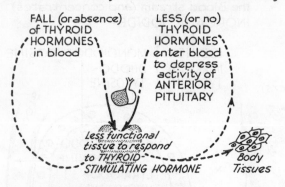

FALL (or absence) of THYROID HORMONES in blood

LESS (or no) THYROID HORMONES enter blood to depress activity of ANTERIOR PITUITARY

Less functional tissue to respond to THYROID-STIMULATING HORMONE

Body Tissues

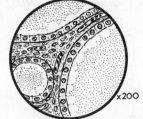

x200

Insufficient HORMONAL SECRETION released to Blood Stream.
TISSUE OXIDATIONS are depressed, *i.e.* Rate at which cells use energy is reduced.
The Basal Metabolic Rate falls.
Less Heat is produced.
Body Temperature falls (and person feels COLD).
Energy stores increase (*e.g.* GLYCOGEN and FAT).

In the <u>ADULT</u>

MYXOEDEMA

<u>SLOWING UP OF ALL BODILY PROCESSES</u>

Blood cholesterol increases.
Appetite is reduced; Weight increases.
Gut movements are sluggish →
Constipation. Heart and Respiratory Rates and Blood Pressure reduced.
Thought processes slow down →
Lethargy; Apathy.
SKIN—Thick, leathery, puffy.
HAIR - Brittle, sparse, dry.

In the <u>CHILD</u> — *e.g. congenital absence of the gland* →

<u>CRETIN</u>

DWARFING

FAILURE of *SKELETAL SEXUAL MENTAL* } GROWTH and DEVELOPMENT

All "milestones" of babyhood are delayed.

THYROID EXTRACT (taken by mouth) restores individuals to normal.

OVERACTIVITY of THYROID

If an *enlarged* THYROID shows increased activity of its secretory cells:-

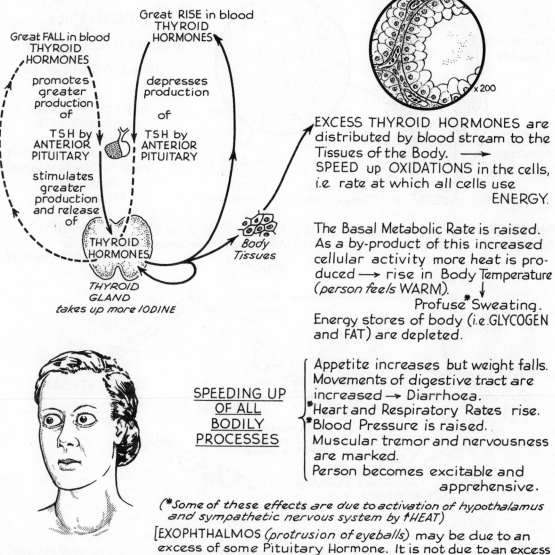

Great FALL in blood
THYROID
HORMONES

Great RISE in blood
THYROID
HORMONES

promotes
greater
production
of

depresses
production

of

TSH by
ANTERIOR
PITUITARY

TSH by
ANTERIOR
PITUITARY

stimulates
greater
production
and release
of

THYROID
HORMONES

*Body
Tissues*

*THYROID
GLAND
takes up more IODINE*

EXCESS THYROID HORMONES are
distributed by blood stream to the
Tissues of the Body. ⟶
SPEED up OXIDATIONS in the cells,
i.e. rate at which all cells use
ENERGY.

The Basal Metabolic Rate is raised.
As a by-product of this increased
cellular activity more heat is produced ⟶ rise in Body Temperature
(*person feels* WARM).
        Profuse*Sweating.
Energy stores of body (i.e.GLYCOGEN
and FAT) are depleted.

SPEEDING UP
OF ALL
BODILY
PROCESSES

{
Appetite increases but weight falls.
Movements of digestive tract are
increased ⟶ Diarrhoea.
*Heart and Respiratory Rates rise.
*Blood Pressure is raised.
Muscular tremor and nervousness
are marked.
Person becomes excitable and
                apprehensive.
}

(*Some of these effects are due to activation of hypothalamus
and sympathetic nervous system by ↑HEAT)

[EXOPHTHALMOS (protrusion of eyeballs) may be due to an
excess of some Pituitary Hormone. It is not due to an excess
of Thyroid Hormones.]

Surgical removal of part of the overactive gland reduces the Thyroid activity.

PARATHYROIDS

FOUR small glands composed of *cords of cells* which secrete Parathyroid
Hormone – **PARATHORMONE**
or PTH

↓

CAPILLARIES

GENERAL CIRCULATION

to ALL TISSUES of the body

But not all tissues are sensitive to it.

It plays an important rôle in CALCIUM and PHOSPHATE METABOLISM

PARATHYROID GLANDS

Situated behind THYROID

OESOPHAGUS

TRACHEA

Each weighs from 20-50mg in Adult

×200

Function of EOSINOPHIL cells is unknown

PARATHORMONE acts on <u>KIDNEY TUBULES</u>, <u>BONE</u> and on <u>GUT</u> to maintain
BLOOD CALCIUM level at 11mg/100ml PLASMA *(necessary for normal neuromuscular excitability)*.

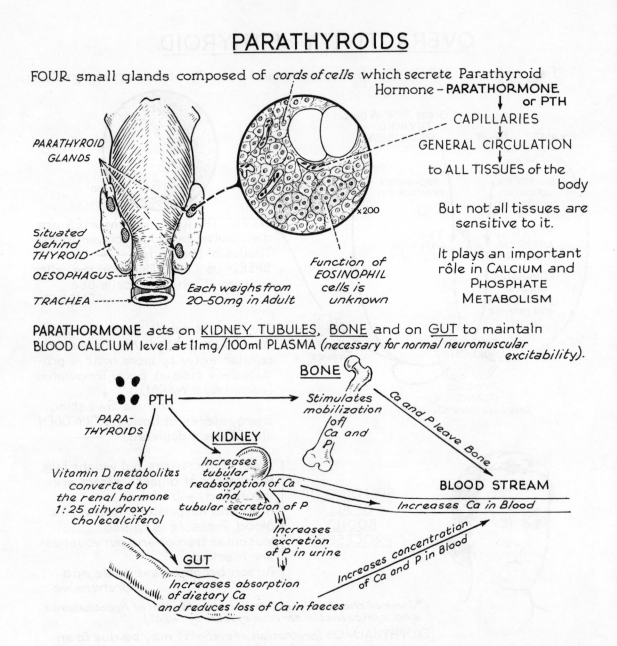

PTH

PARA-THYROIDS

Vitamin D metabolites converted to the renal hormone 1:25 dihydroxy-cholecalciferol

KIDNEY

Increases tubular reabsorption of Ca and tubular secretion of P

Increases excretion of P in urine

<u>BONE</u>

Stimulates mobilization of Ca and P

Ca and P leave Bone

BLOOD STREAM

Increases Ca in Blood

Increases concentration of Ca and P in Blood

GUT

Increases absorption of dietary Ca and reduces loss of Ca in faeces

Alterations in concentration of CALCIUM ions in extracellular fluids control Parathyroid activity. A rise in blood Calcium depresses Parathyroid secretion. A fall in Calcium increases Parathyroid secretion. [Note: This response to the level of plasma Ca differs from thyrocalcitonin's - see 'thyroid'.]

UNDERACTIVITY of PARATHYROIDS

Atrophy or removal of Parathyroid tissue causes a fall in BLOOD CALCIUM level and increased excitability of Neuromuscular tissue. This leads to severe convulsive disorder — TETANY.

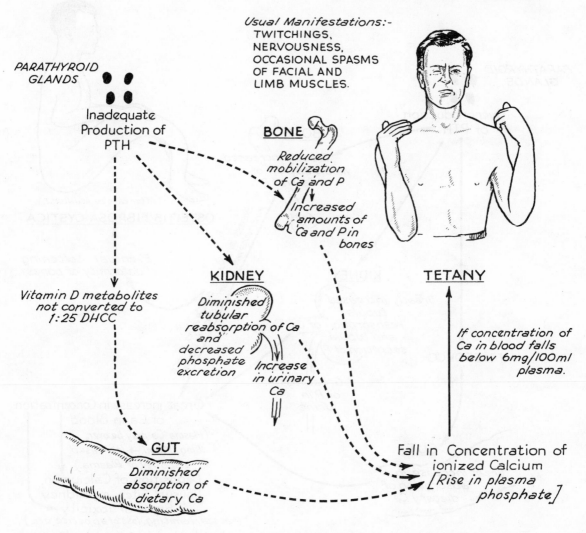

Usual Manifestations:-
TWITCHINGS,
NERVOUSNESS,
OCCASIONAL SPASMS
OF FACIAL AND
LIMB MUSCLES.

PARATHYROID
GLANDS

Inadequate
Production of
PTH

BONE
Reduced
mobilization
of Ca and P
↑↓
Increased
amounts of
Ca and P in
bones

TETANY

Vitamin D metabolites
not converted to
1:25 DHCC

KIDNEY
Diminished
tubular
reabsorption of Ca
and
decreased
phosphate
excretion
Increase
in urinary
Ca

If concentration of
Ca in blood falls
below 6mg/100ml
plasma.

GUT
Diminished
absorption of
dietary Ca

Fall in Concentration of
ionized Calcium
[Rise in plasma
phosphate]

[Note the inverse relationship between plasma calcium and inorganic phosphate]

Symptoms are relieved by injection of Calcium or of Extract of Parathyroid.

155

OVERACTIVITY of PARATHYROIDS

Overactivity of the Parathyroids *(due often to tumour)* leads to rise in BLOOD CALCIUM level and eventually to <u>OSTEITIS FIBROSA CYSTICA</u>.

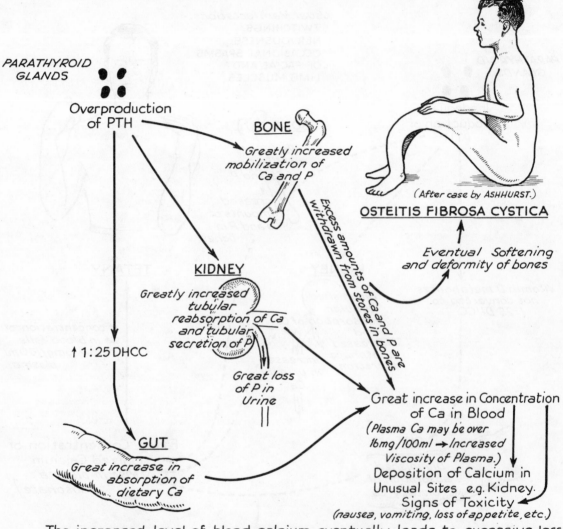

PARATHYROID GLANDS

Overproduction of PTH

↑ 1 : 25 DHCC

BONE
Greatly increased mobilization of Ca and P

Excess amounts of Ca and P are withdrawn from stores in bones

KIDNEY
Greatly increased tubular reabsorption of Ca and tubular secretion of P

Great loss of P in Urine

GUT
Great increase in absorption of dietary Ca

(After case by ASHHURST.)
OSTEITIS FIBROSA CYSTICA

Eventual Softening and deformity of bones

Great increase in Concentration of Ca in Blood
(Plasma Ca may be over 16mg/100ml → Increased Viscosity of Plasma.)
Deposition of Calcium in Unusual Sites e.g. Kidney.
Signs of Toxicity
(nausea, vomiting, loss of appetite, etc.)

The increased level of blood calcium eventually leads to excessive loss of CALCIUM in URINE and also of WATER since the salts are excreted in solution. POLYURIA and THIRST result.

Excision of the overactive Parathyroid tissue abolishes syndrome.

156

SUPRARENAL GLANDS

There are TWO Suprarenal Glands. They lie close to the kidneys.

Each has an outer CORTEX and an inner MEDULLA

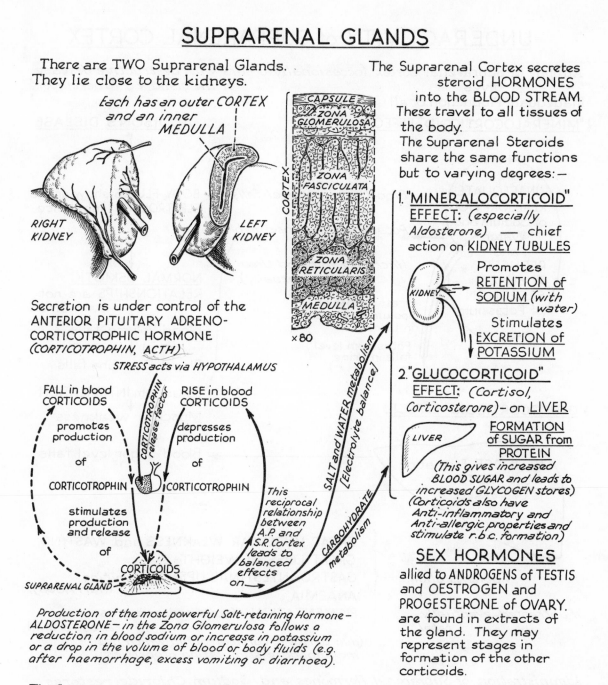

RIGHT KIDNEY

LEFT KIDNEY

CAPSULE
ZONA GLOMERULOSA
CORTEX
ZONA FASCICULATA
ZONA RETICULARIS
MEDULLA

×80

Secretion is under control of the ANTERIOR PITUITARY ADRENO-CORTICOTROPHIC HORMONE (*CORTICOTROPHIN, ACTH*)

STRESS *acts via* HYPOTHALAMUS

CORTICOTROPHIN release factor

FALL in blood CORTICOIDS

RISE in blood CORTICOIDS

promotes production of CORTICOTROPHIN

depresses production of CORTICOTROPHIN

stimulates production and release of CORTICOIDS

CORTICOIDS

SUPRARENAL GLAND

This reciprocal relationship between A.P. and S.R. Cortex leads to balanced effects on →

SALT *and* WATER *metabolism* [Electrolyte balance]

CARBOHYDRATE *metabolism*

Production of the most powerful Salt-retaining Hormone – ALDOSTERONE – in the Zona Glomerulosa follows a reduction in blood sodium or increase in potassium or a drop in the volume of blood or body fluids (e.g. after haemorrhage, excess vomiting or diarrhoea).

The Suprarenal Cortex secretes steroid HORMONES into the BLOOD STREAM. These travel to all tissues of the body.
The Suprarenal Steroids share the same functions but to varying degrees:—

1. "MINERALOCORTICOID" EFFECT: (especially Aldosterone) — chief action on KIDNEY TUBULES

KIDNEY

Promotes RETENTION of SODIUM (with water)
Stimulates EXCRETION of POTASSIUM

2. "GLUCOCORTICOID" EFFECT: (Cortisol, Corticosterone) – on LIVER

LIVER

FORMATION of SUGAR from PROTEIN
(*This gives increased BLOOD SUGAR and leads to increased GLYCOGEN stores*) (*Corticoids also have Anti-inflammatory and Anti-allergic properties and stimulate r.b.c. formation*)

SEX HORMONES

allied to ANDROGENS of TESTIS and OESTROGEN and PROGESTERONE of OVARY. are found in extracts of the gland. They may represent stages in formation of the other corticoids.

The Suprarenal Cortex is essential to life and plays an important rôle in states of stress.

UNDERACTIVITY of SUPRARENAL CORTEX

Atrophy of Suprarenal Cortex *(occasionally occurs with destructive disease of the gland, e.g. Tuberculosis)* gives
Inadequate Production of all CORTICOIDS:-

↓"MINERALOCORTICOID" EFFECT :   ADDISON'S DISEASE

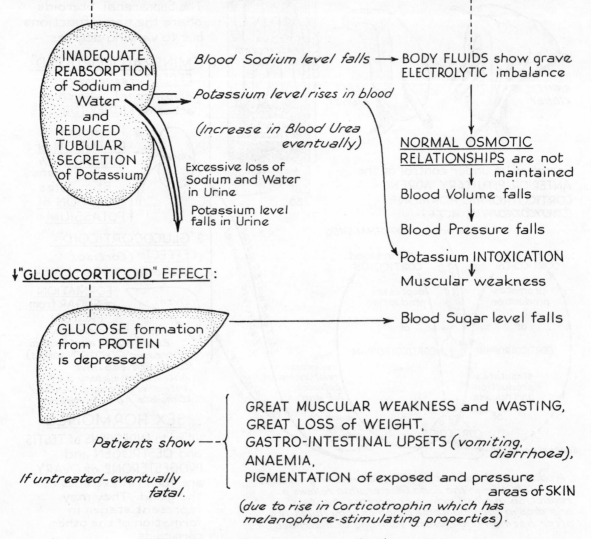

INADEQUATE REABSORPTION of Sodium and Water and REDUCED TUBULAR SECRETION of Potassium

Blood Sodium level falls → BODY FLUIDS show grave ELECTROLYTIC imbalance

Potassium level rises in blood

(Increase in Blood Urea eventually)

NORMAL OSMOTIC RELATIONSHIPS are not maintained

Excessive loss of Sodium and Water in Urine

Blood Volume falls

Potassium level falls in Urine

Blood Pressure falls

Potassium INTOXICATION

Muscular weakness

↓"GLUCOCORTICOID" EFFECT :

GLUCOSE formation from PROTEIN is depressed → Blood Sugar level falls

Patients show ─ ─{
GREAT MUSCULAR WEAKNESS and WASTING,
GREAT LOSS of WEIGHT,
GASTRO-INTESTINAL UPSETS *(vomiting, diarrhoea),*
ANAEMIA,
PIGMENTATION of exposed and pressure areas of SKIN

If untreated - eventually fatal.

(due to rise in Corticotrophin which has melanophore-stimulating properties).

Administration of Suprarenal Hormones and Sodium Chloride restores individual to normal.

158

OVERACTIVITY of SUPRARENAL CORTEX

Overactivity or tumour of Suprarenal Cortex may give
Overproduction of any or all of the CORTICOIDS:-

e.g. <u>ALDOSTERONE</u> ⟶ **PRIMARY ALDOSTERONISM**

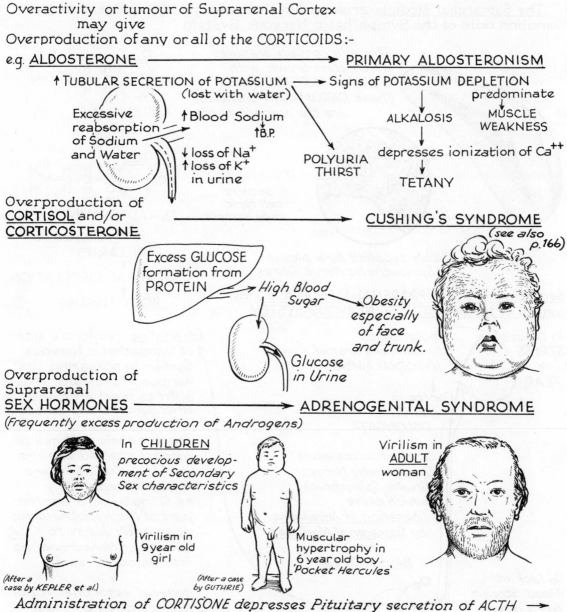

↑TUBULAR SECRETION of POTASSIUM ⟶ Signs of POTASSIUM DEPLETION
(lost with water)

Excessive reabsorption of Sodium and Water

↑Blood Sodium
↑B.P.
↓ loss of Na⁺
↑ loss of K⁺ in urine

predominate

ALKALOSIS   MUSCLE WEAKNESS

depresses ionization of Ca^{++}

POLYURIA THIRST

TETANY

Overproduction of <u>CORTISOL</u> and/or <u>CORTICOSTERONE</u> ⟶ **CUSHING'S SYNDROME**
(see also p.166)

Excess GLUCOSE formation from PROTEIN ⟶ High Blood Sugar ⟶ Obesity especially of face and trunk.

Glucose in Urine

Overproduction of Suprarenal <u>SEX HORMONES</u> ⟶ **ADRENOGENITAL SYNDROME**

(Frequently excess production of Androgens)

In <u>CHILDREN</u> precocious development of Secondary Sex characteristics

Virilism in 9 year old girl

(After a case by KEPLER et al.)

Muscular hypertrophy in 6 year old boy 'Pocket Hercules'

(After a case by GUTHRIE)

Virilism in <u>ADULT</u> woman

Administration of CORTISONE depresses Pituitary secretion of ACTH ⟶ inhibits production of the abnormal steroids.
Removal of the over-secreting tissue or tumour restores individual.

SUPRARENAL MEDULLA

The Suprarenal Medulla arises from the same primitive tissue as the Ganglion cells of the Sympathetic Nervous System.

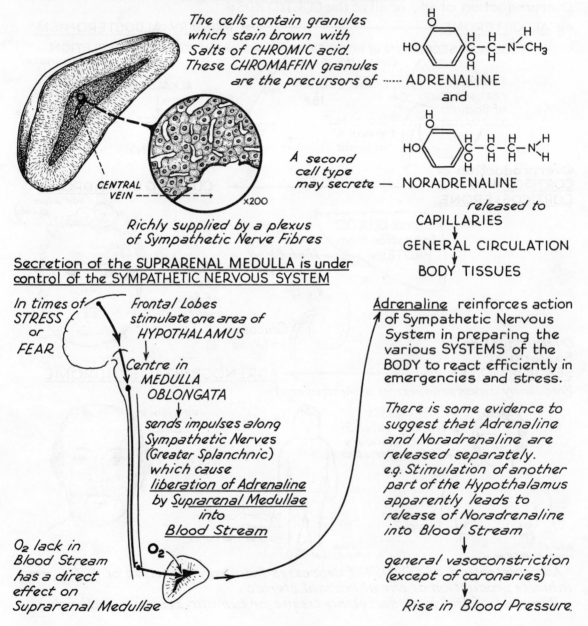

The cells contain granules which stain brown with Salts of CHROMIC acid. These CHROMAFFIN granules are the precursors of ----- ADRENALINE

and

A second cell type may secrete — NORADRENALINE

released to
CAPILLARIES
↓
GENERAL CIRCULATION
↓
BODY TISSUES

CENTRAL VEIN - - - - -→

×200

Richly supplied by a plexus of Sympathetic Nerve Fibres

Secretion of the SUPRARENAL MEDULLA is under control of the SYMPATHETIC NERVOUS SYSTEM

In times of STRESS or FEAR

Frontal Lobes stimulate one area of HYPOTHALAMUS

Centre in MEDULLA OBLONGATA

sends impulses along Sympathetic Nerves (Greater Splanchnic) which cause liberation of Adrenaline by Suprarenal Medullae into Blood Stream

O₂ lack in Blood Stream has a direct effect on Suprarenal Medullae

O₂

Adrenaline reinforces action of Sympathetic Nervous System in preparing the various SYSTEMS of the BODY to react efficiently in emergencies and stress.

There is some evidence to suggest that Adrenaline and Noradrenaline are released separately. e.g. Stimulation of another part of the Hypothalamus apparently leads to release of Noradrenaline into Blood Stream

↓

general vasoconstriction (except of coronaries)

↓

Rise in Blood Pressure.

ADRENALINE

Under quiet resting conditions the blood contains very little adrenaline. During excitement or circumstances which demand special efforts adrenaline is released into the blood stream, and is responsible for the following actions summed up as the "*FIGHT or FLIGHT*" FUNCTION of the Suprarenal medullae.

It _Constricts_ Smooth Muscle of Skin → Hairs 'stand on end'; 'Gooseflesh'.

Dilates Pupil of Eye to admit more light.

Constricts Smooth Muscle of Abdominal Blood Vessels and Cutaneous Blood Vessels → Pallor with Fright.

Dilates Smooth Muscle in Blood Vessels of Heart (Coronaries) and of Skeletal Muscles, i.e. better supply to organs requiring it in emergency.

Increases Heart Rate and Cardiac Output.

Relaxes Smooth Muscle in Wall of Bronchioles → better supply of air to alveoli.

Stimulates Respiration.

Inhibits Movements of Digestive Tract.

Contracts Sphincters of Gut.

Inhibits Wall of Urinary Bladder.

Contracts Ureters and Sphincter of Urinary Bladder.

Mobilizes Muscle and Liver Glycogen → increase in Blood Sugar.

Stimulates Metabolism.

Exerts favourable effect on contracting Skeletal Muscle → Fatigues less readily.

Increases Coagulability of Blood.

Most of these effects can also be produced by stimulating Sympathetic Nerve Fibres.

The Suprarenal Medullae are not essential to life — but without them the body is less able to face emergencies and conditions of stress.

DEVELOPMENT of PITUITARY

The Pituitary Gland consists of ANTERIOR and POSTERIOR parts which differ in ORIGIN, STRUCTURE and FUNCTION.

3rd VENTRICLE

POSTERIOR PART
arises as DOWNGROWTH from FLOOR of PRIMITIVE BRAIN

ANTERIOR PART
arises as UPGROWTH from ROOF of PRIMITIVE MOUTH CAVITY

RATHKE'S POUCH → forms vesicle and connection with mouth cavity disappears

Connection with base of brain persists

The TWO parts meet and fuse

Part of Rathke's Pouch extends upwards to base of brain to form ----→ PARS TUBERALIS

Cavity of Rathke's Pouch is reduced to narrow spaces

Anterior wall of Rathke's Pouch thickens to form ----→ PARS GLANDULARIS (DISTALIS)

Posterior wall of Rathke's Pouch thins down to form ----→ PARS INTERMEDIA

(After GARVEN)

ANTERIOR LOBE

POSTERIOR LOBE

INFUNDIBULAR STALK

PARS NERVOSA

The ADULT PITUITARY (HYPOPHYSIS) is a small oval gland which lies in the SELLA TURCICA — a small cavity in the bone at the base of the skull.

ANTERIOR PITUITARY

This is the MASTER GLAND of the ENDOCRINE SYSTEM.
It regulates the activity of the other Endocrine
Glands, including the GONADS, and influences
ALL METABOLIC PROCESSES
including GROWTH.

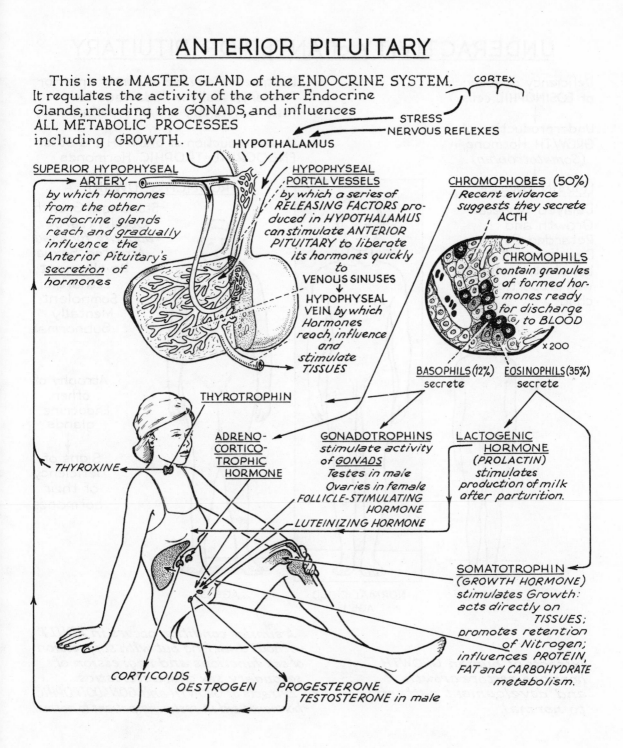

CORTEX

STRESS
NERVOUS REFLEXES

HYPOTHALAMUS

SUPERIOR HYPOPHYSEAL
ARTERY—
by which Hormones
from the other
Endocrine glands
reach and gradually
influence the
Anterior Pituitary's
secretion of
hormones

HYPOPHYSEAL
PORTAL VESSELS
by which a series of
RELEASING FACTORS pro-
duced in HYPOTHALAMUS
can stimulate ANTERIOR
PITUITARY to liberate
its hormones quickly
to
VENOUS SINUSES

HYPOPHYSEAL
VEIN by which
Hormones
reach, influence
and
stimulate
TISSUES

CHROMOPHOBES (50%)
Recent evidence
suggests they secrete
ACTH

CHROMOPHILS
contain granules
of formed hor-
mones ready
for discharge
to BLOOD

x 200

BASOPHILS (12%)   EOSINOPHILS (35%)
secrete   secrete

THYROTROPHIN

THYROXINE

ADRENO-
CORTICO-
TROPHIC
HORMONE

GONADOTROPHINS
stimulate activity
of GONADS
Testes in male
Ovaries in female
FOLLICLE-STIMULATING
HORMONE
LUTEINIZING HORMONE

LACTOGENIC
HORMONE
(PROLACTIN)
stimulates
production of milk
after parturition.

SOMATOTROPHIN
(GROWTH HORMONE)
stimulates Growth:
acts directly on
TISSUES:
promotes retention
of Nitrogen;
influences PROTEIN,
FAT and CARBOHYDRATE
metabolism.

CORTICOIDS   OESTROGEN   PROGESTERONE
TESTOSTERONE in male

UNDERACTIVITY of ANTERIOR PITUITARY

Deficiency or absence of **EOSINOPHIL** cells

↓

Underproduction of GROWTH Hormone (*Somatotrophin*)

Destructive disease of part of Anterior Pituitary (usually with damage to Posterior Pituitary and/or Hypothalamus)

↓

Underproduction of GROWTH and other ENDOCRINE-TROPHIC Hormones

LORAIN DWARF

Delayed Skeletal Growth and Retarded Sexual Development but alert, intelligent, well proportioned child.

FRÖHLICH'S DWARF

Stunting of Growth, Obesity (*Large appetite for sugar*); Arrested Sexual Development; Lethargic; Somnolent; Mentally Subnormal.

Atrophy of other Endocrine glands

↓

Signs of deficiency of their hormones.

AGE 13

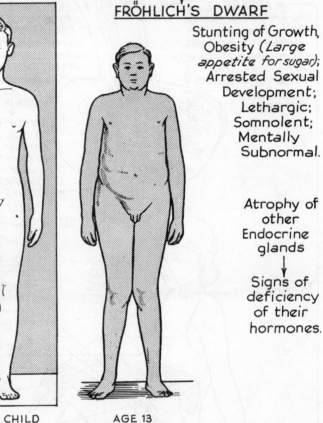

NORMAL CHILD
AGE 13

AGE 13

Extracts of human GROWTH hormone restore growth and development pattern to normal.

A similar condition occurs in ADULT without dwarfing but with suppression of sex functions and regression of secondary sex characteristics.
Extracts of GROWTH and GONADOTROPHIC hormones aid in restoring patient to normal.

OVERACTIVITY of PITUITARY EOSINOPHIL CELLS

Functional overactivity (or tumour) chiefly of the EOSINOPHIL cells of the Anterior Pituitary leads to ⟶ GIANTISM in the CHILD: ACROMEGALY in the ADULT.

Overproduction of GROWTH Hormone

General Circulation

Increases NITROGEN retention. Influences Protein, Carbo-hydrate and Fat metabolism of ALL CELLS of the body.

Overgrowth of all Body Tissues

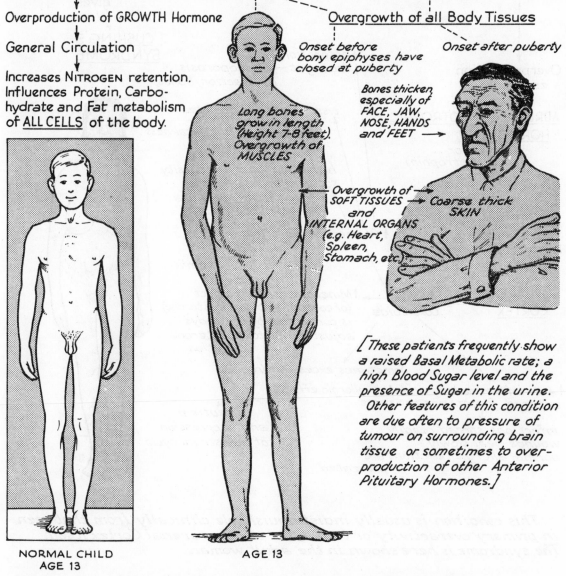

Onset before bony epiphyses have closed at puberty

Onset after puberty

Long bones grow in length (height 7-8 feet). Overgrowth of MUSCLES

Bones thicken, especially of FACE, JAW, NOSE, HANDS and FEET ⟶

Overgrowth of ⟶ SOFT TISSUES and INTERNAL ORGANS (e.g. Heart, Spleen, Stomach, etc.)

Coarse thick SKIN

[These patients frequently show a raised Basal Metabolic rate; a high Blood Sugar level and the presence of Sugar in the urine.
Other features of this condition are due often to pressure of tumour on surrounding brain tissue or sometimes to over-production of other Anterior Pituitary Hormones.]

NORMAL CHILD
AGE 13

AGE 13

Destruction of the overactive tissue — usually by RADIUM therapy — prevents progression of the condition.

OVERACTIVITY of PITUITARY BASOPHIL CELLS

Overactivity (*often due to Tumour*) of the Basophil cells of the Anterior Pituitary
gives

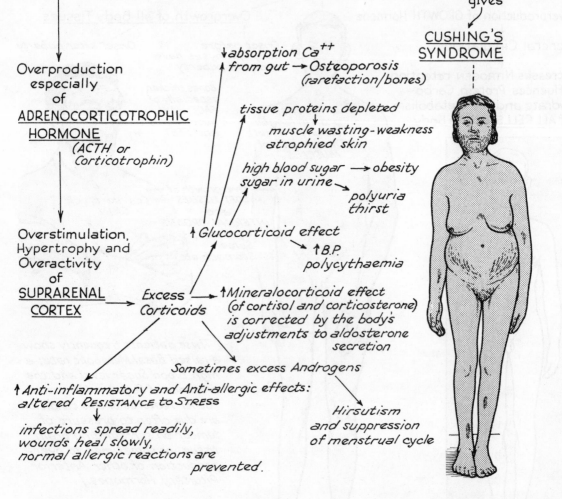

Overproduction
especially
of
ADRENOCORTICOTROPHIC
HORMONE
(*ACTH or
Corticotrophin*)

Overstimulation,
Hypertrophy and
Overactivity
of
SUPRARENAL
CORTEX

Excess
Corticoids

↓absorption Ca++
↑ from gut → Osteoporosis
(rarefaction/bones)

CUSHING'S
SYNDROME

tissue proteins depleted
muscle wasting-weakness
atrophied skin

high blood sugar → obesity
sugar in urine
polyuria
thirst

↑Glucocorticoid effect
→ ↑B.P.
polycythaemia

↑Mineralocorticoid effect
(of cortisol and corticosterone)
is corrected by the body's
adjustments to aldosterone
secretion

Sometimes excess Androgens

↑Anti-inflammatory and Anti-allergic effects:
altered RESISTANCE to STRESS

infections spread readily,
wounds heal slowly,
normal allergic reactions are
prevented.

Hirsutism
and suppression
of menstrual cycle

*This condition is usually indistinguishable clinically from that seen
in primary overactivity or tumour of the Suprarenal Cortex itself.
The syndrome is here shown in the adult woman.*

Overproduction of THYROTROPHIN ⟶ *Overactivity* of THYROID *gland.*

PANHYPOPITUITARISM

Complete Atrophy (or insufficiency) of all secreting cells of Anterior Pituitary
in Adult — SIMMOND'S DISEASE

FAILURE to
PRODUCE
ANY HORMONES ——————→

APPEARANCE of PREMATURE SENILITY

Features usually
associated with
very OLD AGE
|

LACK of
GROWTH HORMONE

↘ *Grave upset in* —————→
  TISSUE metabolism

HAIR *grey, sparse:*
  loss of body hair.
SKIN *dry, sallow,*
  wrinkled.
BODY *emaciated*
  (great loss of weight)
BONES *frail*

LACK of
GONADOTROPHINS ——————→

SEX ORGANS *atrophy.*
  Menstruation ceases.
  Reproductive cycle stops.
  Secondary sex
    characteristics
    gradually regress.

*[This patient is
only 42 years of
age]*
After ZONDEK, *Diseases
of the Endocrine Glands.*

LACK of
ENDOCRINE-TROPHIC ——————→
HORMONES

ALL ENDOCRINES
  *atrophy
  and show
  depressed
  secretion of
  their Hormones*

}——→ Basal metabolism depressed

  Body temperature depressed,
  Heart rate low.
  Blood pressure low.
  Blood sugar low.
  Electrolytic upset.

  *(Mental changes
  supervene)*

*Extracts of Anterior Pituitary may relieve the condition but rarely
succeed in completely restoring the patient to normal.*

POSTERIOR PITUITARY

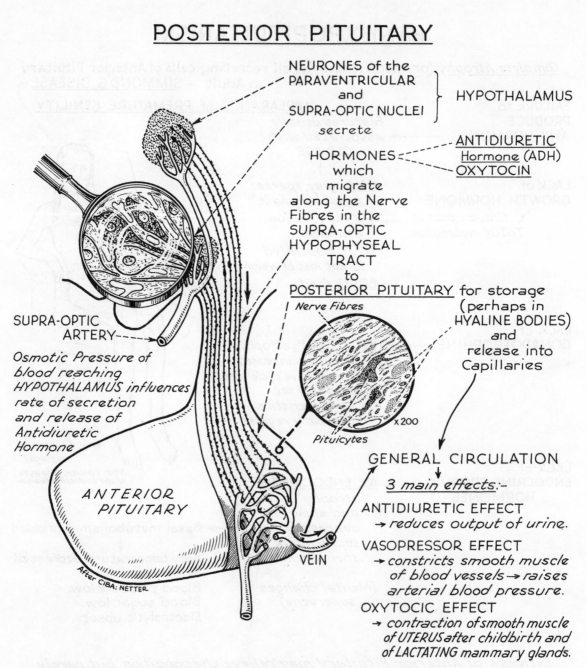

NEURONES of the
PARAVENTRICULAR
and
SUPRA-OPTIC NUCLEI
secrete

HYPOTHALAMUS

HORMONES
which
migrate
along the Nerve
Fibres in the
SUPRA-OPTIC
HYPOPHYSEAL
TRACT
to
POSTERIOR PITUITARY

ANTIDIURETIC
Hormone (ADH)
OXYTOCIN

for storage
(perhaps in
HYALINE BODIES)
and
release into
Capillaries

Nerve Fibres

SUPRA-OPTIC
ARTERY

Osmotic Pressure of
blood reaching
HYPOTHALAMUS influences
rate of secretion
and release of
Antidiuretic
Hormone

Pituicytes

x 200

ANTERIOR
PITUITARY

After CIBA: NETTER.

VEIN

GENERAL CIRCULATION

3 main effects:-

ANTIDIURETIC EFFECT
→ reduces output of urine.

VASOPRESSOR EFFECT
→ constricts smooth muscle
of blood vessels → raises
arterial blood pressure.

OXYTOCIC EFFECT
→ contraction of smooth muscle
of UTERUS after childbirth and
of LACTATING mammary glands.

There are no obvious secreting cells in Posterior Pituitary such as are
seen in other ENDOCRINE GLANDS.

OXYTOCIN

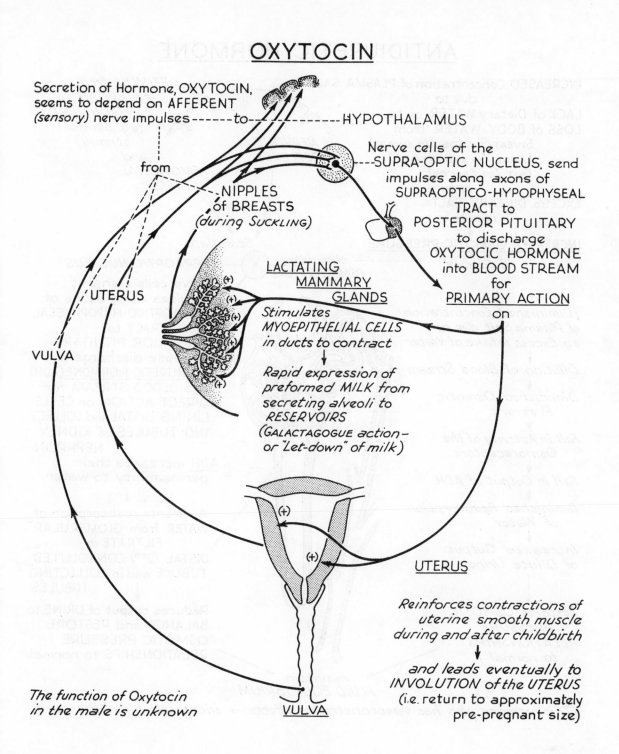

Secretion of Hormone, OXYTOCIN, seems to depend on AFFERENT (sensory) nerve impulses - - - - - - to - - - - - - - - - HYPOTHALAMUS

from

NIPPLES of BREASTS (during Suckling)

Nerve cells of the SUPRA-OPTIC NUCLEUS, send impulses along axons of SUPRAOPTICO-HYPOPHYSEAL TRACT to POSTERIOR PITUITARY to discharge OXYTOCIC HORMONE into BLOOD STREAM for PRIMARY ACTION on

UTERUS

VULVA

LACTATING MAMMARY GLANDS

(+) (+) (+) (+) (+)

Stimulates MYOEPITHELIAL CELLS in ducts to contract

Rapid expression of preformed MILK from secreting alveoli to RESERVOIRS (Galactagogue action - or "Let-down" of milk)

(+)

(+)

UTERUS

Reinforces contractions of uterine smooth muscle during and after childbirth

and leads eventually to INVOLUTION of the UTERUS (i.e. return to approximately pre-pregnant size)

The function of Oxytocin in the male is unknown

VULVA

ANTIDIURETIC HORMONE

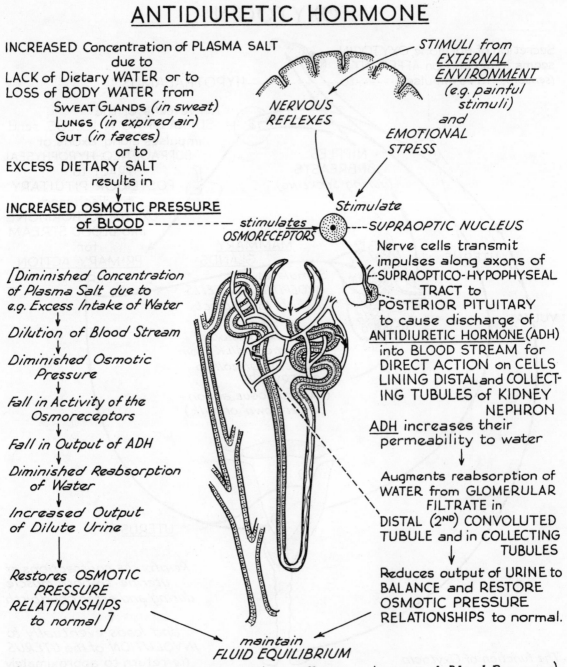

INCREASED Concentration of PLASMA SALT
due to
LACK of Dietary WATER or to
LOSS of BODY WATER from
   Sweat Glands (in sweat)
   Lungs (in expired air)
   Gut (in faeces)
      or to
EXCESS DIETARY SALT
   results in

INCREASED OSMOTIC PRESSURE
   of BLOOD - - - - - - - - - stimulates
                              OSMORECEPTORS

STIMULI from
EXTERNAL
ENVIRONMENT
(e.g. painful
stimuli)
and
EMOTIONAL
STRESS

NERVOUS
REFLEXES

Stimulate

SUPRAOPTIC NUCLEUS
Nerve cells transmit
impulses along axons of
SUPRAOPTICO-HYPOPHYSEAL
   TRACT to
POSTERIOR PITUITARY
to cause discharge of
ANTIDIURETIC HORMONE (ADH)
into BLOOD STREAM for
DIRECT ACTION on CELLS
LINING DISTAL and COLLECT-
ING TUBULES of KIDNEY
   NEPHRON
ADH increases their
   permeability to water

Augments reabsorption of
WATER from GLOMERULAR
   FILTRATE in
DISTAL (2ND) CONVOLUTED
TUBULE and in COLLECTING
   TUBULES

Reduces output of URINE to
BALANCE and RESTORE
OSMOTIC PRESSURE
RELATIONSHIPS to normal.

[Diminished Concentration
of Plasma Salt due to
e.g. Excess Intake of Water

Dilution of Blood Stream

Diminished Osmotic
Pressure

Fall in Activity of the
Osmoreceptors

Fall in Output of ADH

Diminished Reabsorption
of Water

Increased Output
of Dilute Urine

Restores OSMOTIC
PRESSURE
RELATIONSHIPS
to normal]

maintain
FLUID EQUILIBRIUM

(This hormone also has vasoconstrictor effects → increased Blood Pressure)

UNDERACTIVITY of POSTERIOR PITUITARY

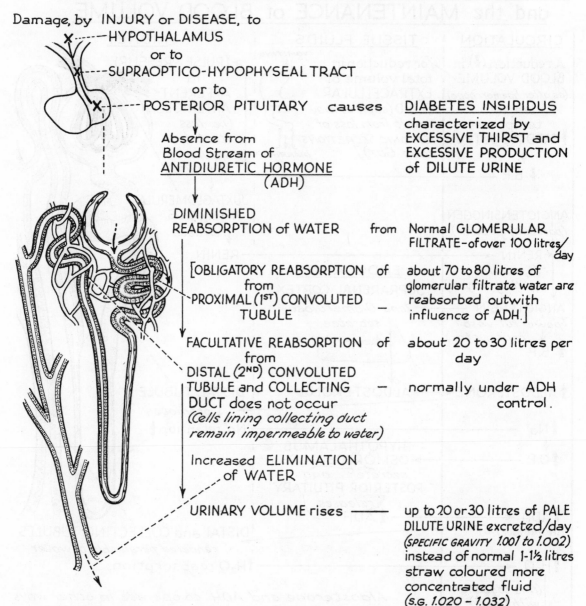

Damage, by INJURY or DISEASE, to

X ----- HYPOTHALAMUS
   or to
X ----- SUPRAOPTICO-HYPOPHYSEAL TRACT
    or to
X ----- POSTERIOR PITUITARY causes **DIABETES INSIPIDUS**

Absence from
Blood Stream of
<u>ANTIDIURETIC HORMONE</u>
(ADH)

characterized by
EXCESSIVE THIRST and
EXCESSIVE PRODUCTION
of DILUTE URINE

DIMINISHED
REABSORPTION of WATER from Normal GLOMERULAR
FILTRATE-of over 100 litres/day

[OBLIGATORY REABSORPTION of
from
PROXIMAL (1ST) CONVOLUTED —
TUBULE

about 70 to 80 litres of
glomerular filtrate water are
reabsorbed outwith
influence of ADH.]

FACULTATIVE REABSORPTION of
from
DISTAL (2ND) CONVOLUTED
TUBULE and COLLECTING —
DUCT does not occur
(*Cells lining collecting duct
remain impermeable to water*)

about 20 to 30 litres per
day

normally under ADH
control.

Increased ELIMINATION
of WATER

URINARY VOLUME rises ————

up to 20 or 30 litres of PALE
DILUTE URINE excreted/day
(*SPECIFIC GRAVITY 1.001 to 1.002*)
instead of normal 1-1½ litres
straw coloured more
concentrated fluid
(*s.G. 1.020 – 1.032*)

*Small amounts of Posterior Pituitary extract absorbed from under the tongue
or given by subcutaneous injection reduce elimination of water to normal.*

ALDOSTERONE and ANTIDIURETIC HORMONE (ADH) and the MAINTENANCE of BLOOD VOLUME

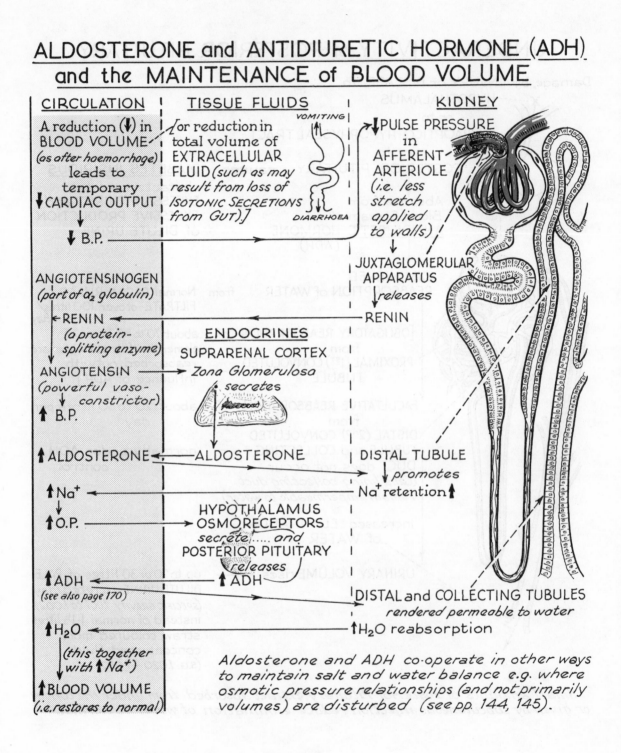

CIRCULATION

A reduction (↓) in BLOOD VOLUME (as after haemorrhage) leads to temporary ↓CARDIAC OUTPUT

↓ B.P.

ANGIOTENSINOGEN (part of α_2 globulin)

← RENIN ← (a protein-splitting enzyme)

ANGIOTENSIN (powerful vaso-constrictor)

↑ B.P.

↑ALDOSTERONE ←

↑Na⁺ ←

↑O.P.

↑ADH ← (see also page 170)

↑H₂O ← (this together with ↑Na⁺)

↑BLOOD VOLUME (i.e. restores to normal)

TISSUE FLUIDS

[or reduction in total volume of EXTRACELLULAR FLUID (such as may result from loss of ISOTONIC SECRETIONS from GUT).]

VOMITING

DIARRHOEA

ENDOCRINES

SUPRARENAL CORTEX

Zona Glomerulosa secretes

ALDOSTERONE

HYPOTHALAMUS OSMORECEPTORS secrete......and POSTERIOR PITUITARY releases

↑ADH

KIDNEY

↓PULSE PRESSURE in AFFERENT ARTERIOLE (i.e. less stretch applied to walls)

JUXTAGLOMERULAR APPARATUS ↓releases

RENIN

DISTAL TUBULE ↓promotes Na⁺retention↑

DISTAL and COLLECTING TUBULES rendered permeable to water

↑H₂O reabsorption

Aldosterone and ADH co-operate in other ways to maintain salt and water balance e.g. where osmotic pressure relationships (and not primarily volumes) are disturbed (see pp. 144, 145).

PANCREAS: ISLETS of LANGERHANS

Islets of Langerhans make up 1-2% of Pancreatic Tissue.

α cells — secrete ⟶ GLUCAGON
(25% of Islets)

BLOOD STREAM

promotes GLYCOGEN breakdown

increases Blood Sugar and
utilization of GLUCOSE by tissues

Its effects are overshadowed by

x150

β cells — secrete ⟶ INSULIN
(50-75% of Islets)   [In cells probably in combination with Zinc]

SECRETION of INSULIN appears to be stimulated
primarily by a RISE in BLOOD SUGAR reaching Islets

BLOOD STREAM

Increases GLUCOSE UPTAKE by all
"GLUCOSE-consuming tissues"– i.e.
It plays important rôle in
CARBOHYDRATE METABOLISM,
probably regulating entrance of
GLUCOSE to cells.

ANTERIOR PITUITARY

GROWTH HORMONE ⎫
                ⎬ oppose action
SUPRARENAL   ⎭ of Insulin
GLUCOCORTICOIDS

CHIEF ACTIONS on:-

KIDNEY   *stimulates reabsorption*
TUBULES   *of GLUCOSE from FILTRATE*

ATROPHY ⟶ *DIABETES*
of ISLETS   *MELLITUS*
             complex
             disorder

Absence of
Insulin ⟶ HIGH BLOOD SUGAR ⟶
           SUGAR in URINE ⟶
           POLYURIA (since sugar lost
                in solution in water)

THIRST ⟶ POLYDIPSIA and POLYPHAGIA

LIVER *reduces GLYCOGEN*   ⎫
       breakdown ⟶ GLUCOSE   ⎬ BLOOD
                                ⎮ SUGAR
MUSCLES *increases GLYCO-*   ⎮ LOWERED
        GEN formation   ⎭
        from GLUCOSE

Muscles are unable to use GLUCOSE efficiently ⟶ Great weakness.

Fat metabolized instead of Carbohydrate ⟨ Fat stores depleted ⟶ Wasting and loss of weight.
                                          Ketone bodies in Blood and Urine.

If untreated ⟶ progressive ⟶ drowsiness ⟶ coma ⟶ death.

Excess Insulin (*HYPERINSULINISM*) ⟶ low blood sugar ⟶ irritability; sweating; hunger.
If untreated ⟶ reduction of metabolism of nervous tissues ⟶ giddiness ⟶ coma ⟶ death.

THYMUS

The Thymus is an irregularly-shaped organ lying behind the breast bone. It is relatively large in the child and reaches its maximum size at puberty. It closely resembles a Lymph Node.

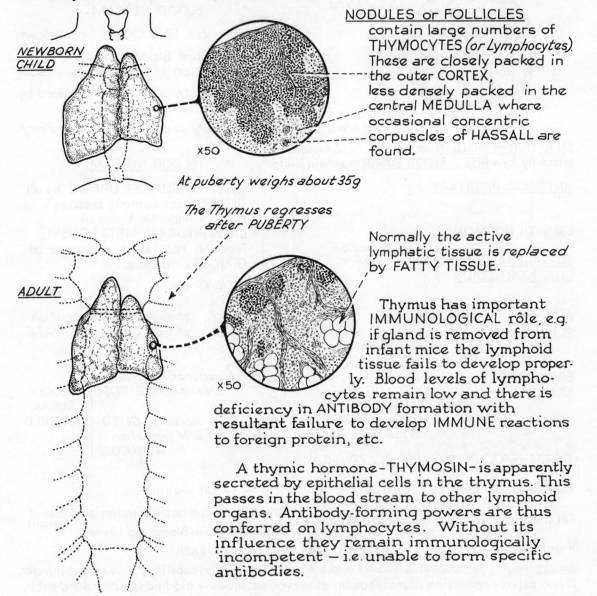

NEWBORN CHILD

x50

At puberty weighs about 35g

NODULES or FOLLICLES contain large numbers of **THYMOCYTES** (or Lymphocytes). These are closely packed in the outer **CORTEX**, less densely packed in the central **MEDULLA** where occasional concentric corpuscles of **HASSALL** are found.

The Thymus regresses after PUBERTY

ADULT

x50

Normally the active lymphatic tissue is *replaced* by **FATTY TISSUE**.

Thymus has important **IMMUNOLOGICAL** rôle, e.g. if gland is removed from infant mice the lymphoid tissue fails to develop properly. Blood levels of lymphocytes remain low and there is deficiency in **ANTIBODY** formation with resultant failure to develop **IMMUNE** reactions to foreign protein, etc.

A thymic hormone – **THYMOSIN** – is apparently secreted by epithelial cells in the thymus. This passes in the blood stream to other lymphoid organs. Antibody-forming powers are thus conferred on lymphocytes. Without its influence they remain immunologically 'incompetent' – i.e. unable to form specific antibodies.

174

CHAPTER 8

REPRODUCTIVE SYSTEM

MALE REPRODUCTIVE SYSTEM

<u>PRIMARY SEX ORGANS</u>   *produce the MALE GERM CELLS — <u>SPERMATOZOA</u>*
       TESTES (Two)   *and the MALE SEX HORMONE —*
                                 TESTOSTERONE

Testosterone is responsible for
development at Puberty of:-
      <u>SECONDARY SEX ORGANS</u>
         EPIDIDYMIS (Two) ———⎤   *transfer Spermatozoa from the Testes.*
         VAS DEFERENS (Two) ——⎦
         SEMINAL VESICLES (Two) -⎤   *secrete fluid medium for transport of Spermatozoa.*
         PROSTATE GLAND ———⎦
         PENIS ——————— *transfers Spermatozoa from male to female.*

   and
appearance of
<u>SECONDARY SEX CHARACTERISTICS</u>
Laryngeal changes → Deep voice.
Pubic, Axillary and Facial hair.
Characteristic Male shape of
                body.

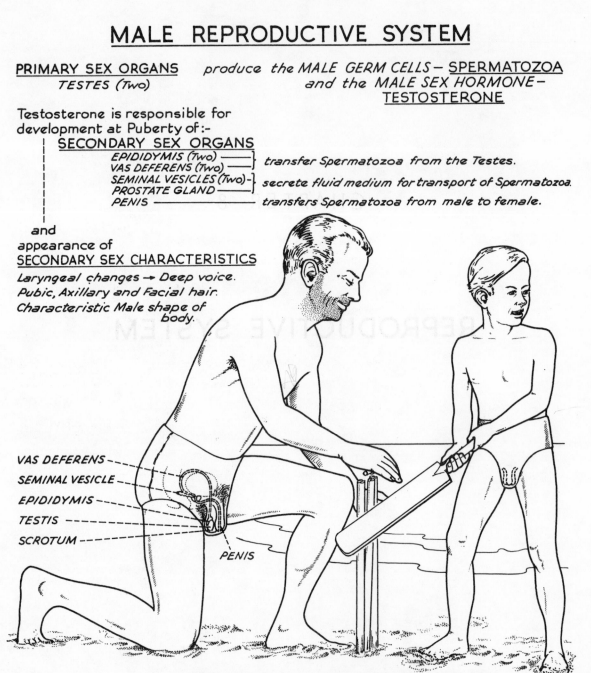

VAS DEFERENS
SEMINAL VESICLE
EPIDIDYMIS
TESTIS
SCROTUM
PENIS

In the male the process of spermatogenesis starts just after
         puberty and is normally continuous until old age.

TESTIS

There are TWO TESTES.
These produce the Male GERM CELLS:—

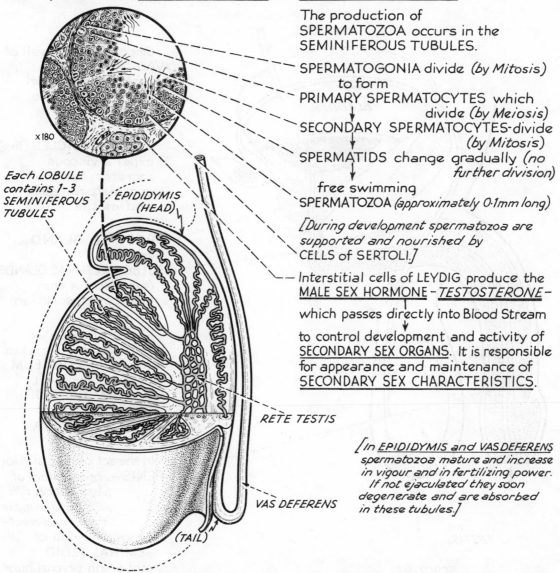

×180

Each LOBULE
contains 1-3
SEMINIFEROUS
TUBULES

EPIDIDYMIS
(HEAD)

RETE TESTIS

VAS DEFERENS

(TAIL)

SPERMATOGENESIS

The production of
SPERMATOZOA occurs in the
SEMINIFEROUS TUBULES.

SPERMATOGONIA divide (by Mitosis)
  to form
PRIMARY SPERMATOCYTES which
  divide (by Meiosis)
SECONDARY SPERMATOCYTES-divide
  (by Mitosis)
SPERMATIDS change gradually (no
  further division)
  free swimming
SPERMATOZOA (approximately 0·1mm long)

[During development spermatozoa are
supported and nourished by
CELLS of SERTOLI.]

Interstitial cells of LEYDIG produce the
MALE SEX HORMONE - TESTOSTERONE -

which passes directly into Blood Stream

to control development and activity of
SECONDARY SEX ORGANS. It is responsible
for appearance and maintenance of
SECONDARY SEX CHARACTERISTICS.

[In EPIDIDYMIS and VASDEFERENS
spermatozoa mature and increase
in vigour and in fertilizing power.
If not ejaculated they soon
degenerate and are absorbed
in these tubules.]

Events occurring in the testes are under CONTROL of HORMONES chiefly
those of ANTERIOR PITUITARY and the HYPOTHALAMUS.

MALE SECONDARY SEX ORGANS

These are the organs adapted for TRANSFER of live SPERMATOZOA from male to female.

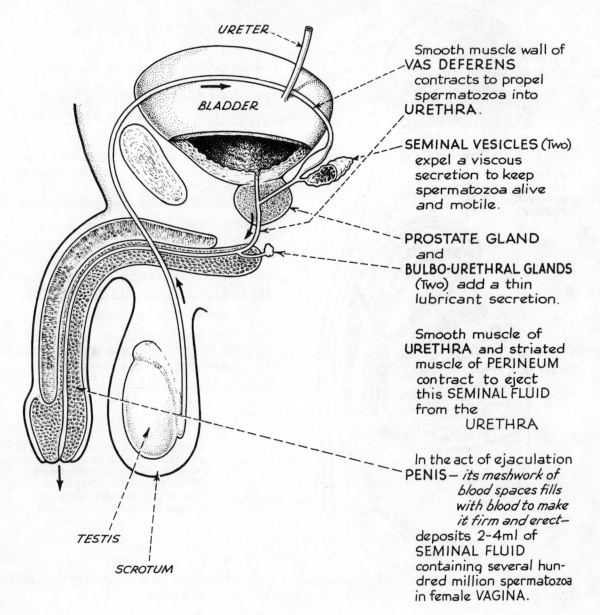

URETER

BLADDER

Smooth muscle wall of **VAS DEFERENS** contracts to propel spermatozoa into **URETHRA.**

SEMINAL VESICLES (Two) expel a viscous secretion to keep spermatozoa alive and motile.

PROSTATE GLAND and **BULBO-URETHRAL GLANDS** (Two) add a thin lubricant secretion.

Smooth muscle of **URETHRA** and striated muscle of PERINEUM contract to eject this SEMINAL FLUID from the URETHRA

In the act of ejaculation **PENIS** — *its meshwork of blood spaces fills with blood to make it firm and erect* — deposits 2-4ml of SEMINAL FLUID containing several hundred million spermatozoa in female VAGINA.

TESTIS

SCROTUM

PUBERTY in MALE

Between the ages of 13 and 16 years Testicular tissue becomes responsive to stimulation by <u>ANTERIOR PITUITARY GONADOTROPHIC HORMONES</u>

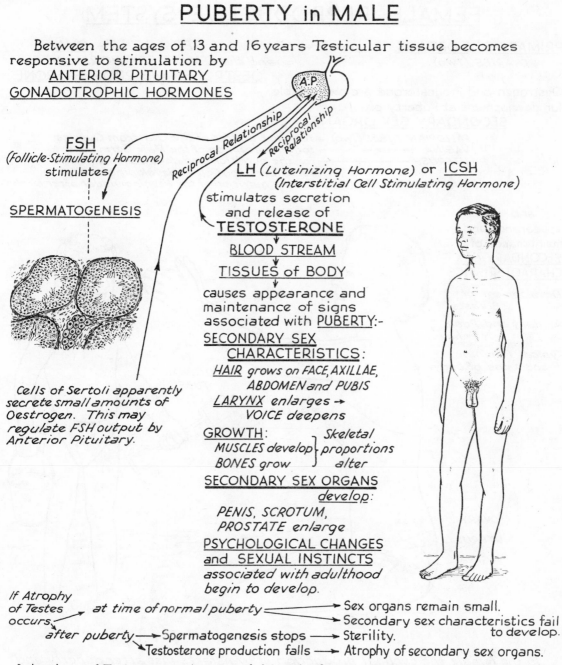

FSH
(Follicle-Stimulating Hormone)
stimulates

<u>SPERMATOGENESIS</u>

Reciprocal Relationship

Reciprocal Relationship

A.P.

<u>LH</u> *(Luteinizing Hormone)* or <u>ICSH</u>
(Interstitial Cell Stimulating Hormone)

stimulates secretion
and release of
TESTOSTERONE

<u>BLOOD STREAM</u>

<u>TISSUES of BODY</u>

causes appearance and
maintenance of signs
associated with <u>PUBERTY</u>:-
<u>SECONDARY SEX</u>
<u>CHARACTERISTICS</u>:

<u>HAIR</u> grows on *FACE, AXILLAE,
ABDOMEN and PUBIS*
<u>LARYNX</u> enlarges →
VOICE deepens

<u>GROWTH</u>:   Skeletal
MUSCLES develop ⎱ proportions
BONES grow   ⎰ alter

<u>SECONDARY SEX ORGANS</u>
<u>develop</u>:

*PENIS, SCROTUM,
PROSTATE enlarge*

<u>PSYCHOLOGICAL CHANGES</u>
and SEXUAL INSTINCTS
*associated with adulthood
begin to develop.*

*Cells of Sertoli apparently
secrete small amounts of
Oestrogen. This may
regulate FSH output by
Anterior Pituitary.*

*If Atrophy
of Testes
occurs,* → *at time of normal puberty* → Sex organs remain small.
↳ Secondary sex characteristics fail
after puberty → Spermatogenesis stops → Sterility.   to develop.
↳ Testosterone production falls → Atrophy of secondary sex organs.

Injections of Testosterone in cases of delayed puberty → changes associated with puberty.

FEMALE REPRODUCTIVE SYSTEM

PRIMARY SEX ORGANS   produce the FEMALE GERM CELLS — OVA
    OVARIES (Two)   and the FEMALE SEX HORMONES —
                                             OESTROGEN and PROGESTERONE

Oestrogen and Progesterone are responsible
for development at Puberty of:-
     SECONDARY SEX ORGANS
        FALLOPIAN TUBES (Two) — for the transfer of the Ova from Ovaries.
        VAGINA ———————————— for the reception of the Male Germ Cells.
        UTERUS ———————————— for the nutrition and development of the
                                fertilized Egg Cell → developing embryo.
        MAMMARY GLANDS (Two)- for the nutrition of the New Individual after birth.

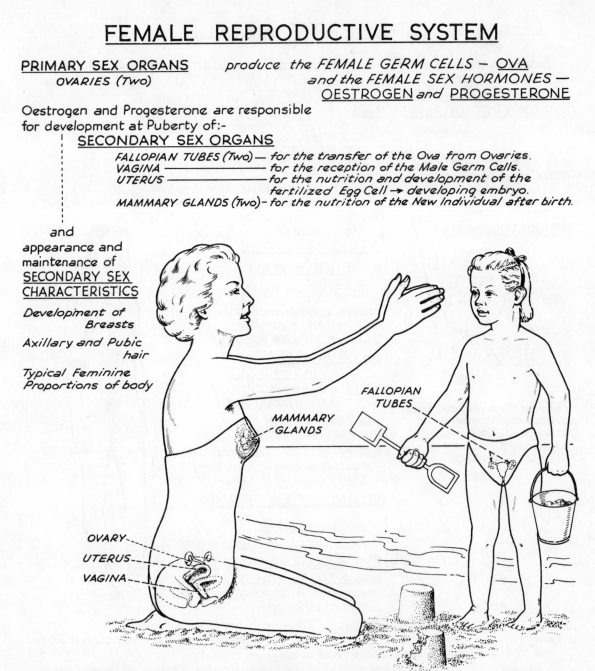

     and
appearance and
maintenance of
**SECONDARY SEX
CHARACTERISTICS**

*Development of
    Breasts*

*Axillary and Pubic
    hair*

*Typical Feminine
Proportions of body*

MAMMARY
GLANDS

FALLOPIAN
TUBES

OVARY

UTERUS

VAGINA

*In the female the cyclical production of ova starts just after puberty and
continues (unless interrupted by pregnancy or disease) until the menopause.*

ADULT PELVIC SEX ORGANS
in ORDINARY FEMALE CYCLE

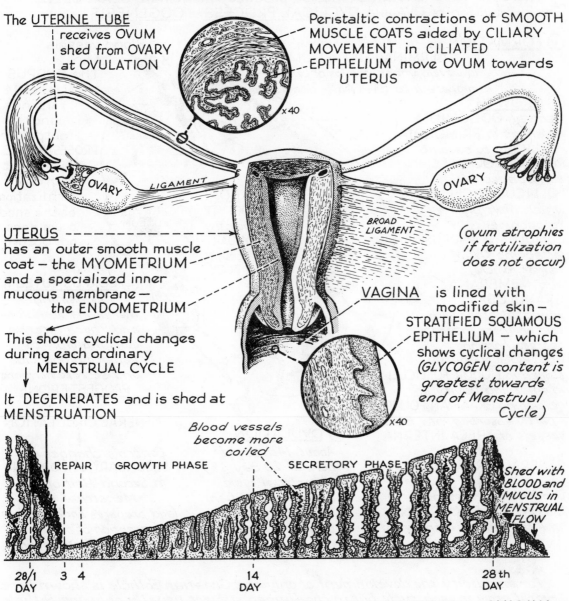

The UTERINE TUBE receives OVUM shed from OVARY at OVULATION

Peristaltic contractions of SMOOTH MUSCLE COATS aided by CILIARY MOVEMENT in CILIATED EPITHELIUM move OVUM towards UTERUS

x 40

OVARY

LIGAMENT

OVARY

BROAD LIGAMENT

(ovum atrophies if fertilization does not occur)

UTERUS has an outer smooth muscle coat – the MYOMETRIUM and a specialized inner mucous membrane – the ENDOMETRIUM

This shows cyclical changes during each ordinary MENSTRUAL CYCLE

It DEGENERATES and is shed at MENSTRUATION

VAGINA is lined with modified skin – STRATIFIED SQUAMOUS EPITHELIUM – which shows cyclical changes (GLYCOGEN content is greatest towards end of Menstrual Cycle)

x 40

Blood vessels become more coiled

REPAIR   GROWTH PHASE   SECRETORY PHASE

Shed with BLOOD and MUCUS in MENSTRUAL FLOW

28/1 DAY   3 4   14 DAY   28 th DAY

Rhythmical changes occur in UTERUS, UTERINE TUBES and VAGINA under action of OVARIAN Hormones.

OVARY in ORDINARY ADULT CYCLE

There are TWO OVARIES. These produce the Female GERM CELLS.
The production of OVA is a CYCLICAL PROCESS — OOGENESIS

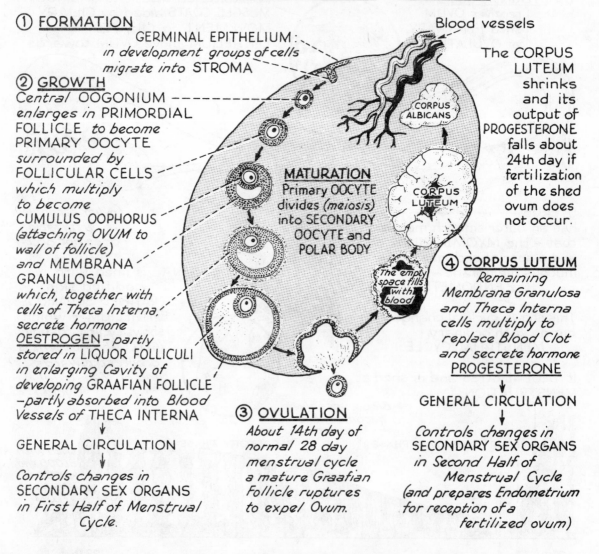

① **FORMATION**

GERMINAL EPITHELIUM:
in development groups of cells
migrate into STROMA

Blood vessels

② **GROWTH**
Central OOGONIUM
enlarges in PRIMORDIAL
FOLLICLE to become
PRIMARY OOCYTE
surrounded by
FOLLICULAR CELLS
which multiply
to become
CUMULUS OOPHORUS
(attaching OVUM to
wall of follicle)
and MEMBRANA
GRANULOSA
which, together with
cells of Theca Interna,
secrete hormone
OESTROGEN — partly
stored in LIQUOR FOLLICULI
in enlarging Cavity of
developing GRAAFIAN FOLLICLE
— partly absorbed into Blood
Vessels of THECA INTERNA

↓

GENERAL CIRCULATION

↓

Controls changes in
SECONDARY SEX ORGANS
in First Half of Menstrual
Cycle.

MATURATION
Primary OOCYTE
divides (meiosis)
into SECONDARY
OOCYTE and
POLAR BODY

The empty space fills with blood

③ **OVULATION**
About 14th day of
normal 28 day
menstrual cycle
a mature Graafian
Follicle ruptures
to expel Ovum.

CORPUS
ALBICANS

CORPUS
LUTEUM

The CORPUS
LUTEUM
shrinks
and its
output of
PROGESTERONE
falls about
24th day if
fertilization
of the shed
ovum does
not occur.

④ **CORPUS LUTEUM**
Remaining
Membrana Granulosa
and Theca Interna
cells multiply to
replace Blood Clot
and secrete hormone
PROGESTERONE

↓

GENERAL CIRCULATION

↓

Controls changes in
SECONDARY SEX ORGANS
in Second Half of
Menstrual Cycle
(and prepares Endometrium
for reception of a
fertilized ovum)

For simplicity the development of only one Graafian Follicle is shown here.
Several grow in each cycle but in the Human subject usually only one Follicle
ruptures. The others atrophy: i.e. ONE MATURE OVUM is shed each month.
Events in the OVARY are under control of ANTERIOR PITUITARY HORMONES.

OVARY in PREGNANCY

When pregnancy occurs the ordinary ovarian cycle is suspended.

The CORPUS LUTEUM continues to grow until it may come to occupy 30% to 50% of the total volume of the OVARY.

The large amount of PROGESTERONE

helps to maintain the PREGNANCY in its early stages and is essential for development of the PLACENTA – the special structure through which the child receives its nourishment from the mother.
It also produces PROGESTERONE which gradually takes over from the

CORPUS LUTEUM —

reaches its peak at about 6 weeks after conception

falls off about 2nd. month

ceases to be active about 4th. month

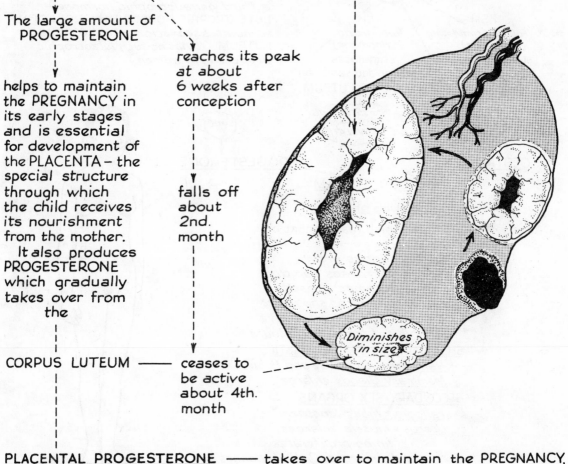

Diminishes in size

PLACENTAL PROGESTERONE —— takes over to maintain the PREGNANCY, and to help to prepare the mammary glands for Lactation.

PUBERTY in FEMALE

Between the ages of 10 and 14 years Ovarian tissue usually begins to respond to stimulation by <u>ANTERIOR PITUITARY GONADOTROPHIC HORMONES</u>.

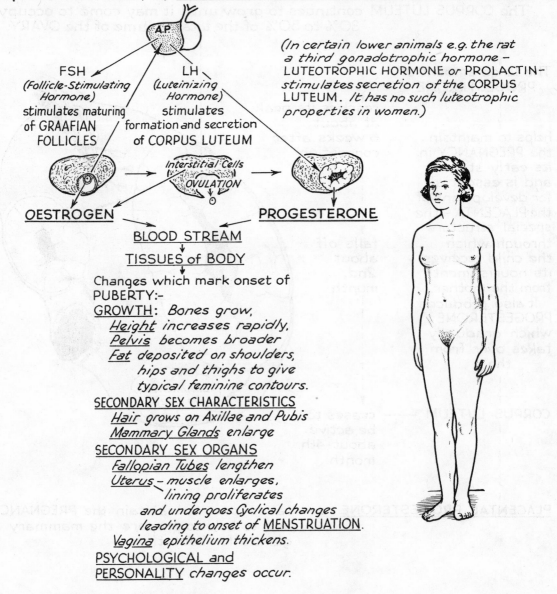

A.P.

FSH
(Follicle-Stimulating Hormone)
stimulates maturing
of GRAAFIAN
FOLLICLES

LH
(Luteinizing Hormone)
stimulates
formation and secretion
of CORPUS LUTEUM

*(In certain lower animals e.g. the rat a third gonadotrophic hormone –
LUTEOTROPHIC HORMONE or PROLACTIN-stimulates secretion of the CORPUS LUTEUM. It has no such luteotrophic properties in women.)*

Interstitial Cells

OVULATION

<u>OESTROGEN</u>

<u>PROGESTERONE</u>

<u>BLOOD STREAM</u>

<u>TISSUES of BODY</u>

Changes which mark onset of
PUBERTY:-

<u>GROWTH</u>: *Bones grow,*
  <u>Height</u> increases rapidly,
  <u>Pelvis</u> becomes broader
  <u>Fat</u> deposited on shoulders,
    hips and thighs to give
    typical feminine contours.

<u>SECONDARY SEX CHARACTERISTICS</u>
  <u>Hair</u> grows on Axillae and Pubis
  <u>Mammary Glands</u> enlarge

<u>SECONDARY SEX ORGANS</u>
  <u>Fallopian Tubes</u> lengthen
  <u>Uterus</u> – muscle enlarges,
    lining proliferates
    and undergoes Cyclical changes
    leading to onset of <u>MENSTRUATION</u>.
  <u>Vagina</u> epithelium thickens.

<u>PSYCHOLOGICAL</u> and
<u>PERSONALITY</u> *changes occur.*

i.e. PUBERTY changes GIRL CHILD into WOMAN able to bear children.

OVARIAN HORMONES

The Ovarian Cycle is repeated monthly from PUBERTY to the MENOPAUSE unless interrupted by PREGNANCY or DISEASE.

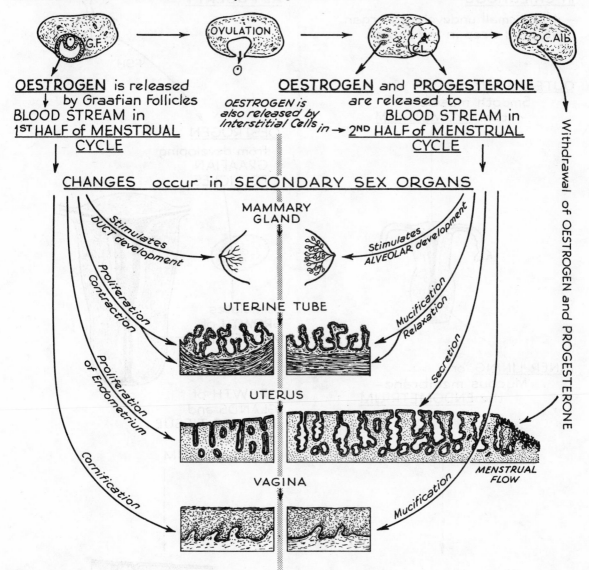

OVARIAN HORMONES are therefore directly responsible for regular cycle of events in SECONDARY SEX ORGANS.

UTERUS and UTERINE TUBES

The UTERUS (or Womb) is the organ which bears the developing child till birth.

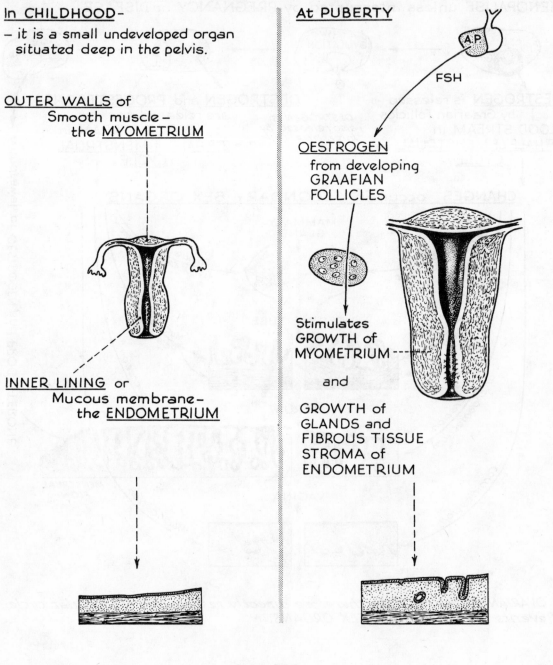

In <u>CHILDHOOD</u>-

- it is a small undeveloped organ situated deep in the pelvis.

<u>OUTER WALLS</u> of
  Smooth muscle –
    the <u>MYOMETRIUM</u>

<u>INNER LINING</u> or
  Mucous membrane –
    the <u>ENDOMETRIUM</u>

At <u>PUBERTY</u>

A.P.

FSH

<u>OESTROGEN</u>
  from developing
  GRAAFIAN
  FOLLICLES

Stimulates
GROWTH of
MYOMETRIUM - - - -

and

GROWTH of
GLANDS and
FIBROUS TISSUE
STROMA of
ENDOMETRIUM

186

UTERUS and UTERINE TUBES

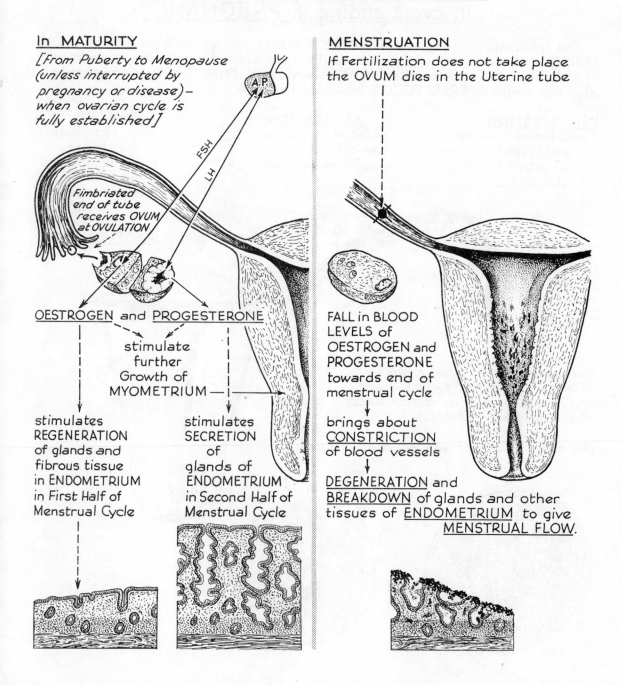

In MATURITY

[From Puberty to Menopause (unless interrupted by pregnancy or disease) – when ovarian cycle is fully established]

FSH

LH

A.P.

Fimbriated end of tube receives OVUM at OVULATION

OESTROGEN and PROGESTERONE

stimulate further Growth of MYOMETRIUM —

stimulates REGENERATION of glands and fibrous tissue in ENDOMETRIUM in First Half of Menstrual Cycle

stimulates SECRETION of glands of ENDOMETRIUM in Second Half of Menstrual Cycle

MENSTRUATION

If Fertilization does not take place the OVUM dies in the Uterine tube

FALL in BLOOD LEVELS of OESTROGEN and PROGESTERONE towards end of menstrual cycle

brings about CONSTRICTION of blood vessels

DEGENERATION and BREAKDOWN of glands and other tissues of ENDOMETRIUM to give MENSTRUAL FLOW.

UTERINE TUBES
in cycle ending in PREGNANCY

The fimbriated end of the UTERINE TUBE receives the OVUM at OVULATION. PERISTALTIC contractions of the muscular tube aided by ciliary movements of its lining cells transfer the OVUM towards the UTERUS. The uterine tube also transmits SPERMATOZOA towards the OVA.

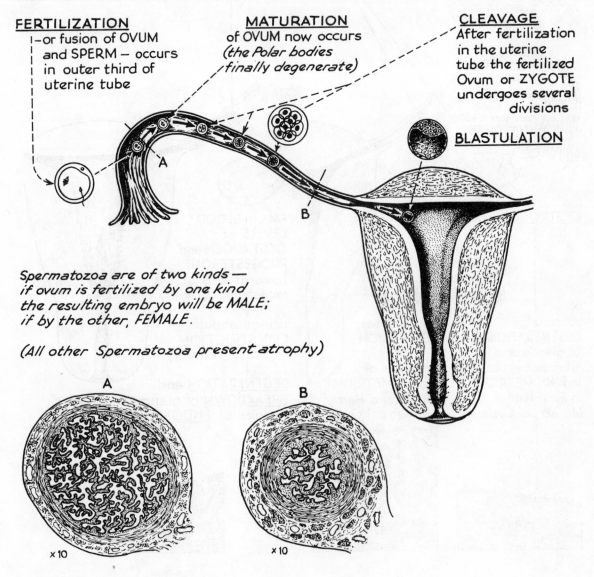

FERTILIZATION
—or fusion of OVUM and SPERM — occurs in outer third of uterine tube

MATURATION
of OVUM now occurs
(the Polar bodies finally degenerate)

CLEAVAGE
After fertilization in the uterine tube the fertilized Ovum or ZYGOTE undergoes several divisions

BLASTULATION

*Spermatozoa are of two kinds —
if ovum is fertilized by one kind
the resulting embryo will be MALE;
if by the other, FEMALE.*

(All other Spermatozoa present atrophy)

A

× 10

B

× 10

UTERUS

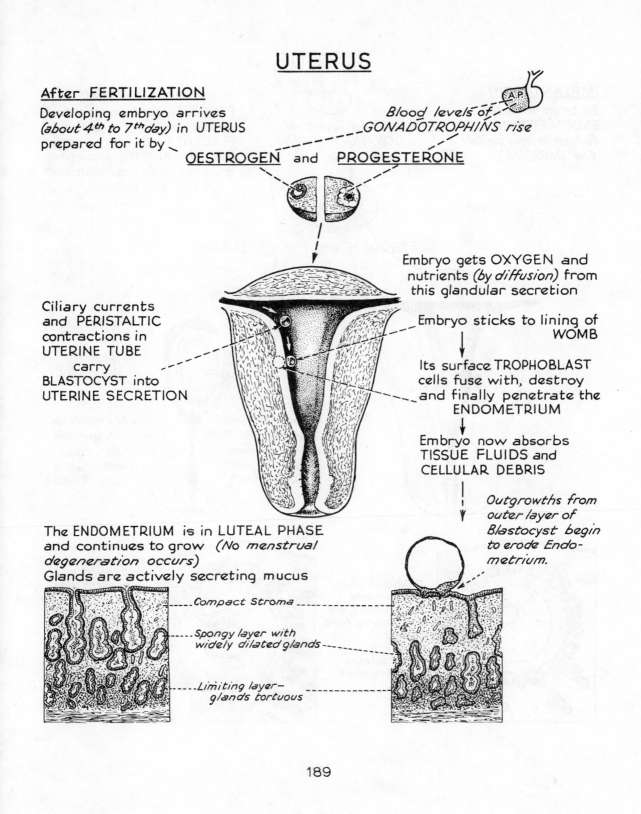

After FERTILIZATION

Developing embryo arrives
(about 4th to 7th day) in UTERUS
prepared for it by

Blood levels of
GONADOTROPHINS rise

A.P.

OESTROGEN and PROGESTERONE

Embryo gets OXYGEN and
nutrients *(by diffusion)* from
this glandular secretion

Ciliary currents
and PERISTALTIC
contractions in
UTERINE TUBE
carry
BLASTOCYST into
UTERINE SECRETION

Embryo sticks to lining of
WOMB

Its surface TROPHOBLAST
cells fuse with, destroy
and finally penetrate the
ENDOMETRIUM

Embryo now absorbs
TISSUE FLUIDS and
CELLULAR DEBRIS

*Outgrowths from
outer layer of
Blastocyst begin
to erode Endo-
metrium.*

The ENDOMETRIUM is in LUTEAL PHASE
and continues to grow *(No menstrual
degeneration occurs)*
Glands are actively secreting mucus

Compact Stroma

*Spongy layer with
widely dilated glands*

*Limiting layer –
glands tortuous*

UTERUS

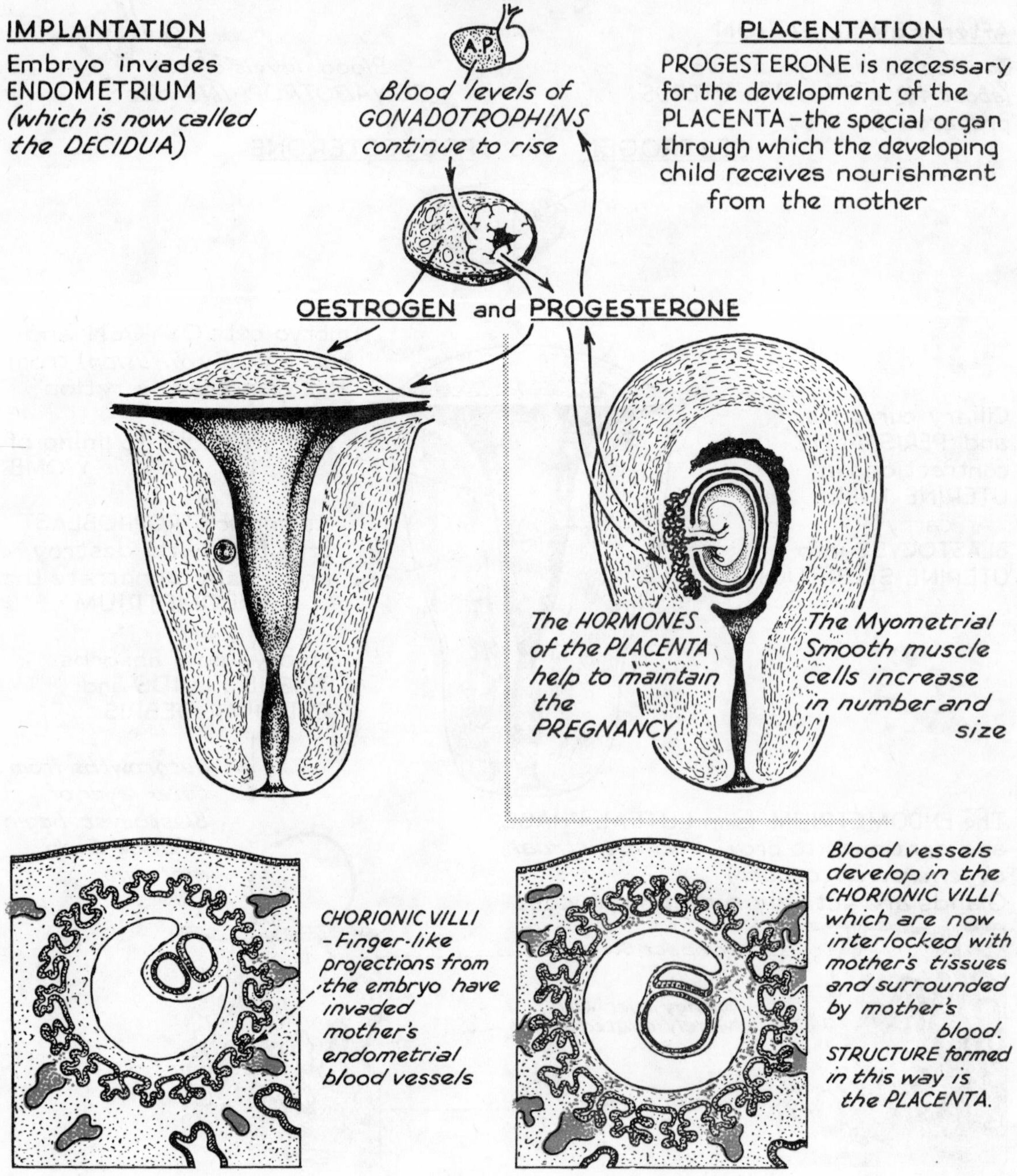

IMPLANTATION

Embryo invades
ENDOMETRIUM
*(which is now called
the DECIDUA)*

*Blood levels of
GONADOTROPHINS
continue to rise*

PLACENTATION

PROGESTERONE is necessary
for the development of the
PLACENTA – the special organ
through which the developing
child receives nourishment
from the mother

OESTROGEN and PROGESTERONE

*The HORMONES
of the PLACENTA
help to maintain
the
PREGNANCY*

*The Myometrial
Smooth muscle
cells increase
in number and
size*

*CHORIONIC VILLI
– Finger-like
projections from
the embryo have
invaded
mother's
endometrial
blood vessels*

*Blood vessels
develop in the
CHORIONIC VILLI
which are now
interlocked with
mother's tissues
and surrounded
by mother's
blood.
STRUCTURE formed
in this way is
the PLACENTA.*

PLACENTA

The PLACENTA in WOMAN is HAEMOCHORIAL – *i.e. chorionic villi dip directly into maternal blood.*
It increases in weight throughout Pregnancy.

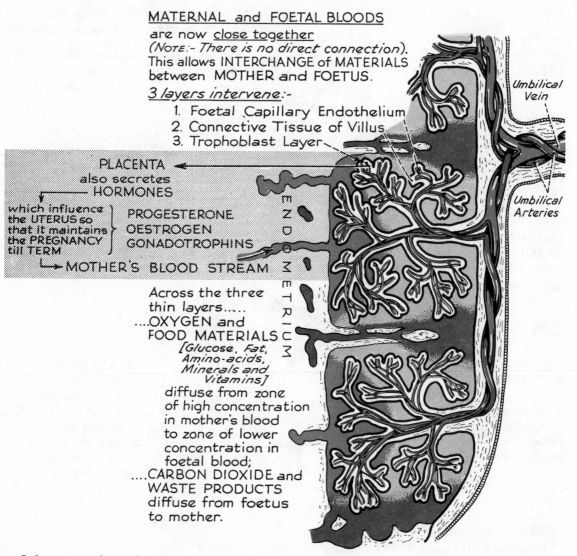

MATERNAL and FOETAL BLOODS
are now <u>close together</u>
(NOTE:- *There is no direct connection*).
This allows INTERCHANGE of MATERIALS
between MOTHER and FOETUS.

3 layers intervene:-
1. Foetal Capillary Endothelium
2. Connective Tissue of Villus
3. Trophoblast Layer

Umbilical
Vein

Umbilical
Arteries

PLACENTA
also secretes
HORMONES
which influence
the UTERUS so
that it maintains
the PREGNANCY
till TERM
PROGESTERONE
OESTROGEN
GONADOTROPHINS
→ MOTHER'S BLOOD STREAM

ENDOMETRIUM

Across the three
thin layers.....
....OXYGEN and
FOOD MATERIALS
*[Glucose, Fat,
Amino-acids,
Minerals and
Vitamins]*
diffuse from zone
of high concentration
in mother's blood
to zone of lower
concentration in
foetal blood;
....CARBON DIOXIDE and
WASTE PRODUCTS
diffuse from foetus
to mother.

*Substances of small molecular weight usually pass in either direction by diffusion.
Larger molecules are probably transported by special carrier systems involving enzymes.*

UTERUS

UTERUS

As <u>PREGNANCY ADVANCES</u>

? Role of Maternal Pituitary Gonado-trophins in maintaining pregnancy.

Foetus grows larger and comes to fill UTERINE CAVITY

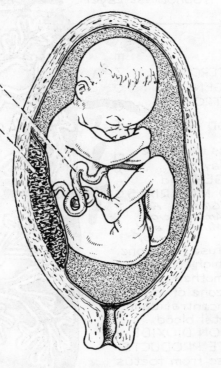

Corpus Luteum ceases to function after 3RD or 4TH month.

Foetus is attached by UMBILICAL CORD to PLACENTA

Foetus is bathed by AMNIOTIC FLUID (within Amniotic and Chorionic membranes)

GROWTH of MYOMETRIUM —increase in number and size of smooth muscle cells and of the blood vessels

PLACENTAL OESTROGEN and PROGESTERONE GONADOTROPHINS in blood stream condition the maintenance of pregnancy

STRETCHING of MYOMETRIUM

192

FOETAL CIRCULATION

For the foetus the PLACENTA acts as the organ of transfer for Oxygen, Nutritives and Waste Products. Only a small volume of blood passes through the foetal lungs.

BLOOD RETURNING TO HEART

...To RIGHT ATRIUM

Small amount from heart, head, neck and arms →S.V.C.

LARGE AMOUNT via UMBILICAL VEINS

through LIVER – short circuits to I.V.C. via

DUCTUS VENOSUS.

Some of this passes to Right Atrium
Small amount from abdominal cavity and legs.

...To LEFT ATRIUM

Small amount from 2 lungs

LARGE AMOUNT from INFERIOR VENA CAVA *through*

FORAMEN OVALE.

(thus by-passing pulmonary circulation)

BLOOD LEAVING HEART

...From RIGHT VENTRICLE

Small amount to 2 lungs

LARGE AMOUNT to AORTA

through

DUCTUS ARTERIOSUS

(thus by-passing pulmonary circulation) joins

OUTPUT from LEFT VENTRICLE

↓

Small amount to heart, head, neck and arms.
LARGE AMOUNT to PLACENTA
through
UMBILICAL ARTERIES
Small amount to abdominal cavity and legs.

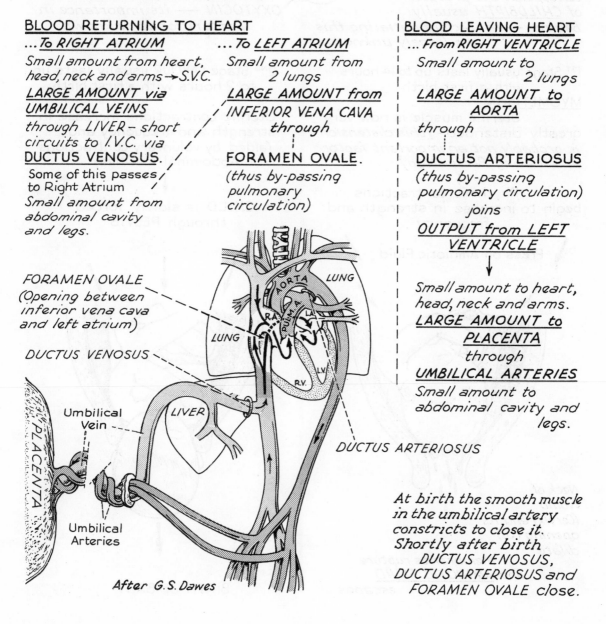

FORAMEN OVALE (Opening between inferior vena cava and left atrium)

DUCTUS VENOSUS

Umbilical Vein

LIVER

Umbilical Arteries

After G.S. Dawes

DUCTUS ARTERIOSUS

At birth the smooth muscle in the umbilical artery constricts to close it. Shortly after birth DUCTUS VENOSUS, DUCTUS ARTERIOSUS and FORAMEN OVALE close.

UTERUS

PARTURITION

About 40 weeks after conception the process of _CHILDBIRTH_ usually begins —— Factors initiating this are largely unknown.

After 32ND week of Pregnancy the uterus becomes sensitive to OXYTOCIN —— its importance in Parturition is unknown.

1ST Stage usually lasts up to 14 hours with a first birth

2ND Stage LABOUR usually lasts up to 2 hours with a first birth

MYOMETRIUM

Uterine muscle is now very greatly distended – [this distension is probably not an important factor in initiating Parturition].

Uterine contractions increase in strength and frequency and (aided by voluntary contractions of Abdominal muscles)

Rhythmic contractions begin to increase in strength and frequency

CHILD is slowly forced through PELVIS

Press on Amniotic Fluid

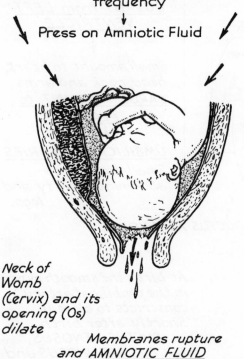

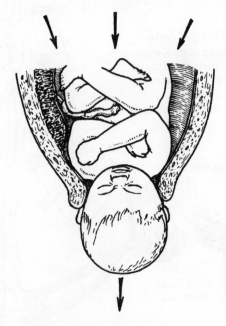

Neck of Womb (Cervix) and its opening (Os) dilate

Membranes rupture and _AMNIOTIC FLUID_ escapes

BIRTH of BABY

UTERUS

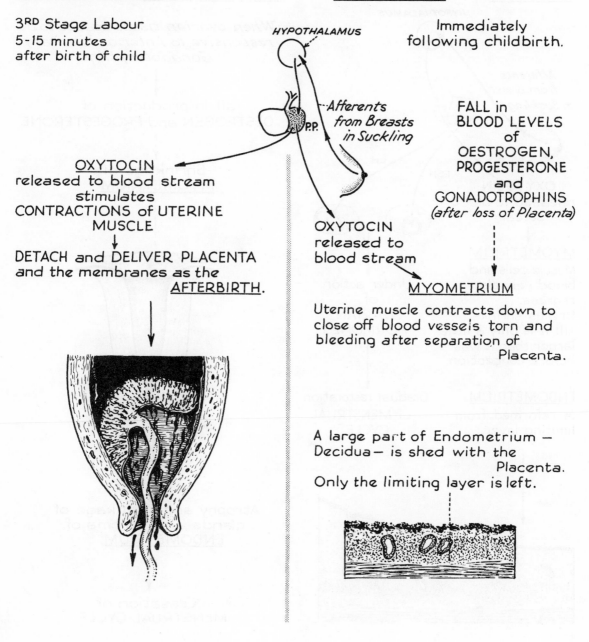

After PARTURITION

3RD Stage Labour
5-15 minutes
after birth of child

HYPOTHALAMUS

P.P.

--Afferents
from Breasts
in Suckling

OXYTOCIN
released to blood stream
stimulates
CONTRACTIONS of UTERINE
MUSCLE

↓

DETACH and DELIVER PLACENTA
and the membranes as the
AFTERBIRTH.

In PUERPERIUM

Immediately
following childbirth.

FALL in
BLOOD LEVELS
of
OESTROGEN,
PROGESTERONE
and
GONADOTROPHINS
(after loss of Placenta)

OXYTOCIN
released to
blood stream

MYOMETRIUM

Uterine muscle contracts down to
close off blood vessels torn and
bleeding after separation of
Placenta.

A large part of Endometrium —
Decidua— is shed with the
Placenta.
Only the limiting layer is left.

UTERUS

INVOLUTION

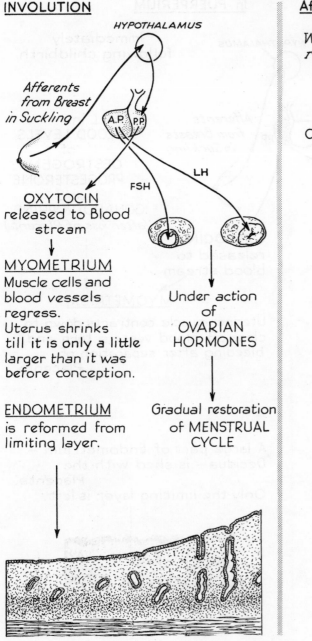

HYPOTHALAMUS

Afferents from Breast in Suckling

A.P. P.P

FSH LH

OXYTOCIN
released to Blood stream

MYOMETRIUM
Muscle cells and blood vessels regress.
Uterus shrinks till it is only a little larger than it was before conception.

Under action of **OVARIAN HORMONES**

ENDOMETRIUM
is reformed from limiting layer.

Gradual restoration of MENSTRUAL CYCLE

After MENOPAUSE

When ovarian tissue ceases to be responsive to Anterior Pituitary Gonadotrophins

↓

Fall in production of OESTROGEN and PROGESTERONE

↓

Shrinkage of **MYOMETRIUM**

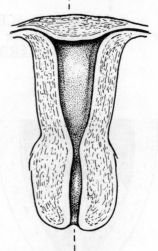

Atrophy and shrinkage of glands and stroma of **ENDOMETRIUM**

↓

Cessation of MENSTRUAL CYCLE

MAMMARY GLANDS

There are 2 mammary glands.

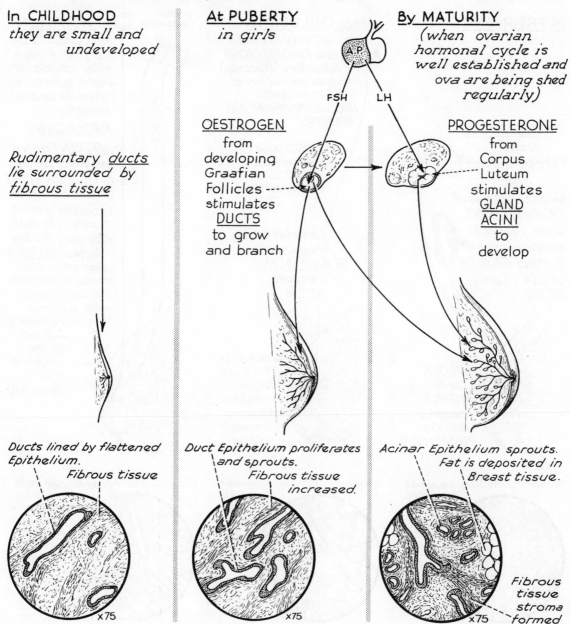

In CHILDHOOD
*they are small and
undeveloped*

*Rudimentary <u>ducts</u>
lie surrounded by
<u>fibrous tissue</u>*

At PUBERTY
in girls

FSH   LH

<u>OESTROGEN</u>
from
developing
Graafian
Follicles
stimulates
<u>DUCTS</u>
to grow
and branch

By MATURITY
*(when ovarian
hormonal cycle is
well established and
ova are being shed
regularly)*

<u>PROGESTERONE</u>
from
Corpus
Luteum
stimulates
<u>GLAND
ACINI</u>
to
develop

*Ducts lined by flattened
Epithelium.
Fibrous tissue*

x75

*Duct Epithelium proliferates
and sprouts.
Fibrous tissue
increased.*

x75

*Acinar Epithelium sprouts.
Fat is deposited in
Breast tissue.*

*Fibrous
tissue
stroma
formed*

x75

197

MAMMARY GLANDS

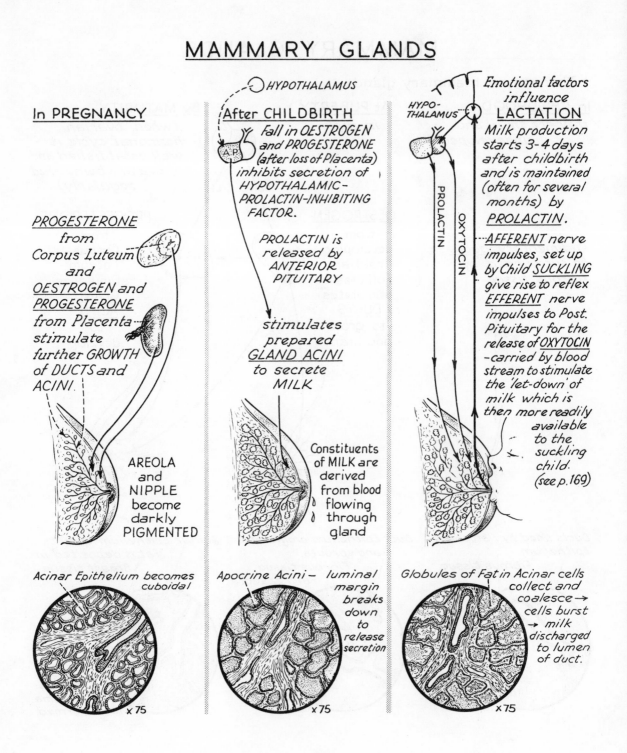

In PREGNANCY

○ HYPOTHALAMUS

After CHILDBIRTH

Emotional factors influence LACTATION

PROGESTERONE from Corpus Luteum and OESTROGEN and PROGESTERONE from Placenta stimulate further GROWTH of DUCTS and ACINI.

Fall in OESTROGEN and PROGESTERONE (after loss of Placenta) inhibits secretion of HYPOTHALAMIC-PROLACTIN-INHIBITING FACTOR.

PROLACTIN is released by ANTERIOR PITUITARY

HYPO-THALAMUS

Milk production starts 3-4 days after childbirth and is maintained (often for several months) by PROLACTIN.

PROLACTIN

OXYTOCIN

AFFERENT nerve impulses, set up by Child SUCKLING give rise to reflex EFFERENT nerve impulses to Post. Pituitary for the release of OXYTOCIN -carried by blood stream to stimulate the 'let-down' of milk which is then more readily available to the suckling child. (see p.169)

stimulates prepared GLAND ACINI to secrete MILK

AREOLA and NIPPLE become darkly PIGMENTED

Constituents of MILK are derived from blood flowing through gland

Acinar Epithelium becomes cuboidal

× 75

Apocrine Acini – luminal margin breaks down to release secretion

× 75

Globules of Fat in Acinar cells collect and coalesce → cells burst → milk discharged to lumen of duct.

× 75

198

MAMMARY GLANDS

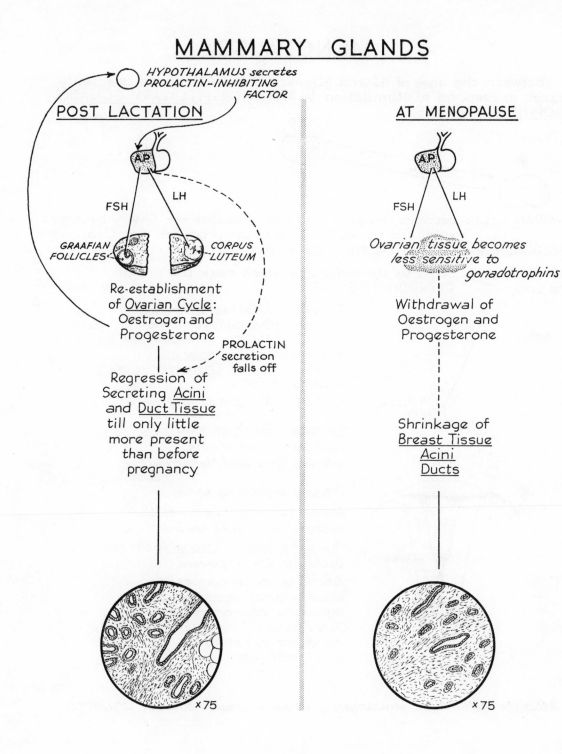

HYPOTHALAMUS *secretes*
PROLACTIN-INHIBITING
FACTOR

POST LACTATION

AP

FSH   LH

GRAAFIAN
FOLLICLES
CORPUS
LUTEUM

Re-establishment
of <u>Ovarian Cycle</u>:
Oestrogen and
Progesterone

PROLACTIN
secretion
falls off

Regression of
Secreting <u>Acini</u>
and <u>Duct Tissue</u>
till only little
more present
than before
pregnancy

×75

AT MENOPAUSE

AP

FSH   LH

*Ovarian tissue becomes
less sensitive to
gonadotrophins*

Withdrawal of
Oestrogen and
Progesterone

Shrinkage of
<u>Breast Tissue</u>
<u>Acini</u>
<u>Ducts</u>

×75

MENOPAUSE

Between the ages of 42 and 50 years OVARIAN tissue gradually ceases to respond to stimulation by <u>ANTERIOR PITUITARY GONADO-TROPHIC HORMONES.</u>

FSH
LH
A.P.

OVARIAN CYCLE becomes irregular and finally ceases ⟶ Ovary becomes small and fibrosed and no longer produces ripe Ova.

OESTROGEN and PROGESTERONE levels in Blood stream fall.

TISSUES of the body ⟶ begin to show changes which mark the end of REPRODUCTIVE LIFE.

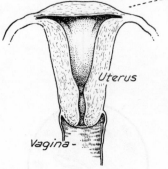

Ducts ⎱
Acini ⎰ Atrophy

Uterus

Vagina -

Sometimes final redistribution of fat → *less typically feminine distribution.*

Regression of <u>Secondary Sex Characteristics</u>.

Breasts shrink.

Hair becomes sparse in axillae and on pubis.

<u>*Secondary Sex Organs*</u> *atrophy.*

Fallopian tubes shrink.

Uterine Cycle and Menstruation cease.

(Muscle and lining shrink).

Vaginal epithelium becomes thin.

External Genitalia shrink.

<u>*Psychological and Personality*</u> *changes.*

Decline in Sexual powers.

Emotional disturbances may occur – often accompanied by Vasomotor phenomena such as "Hot Flushes" (vasodilatation), excessive sweating and giddiness.

After the MENOPAUSE a woman is usually unable to bear children.

PITUITARY, OVARIAN and ENDOMETRIAL CYCLES

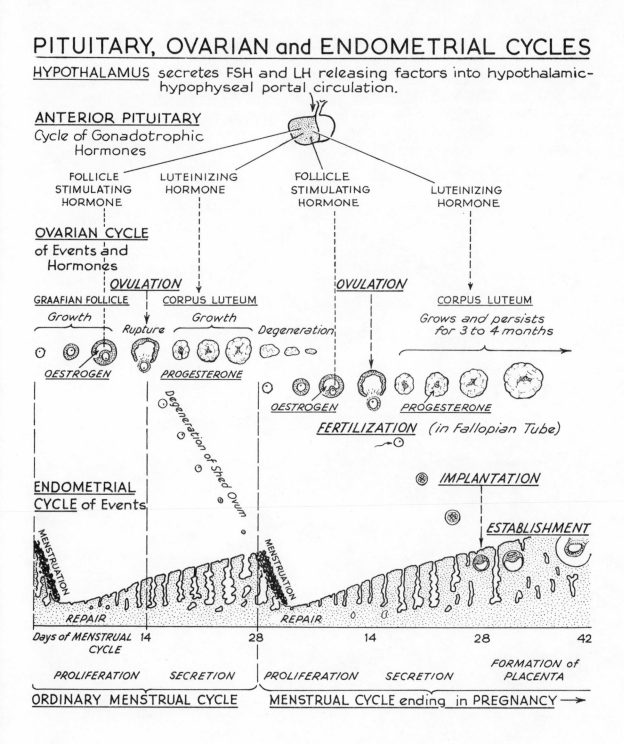

HYPOTHALAMUS secretes FSH and LH releasing factors into hypothalamic-hypophyseal portal circulation.

ANTERIOR PITUITARY
Cycle of Gonadotrophic
    Hormones

FOLLICLE STIMULATING HORMONE   LUTEINIZING HORMONE   FOLLICLE STIMULATING HORMONE   LUTEINIZING HORMONE

OVARIAN CYCLE
of Events and
Hormones

OVULATION   OVULATION

GRAAFIAN FOLLICLE   CORPUS LUTEUM
Growth   Rupture   Growth   Degeneration   CORPUS LUTEUM
Grows and persists
for 3 to 4 months

OESTROGEN   PROGESTERONE   OESTROGEN   PROGESTERONE

Degeneration of Shed Ovum

FERTILIZATION (in Fallopian Tube)

IMPLANTATION

ENDOMETRIAL
CYCLE of Events

ESTABLISHMENT

MENSTRUATION   MENSTRUATION

REPAIR   REPAIR

Days of MENSTRUAL   14   28   14   28   42
CYCLE

FORMATION of
PLACENTA

PROLIFERATION   SECRETION   PROLIFERATION   SECRETION

ORDINARY MENSTRUAL CYCLE   MENSTRUAL CYCLE ending in PREGNANCY ⟶

201

CHAPTER 9

"MASTER" TISSUES:

CENTRAL NERVOUS SYSTEM
LOCOMOTOR SYSTEM

NERVOUS SYSTEM

The Nervous System is concerned with the INTEGRATION and CONTROL of all bodily functions.

It has specialized in IRRITABILITY— *the ability to receive and respond to messages from the external and internal environments*

and also in CONDUCTION – *the ability to transmit messages to and from CO-ORDINATING CENTRES.*

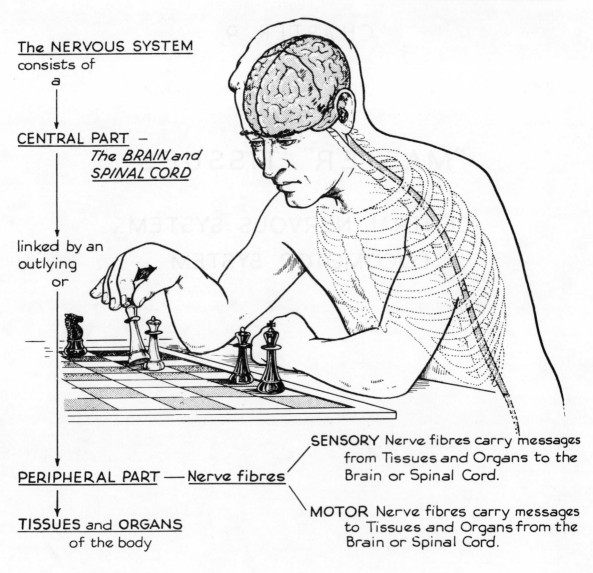

The NERVOUS SYSTEM
consists of
a

↓

CENTRAL PART –
The BRAIN and SPINAL CORD

↓

linked by an
outlying
or

↓

PERIPHERAL PART — Nerve fibres

↓

TISSUES and ORGANS
of the body

SENSORY Nerve fibres carry messages from Tissues and Organs to the Brain or Spinal Cord.

MOTOR Nerve fibres carry messages to Tissues and Organs from the Brain or Spinal Cord.

DEVELOPMENT of the NERVOUS SYSTEM

The Nervous System develops in the embryo from a simple tube of ECTODERM:- The PRIMITIVE NEURAL TUBE.

The CELLS lining it become the NERVOUS tissue of the BRAIN and SPINAL CORD. The CANAL becomes distended to form the VENTRICLES of the BRAIN and CENTRAL CANAL of the SPINAL CORD:-

Prosencephalon   Mesencephalon   Rhombencephalon

Each of these swellings with the cells lining them become more complicated:-

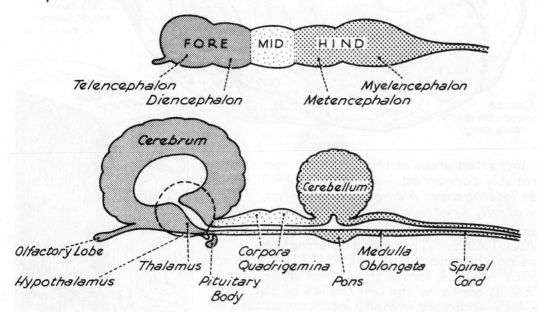

Telencephalon
Diencephalon
Myelencephalon
Metencephalon

Cerebrum

Cerebellum

Olfactory Lobe
Hypothalamus
Thalamus
Pituitary Body
Corpora Quadrigemina
Pons
Medulla Oblongata
Spinal Cord

There are now many layers of cells forming the Brain and Spinal Cord. The Ventricles of the Brain and the Central Canal of the Spinal Cord are filled with CEREBRO-SPINAL FLUID.

CEREBRUM

The largest part of the human brain is the CEREBRUM — made up of 2 CEREBRAL HEMISPHERES. Each of these is divided into LOBES.

INITIATING CENTRES for OUTGOING messages

RECEIVING CENTRES for INCOMING information

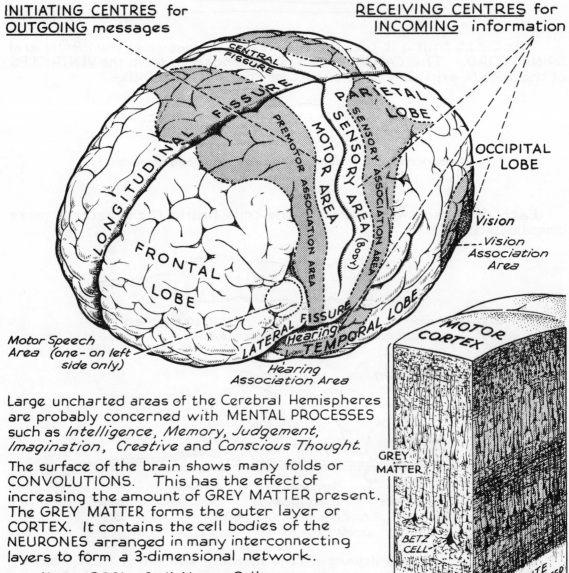

CENTRAL FISSURE

LONGITUDINAL FISSURE

PARIETAL LOBE

SENSORY ASSOCIATION AREA

SENSORY AREA (Body)

MOTOR AREA

PREMOTOR ASSOCIATION AREA

OCCIPITAL LOBE

Vision

Vision Association Area

FRONTAL LOBE

LATERAL FISSURE

Hearing

TEMPORAL LOBE

Motor Speech Area (one- on left side only)

Hearing Association Area

MOTOR CORTEX

GREY MATTER

BETZ CELL

WHITE MATTER

Large uncharted areas of the Cerebral Hemispheres are probably concerned with MENTAL PROCESSES such as *Intelligence, Memory, Judgement, Imagination, Creative* and *Conscious Thought.*

The surface of the brain shows many folds or CONVOLUTIONS. This has the effect of increasing the amount of GREY MATTER present. The GREY MATTER forms the outer layer or CORTEX. It contains the cell bodies of the NEURONES arranged in many interconnecting layers to form a 3-dimensional network.

About 90% of all Nerve Cells are in the Cerebral Cortex.

HORIZONTAL SECTION through BRAIN

This view shows surface GREY MATTER containing Nerve Cells and inner WHITE MATTER made up of Nerve Fibres.

Deep in the substance of the Cerebral Hemispheres there are additional masses of GREY MATTER :-

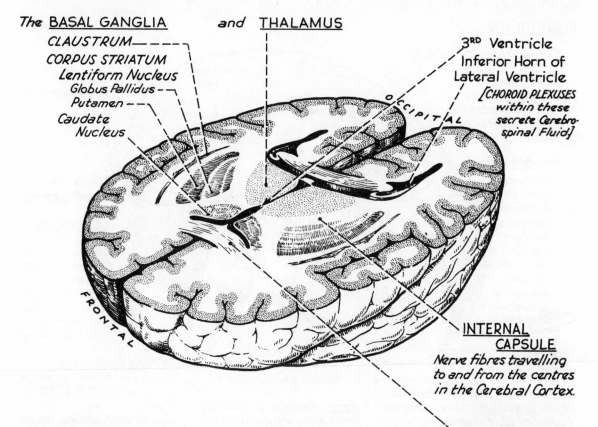

The BASAL GANGLIA   *and* THALAMUS

CLAUSTRUM

CORPUS STRIATUM
   Lentiform Nucleus
   Globus Pallidus
   Putamen
Caudate Nucleus

3RD Ventricle
Inferior Horn of
Lateral Ventricle
[CHOROID PLEXUSES
within these
secrete Cerebro-
spinal Fluid]

OCCIPITAL

FRONTAL

INTERNAL
CAPSULE
*Nerve fibres travelling
to and from the centres
in the Cerebral Cortex.*

ANTERIOR
COMMISSURE
*Nerve fibres linking
the two hemispheres.*

The Basal Ganglia
   are concerned with
   modifying and
   co-ordinating
   VOLUNTARY MUSCLE
   MOVEMENT.

The Thalamus
   is an important
   relay centre for
   SENSORY fibres
   on their way to
   CEREBRAL CORTEX.
   'Crude' Sensation
   and PAIN may be
   appreciated here.

207

VERTICAL SECTION through BRAIN

This is a Vertical Section through the LONGITUDINAL FISSURE — a deep cleft which separates the two Cerebral Hemispheres. At the bottom of this cleft are tracts of nerve fibres which link up the different LOBES of each hemisphere and also link the two hemispheres with each other — the CORPUS CALLOSUM.

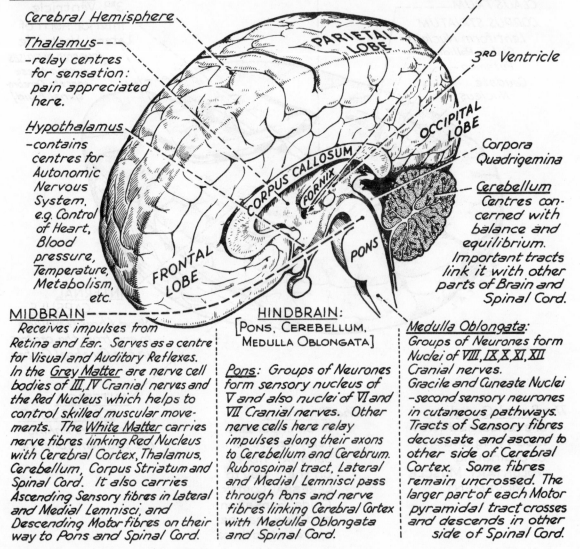

FOREBRAIN

Cerebral Hemisphere

Thalamus---
-relay centres for sensation: pain appreciated here.

Hypothalamus
-contains centres for Autonomic Nervous System. e.g. Control of Heart, Blood pressure, Temperature, Metabolism, etc.

Opening of Lateral Ventricle

3RD Ventricle

PARIETAL LOBE

OCCIPITAL LOBE

CORPUS CALLOSUM

FORNIX

FRONTAL LOBE

PONS

Corpora Quadrigemina

Cerebellum Centres concerned with balance and equilibrium. Important tracts link it with other parts of Brain and Spinal Cord.

MIDBRAIN---
Receives impulses from Retina and Ear. Serves as a centre for Visual and Auditory Reflexes. In the <u>Grey Matter</u> are nerve cell bodies of III, IV Cranial nerves and the Red Nucleus which helps to control skilled muscular movements. The <u>White Matter</u> carries nerve fibres linking Red Nucleus with Cerebral Cortex, Thalamus, Cerebellum, Corpus Striatum and Spinal Cord. It also carries Ascending Sensory fibres in Lateral and Medial Lemnisci, and Descending Motor fibres on their way to Pons and Spinal Cord.

HINDBRAIN:
[PONS, CEREBELLUM, MEDULLA OBLONGATA]

<u>Pons</u>: Groups of Neurones form sensory nucleus of V and also nuclei of VI and VII Cranial nerves. Other nerve cells here relay impulses along their axons to Cerebellum and Cerebrum. Rubrospinal tract, Lateral and Medial Lemnisci pass through Pons and nerve fibres linking Cerebral Cortex with Medulla Oblongata and Spinal Cord.

<u>Medulla Oblongata</u>:
Groups of Neurones form Nuclei of VIII, IX, X, XI, XII Cranial nerves. Gracile and Cuneate Nuclei -second sensory neurones in cutaneous pathways. Tracts of Sensory fibres decussate and ascend to other side of Cerebral Cortex. Some fibres remain uncrossed. The larger part of each Motor pyramidal tract crosses and descends in other side of Spinal Cord.

CORONAL SECTION through BRAIN

This is a section through the TRANSVERSE (CENTRAL) FISSURE.
It shows each of the major developments of the BRAIN –

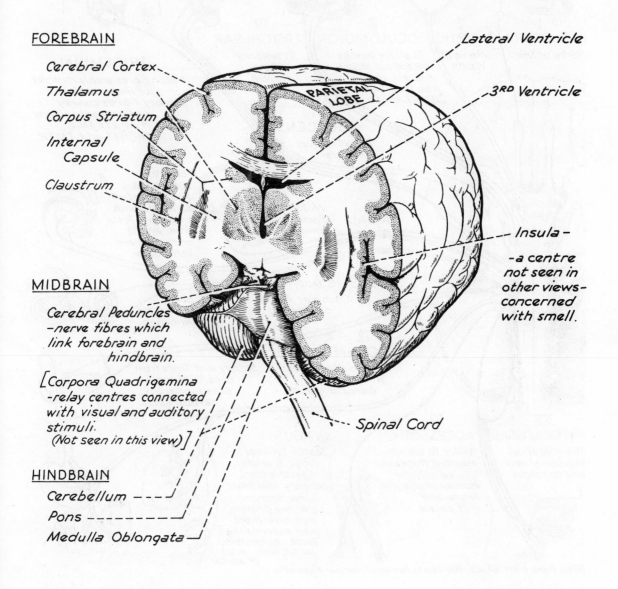

FOREBRAIN

Cerebral Cortex

Thalamus

Corpus Striatum

Internal Capsule

Claustrum

PARIETAL LOBE

Lateral Ventricle

3ʳᵈ Ventricle

Insula –

–a centre not seen in other views– concerned with smell.

MIDBRAIN

Cerebral Peduncles –nerve fibres which link forebrain and hindbrain.

[Corpora Quadrigemina –relay centres connected with visual and auditory stimuli. (Not seen in this view)]

Spinal Cord

HINDBRAIN

Cerebellum – – – –

Pons – – – – – – –

Medulla Oblongata

CRANIAL NERVES

Twelve pairs of nerves arise directly from the undersurface of the Brain to supply Head and Neck and most of the viscera.

I OLFACTORY
Nerve of Smell.

II OPTIC
Nerve of Vision.

III OCULOMOTOR
To all Eye muscles except sup. oblique and ext. rectus. Also to IRIS and CILIARY muscle.

IV TROCHLEAR
To superior oblique muscle.

V TRIGEMINAL
Motor fibres supply muscles of mastication. Sensory fibres convey ordinary sensations from EYE, FACE, SINUSES and TEETH.

VI ABDUCENS
To ext. rectus muscle.

VII FACIAL
Motor to Facial muscles.

Motor to submaxillary and sublingual salivary glands. Sensory and Taste from ant. 2/3 of tongue and soft palate.

VIII ACOUSTIC (AUDITORY)
Cochlear Nerve of Hearing.

Vestibular Nerve for sense of Equilibrium.

IX GLOSSOPHARYNGEAL
Motor to Pharyngeal muscles and parotid gland.

Sensory and Taste from post. 1/3 of tongue, tonsil, pharynx, carotid sinus, carotid body

XII HYPOGLOSSAL
Motor to strap muscles of neck and to tongue.

XI ACCESSORY
Motor to sterno-mastoid, trapezius, constrictor muscles of pharynx, larynx and soft palate.

X VAGUS
Motor to heart, lungs, bronchi, digestive tract, Sensory from heart, lungs, bronchi, trachea, pharynx, digestive tract and external ear. Taste - epiglottis, Aortic body, arch of aorta.

(After Frank H. NETTER, M.D., The Ciba Collection of Medical Illustrations)

SPINAL CORD

The SPINAL CORD lies within the Vertebral Canal. It is continuous above with the Medulla Oblongata.

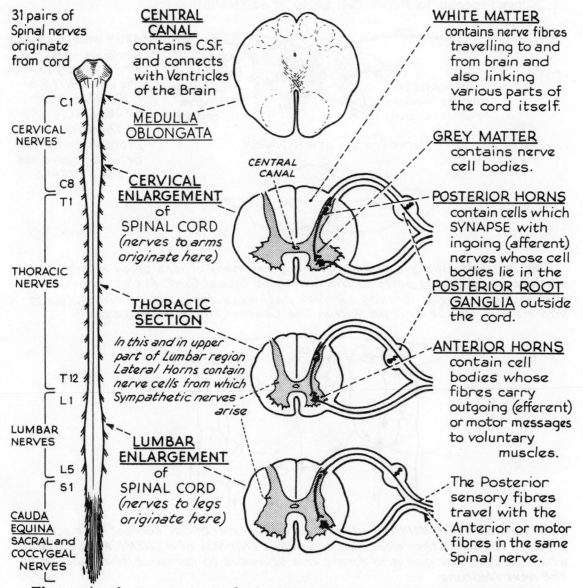

31 pairs of Spinal nerves originate from cord

C1

CERVICAL NERVES

C8
T1

THORACIC NERVES

T 12
L1

LUMBAR NERVES

L5
S1

<u>CAUDA EQUINA</u> SACRAL and COCCYGEAL NERVES

<u>CENTRAL CANAL</u> contains C.S.F. and connects with Ventricles of the Brain

<u>MEDULLA OBLONGATA</u>

CENTRAL CANAL

<u>CERVICAL ENLARGEMENT</u> of SPINAL CORD *(nerves to arms originate here)*

<u>THORACIC SECTION</u>

In this and in upper part of Lumbar region Lateral Horns contain nerve cells from which Sympathetic nerves arise

<u>LUMBAR ENLARGEMENT</u> of SPINAL CORD *(nerves to legs originate here)*

<u>WHITE MATTER</u> contains nerve fibres travelling to and from brain and also linking various parts of the cord itself.

<u>GREY MATTER</u> contains nerve cell bodies.

<u>POSTERIOR HORNS</u> contain cells which SYNAPSE with ingoing (afferent) nerves whose cell bodies lie in the **<u>POSTERIOR ROOT GANGLIA</u>** outside the cord.

<u>ANTERIOR HORNS</u> contain cell bodies whose fibres carry outgoing (efferent) or motor messages to voluntary muscles.

The Posterior sensory fibres travel with the Anterior or motor fibres in the same Spinal nerve.

The spinal nerves travel to all parts of the trunk and limbs.

SYNAPSE

The structural unit of the Nervous System is the NEURONE.
Neurones are linked together in the Nervous System.....

..... Nerve Process to Nerve Cell body at a SYNAPSE

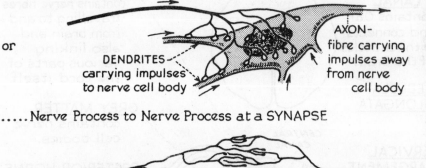

or

DENDRITES
carrying impulses
to nerve cell body

AXON-
fibre carrying
impulses away
from nerve
cell body

*AXON ends in small
swellings – END FEET-
which merely touch
the DENDRITES or
BODY of another
NERVE CELL.
i.e. There is no direct
protoplasmic union
between neurones
at the SYNAPSE.*

..... Nerve Process to Nerve Process at a SYNAPSE

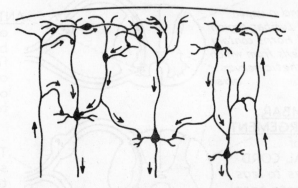

*One neurone usually connects with a great many others often widely
scattered in different parts of the Brain and Spinal Cord. In this way intricate
chains of nerve cells forming complex pathways for <u>incoming</u> and <u>outgoing</u>
information can be built up within the Central Nervous System.*

*When the NERVE IMPULSE – like a very small electrical current — reaches a
SYNAPSE it causes the release at the NERVE ENDINGS of a CHEMICAL SUBSTANCE
which bridges the gap and forms the stimulus to conduct the Impulse to
the next Neurone.*

A Synapse permits transmission of the impulse in one direction only.

212

NERVE IMPULSE

Neurones specialize in IRRITABILITY and in CONDUCTION of IMPULSES.

In RESTING (INACTIVE) NERVE FIBRE:

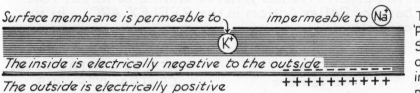

Surface membrane is permeable to   impermeable to Na
K⁺

The inside is electrically negative to the outside ‾ ‾ ‾ ‾ ‾

The outside is electrically positive   + + + + + + + + + +

This is the so-called 'POLARIZED' or resting STATE. It is the result of chemical processes in the cells and is a potential source of energy.

When *NERVE FIBRE is STIMULATED* (e.g. electrically or chemically) ELECTRO-CHEMICAL CHANGES occur at point of stimulation. The resting potential is abolished.

The membrane permeability is altered

The membrane is said to be 'DEPOLARIZED'.

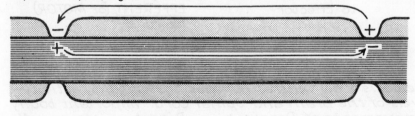

The electrical state is reversed. The inside becomes electrically positive and the outside becomes electrically negative.

Current flows ahead from the stimulated or active part to the inactive part – carried largely by the Potassium and Sodium ions.

Current passing through the membrane of the inactive part in turn depolarizes it. This then becomes the new active area to excite the next segment.

Direction of Impulse ⟶

In MYELINATED NERVE FIBRE:

The active region still 'triggers' the resting part ahead of it by causing an outward flow of current but this occurs only at the Nodes of Ranvier. i.e. The Nerve Impulse is propagated from Node to Node.

This increases speed of CONDUCTION along these large fibres.

After passage of nerve impulse 'REPOLARIZATION' and IONIC BALANCE are quickly restored.

A nerve impulse is the same whether it is in a sensory nerve taking messages to the brain or in a motor nerve taking impulses from the brain to the muscles, etc.

REFLEX ACTION

The NEURONE is the ANATOMICAL or STRUCTURAL UNIT of the Nervous System: the Nervous REFLEX is the PHYSIOLOGICAL or FUNCTIONAL UNIT.

A *Nervous Reflex* is an INVOLUNTARY ACTION caused by the STIMULATION of an AFFERENT (*sensory*) nerve ending or RECEPTOR.

The structural basis of Reflex Action is the REFLEX ARC.   In its simplest form this consists of:-

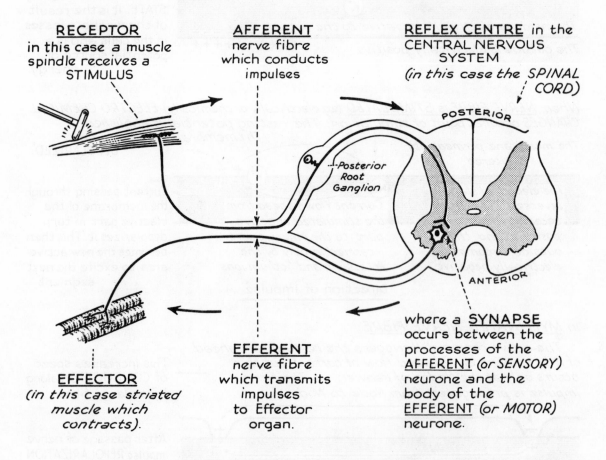

RECEPTOR
in this case a muscle spindle receives a STIMULUS

AFFERENT
nerve fibre which conducts impulses

REFLEX CENTRE in the CENTRAL NERVOUS SYSTEM
(in this case the SPINAL CORD)

POSTERIOR

--Posterior
Root
Ganglion

ANTERIOR

EFFECTOR
(in this case striated muscle which contracts).

EFFERENT
nerve fibre which transmits impulses to Effector organ.

where a **SYNAPSE** occurs between the processes of the **AFFERENT** (*or SENSORY*) neurone and the body of the **EFFERENT** (*or MOTOR*) neurone.

Reflexes form the basis of all Central Nervous System (C.N.S.) activity. They occur at all levels of the Brain and Spinal Cord. Important bodily functions such as movements of Respiration, Digestion, etc, are all controlled through Reflexes. We are made aware of some reflex acts: others occur without our knowledge.

STRETCH REFLEXES

In man a very few REFLEX ARCS involve 2 NEURONES only. Two examples elicited by doctors when testing the Nervous System are:—

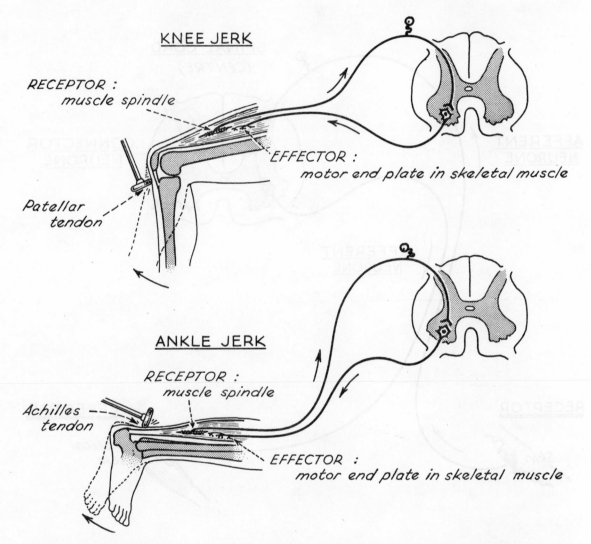

KNEE JERK

RECEPTOR : muscle spindle

EFFECTOR : motor end plate in skeletal muscle

Patellar tendon

ANKLE JERK

RECEPTOR : muscle spindle

Achilles tendon

EFFECTOR : motor end plate in skeletal muscle

When the tendon is sharply tapped the muscle is stretched. (N.B. The stimulus is by stretch of the muscle spindle.) Nerve messages pass into the spinal cord — and out to the muscle which then contracts.

SPINAL REFLEXES

In most REFLEX ARCS in man AFFERENT and EFFERENT
NEURONES are linked by at least 1 CONNECTOR NEURONE.

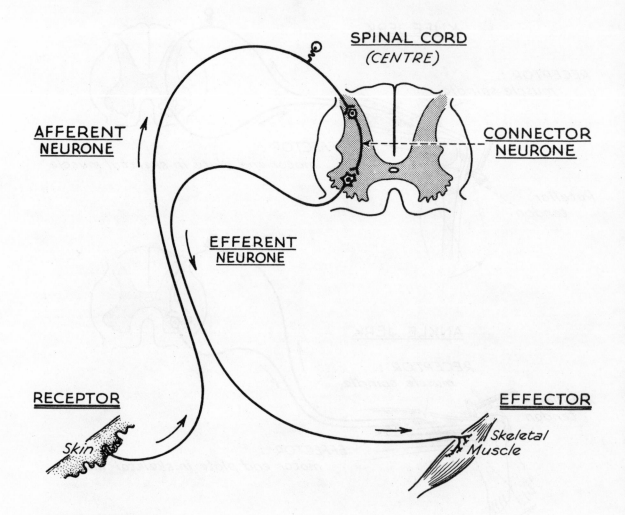

A chain of many connector neurones is frequently found.

"EDIFICE" of the C.N.S. *(After R.C. Garry)*

In the majority of Reflex arcs in man a chain of many connector neurones is found. There may be link-ups with various levels of the Brain and Spinal Cord.

This diagram gives a highly simplified concept of the type of LINK-UP which can occur between different levels of the Central Nervous System.

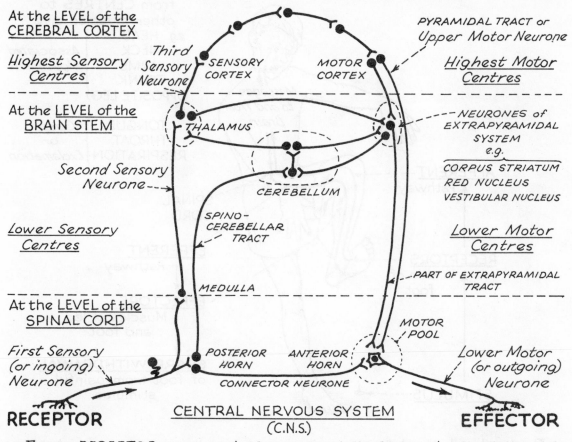

At the LEVEL of the CEREBRAL CORTEX

Highest Sensory Centres

Third Sensory Neurone

SENSORY CORTEX

MOTOR CORTEX

PYRAMIDAL TRACT or Upper Motor Neurone

Highest Motor Centres

At the LEVEL of the BRAIN STEM

THALAMUS

Second Sensory Neurone

CEREBELLUM

NEURONES of EXTRAPYRAMIDAL SYSTEM e.g.

CORPUS STRIATUM
RED NUCLEUS
VESTIBULAR NUCLEUS

Lower Sensory Centres

SPINO-CEREBELLAR TRACT

Lower Motor Centres

MEDULLA

PART OF EXTRAPYRAMIDAL TRACT

At the LEVEL of the SPINAL CORD

First Sensory (or ingoing) Neurone

POSTERIOR HORN

ANTERIOR HORN

MOTOR POOL

Lower Motor (or outgoing) Neurone

CONNECTOR NEURONE

RECEPTOR

CENTRAL NERVOUS SYSTEM (C.N.S.)

EFFECTOR

Every RECEPTOR neurone is thus potentially linked in the C.N.S. with a large number of EFFECTOR organs all over the body and every EFFECTOR neurone is similarly in communication with RECEPTORS all over the body.

Centres in the BRAIN and BRAIN STEM can thus modify REFLEX ACTS which occur through the SPINAL CORD. These centres can send "suppressing" or "facilitating" impulses along their pathways to the cells in the SPINAL CORD.

REFLEX ACTION

Most REFLEX ACTIONS in man involve a great many REFLEX ARCS.

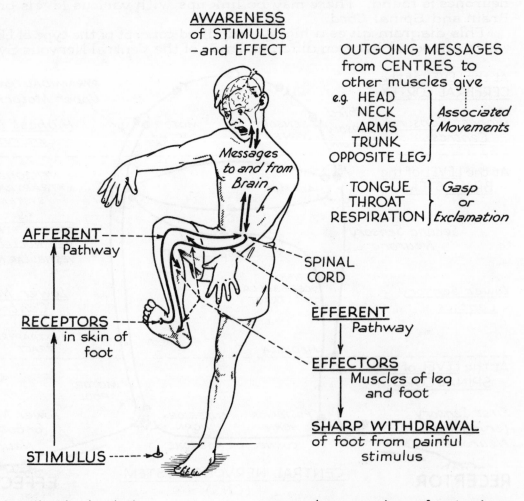

AWARENESS
of STIMULUS
– and EFFECT

OUTGOING MESSAGES
from CENTRES to
other muscles give
e.g. HEAD
NECK
ARMS   *Associated*
TRUNK   *Movements*
OPPOSITE LEG

TONGUE   *Gasp*
THROAT   *or*
RESPIRATION *Exclamation*

Messages to and from Brain

AFFERENT
Pathway

SPINAL
CORD

RECEPTORS
in skin of
foot

EFFERENT
Pathway

EFFECTORS
Muscles of leg
and foot

SHARP WITHDRAWAL
of foot from painful
stimulus

STIMULUS

The localized stimulation of a very few RECEPTORS – sending messages along their AFFERENT NEURONES to SPINAL CORD and to BRAIN – has led to a very large number of outgoing impulses in many EFFECTOR NEURONES to a large number of MUSCLES to give a very widespread and generalized REFLEX RESPONSE.

This is possible because each receptor neurone is potentially connected within the Central Nervous System with all effector neurones.

ARRANGEMENT of NEURONES

Some of the ways in which neurones can be linked are indicated here:-

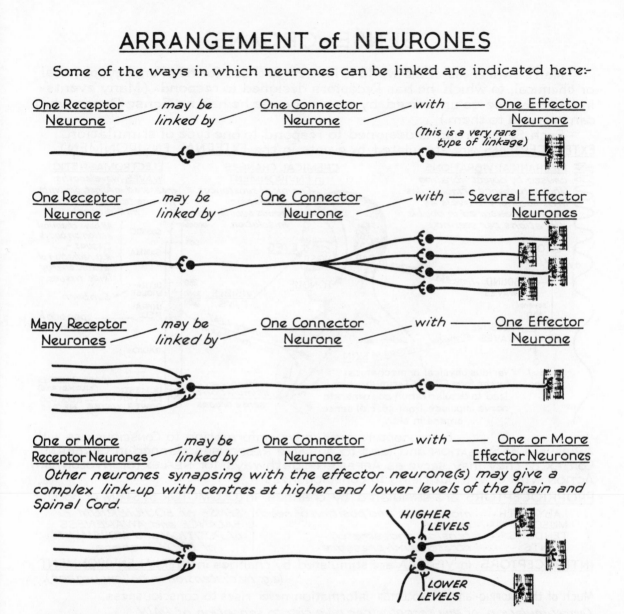

One Receptor Neurone — may be linked by — One Connector Neurone — with — One Effector Neurone
(This is a very rare type of linkage)

One Receptor Neurone — may be linked by — One Connector Neurone — with — Several Effector Neurones

Many Receptor Neurones — may be linked by — One Connector Neurone — with — One Effector Neurone

One or More Receptor Neurones — may be linked by — One Connector Neurone — with — One or More Effector Neurones

Other neurones synapsing with the effector neurone(s) may give a complex link-up with centres at higher and lower levels of the Brain and Spinal Cord.

HIGHER LEVELS

LOWER LEVELS

Through such 'functional' link-ups, neurones in different parts of the Central Nervous System, when active, can influence each other. This makes it possible for 'CONDITIONED' REFLEXES to become established (for simple example see page 55).

(for simple example see page 55)

Such reflexes probably form the basis of all training so that it becomes difficult to say where REFLEX (or INVOLUNTARY) behaviour ends and purely VOLUNTARY behaviour begins.

SENSE ORGANS

Man's awareness of the world is limited to those forms of energy, physical or chemical, to which he has Receptors designed to respond. (Many "events" in the universe go unnoticed by man because he has no Sense Organ which can respond to them.)

Each Sense Organ is designed to respond to one type of stimulation. EXTEROCEPTORS are stimulated by events in the EXTERNAL ENVIRONMENT.

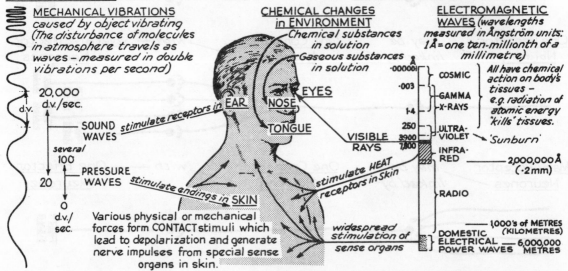

MECHANICAL VIBRATIONS caused by object vibrating (The disturbance of molecules in atmosphere travels as waves - measured in double vibrations per second)

20,000 d.v./sec.

SOUND WAVES *stimulate receptors in EAR*

several 100

20

PRESSURE WAVES *stimulate endings in SKIN*

0 d.v./sec.

Various physical or mechanical forces form CONTACT stimuli which lead to depolarization and generate nerve impulses from special sense organs in skin.

CHEMICAL CHANGES in ENVIRONMENT
- Chemical substances in solution
- Gaseous substances in solution

EYES
NOSE
TONGUE

stimulate HEAT receptors in Skin

widespread stimulation of sense organs

ELECTROMAGNETIC WAVES (wavelengths measured in Ångström units: 1Å = one ten-millionth of a millimetre)

·00001
·003
1·4
250
VISIBLE RAYS 3900
7,800

COSMIC
GAMMA
X-RAYS
ULTRA-VIOLET
INFRA-RED

All have chemical action on body's tissues - e.g. radiation of atomic energy 'kills' tissues.

'Sunburn'

2,000,000Å (·2mm)

RADIO

1,000's of METRES (KILOMETRES)

DOMESTIC ELECTRICAL POWER WAVES — 6,000,000 METRES

Exteroceptors may convey information to CONSCIOUSNESS with AWARENESS or SENSATION and lead to suitable RESPONSES planned in CEREBRAL CORTEX or they may serve as AFFERENT pathways for REFLEX (or INVOLUNTARY) ACTION with or without rising to consciousness.

PROPRIOCEPTORS are stimulated by changes in LOCOMOTOR SYSTEM of body

LABYRINTH ----- *movements and position of head* ⎫ SENSE of EQUILIBRIUM or
MUSCLES ------- *stretch*   ⎬ BALANCE and AWARENESS
TENDONS ------- *tension and stretch*   ⎪ of POSITION and MOVEMENT
JOINTS --------- *stretch and pressure*   ⎭ of BODY in space.

INTEROCEPTORS in VISCERA are stimulated by changes in INTERNAL ENVIRONMENT (e.g. by distension in hollow organs).

Much of the PROPRIO- and INTEROCEPTOR information never rises to consciousness.

Overstimulation of *any receptor* can give rise to sensation of *PAIN*.
Most receptors show ADAPTATION — if continuously stimulated they send reduced numbers of impulses to the brain.

PERCEPTION is awareness of the source of the stimulus.
Information from one Sense Organ is correlated with other information (*past or present*).
APPERCEPTION is recognition or identification of the source of the stimulus and depends on association with past experience.

SMELL

Smell is a CHEMICAL SENSE, i.e. the receptors respond to CHEMICAL STIMULI. To arouse the sensation a substance must first be in a GASEOUS STATE then go into SOLUTION.

The ORGAN of SMELL is the NOSE — *Also serves as the main air passage to Respiratory System.*

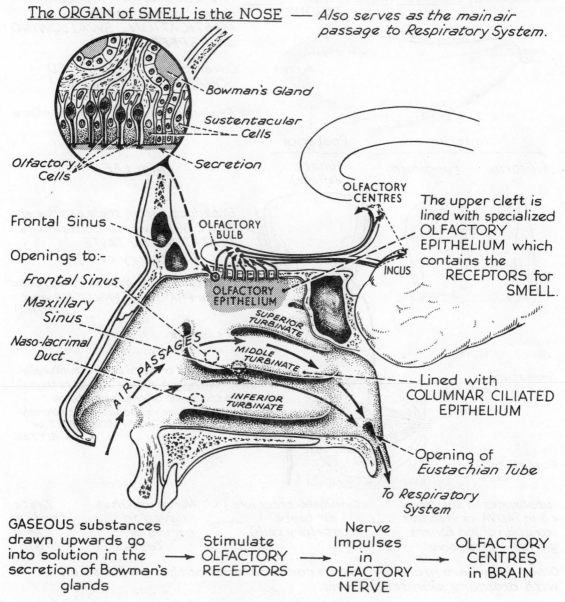

Bowman's Gland

Sustentacular Cells

Secretion

Olfactory Cells

OLFACTORY CENTRES

The upper cleft is lined with specialized OLFACTORY EPITHELIUM which contains the RECEPTORS for SMELL.

Frontal Sinus

OLFACTORY BULB

INCUS

Openings to:-

Frontal Sinus

Maxillary Sinus

Naso-lacrimal Duct

OLFACTORY EPITHELIUM

SUPERIOR TURBINATE

AIR PASSAGES

MIDDLE TURBINATE

INFERIOR TURBINATE

— Lined with COLUMNAR CILIATED EPITHELIUM

Opening of *Eustachian Tube*

To Respiratory System

GASEOUS substances drawn upwards go into solution in the secretion of Bowman's glands → Stimulate OLFACTORY RECEPTORS → Nerve Impulses in OLFACTORY NERVE → OLFACTORY CENTRES in BRAIN

TASTE

Taste is a CHEMICAL SENSE, *i.e.* receptors respond to CHEMICAL STIMULI. To arouse the sensation a substance must be in SOLUTION.

The essential <u>ORGAN of TASTE</u> is the TONGUE

The VOLUNTARY muscular organ concerned also in MASTICATION, SWALLOWING and SPEECH.

Covered with STRATIFIED SQUAMOUS EPITHELIUM.

Projections on its upper surface are called

<u>PAPILLAE</u>

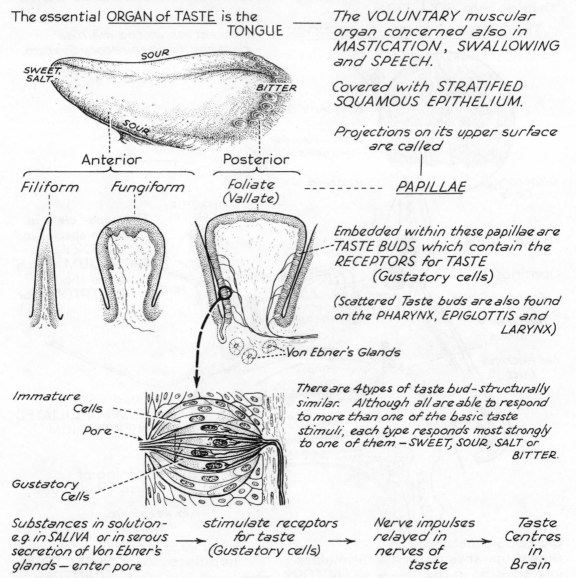

SOUR
SWEET, SALT
BITTER
SOUR

Anterior | Posterior
Filiform | Fungiform | Foliate (Vallate)

Von Ebner's Glands

Immature Cells
Pore
Gustatory Cells

Embedded within these papillae are TASTE BUDS which contain the RECEPTORS for TASTE (Gustatory cells)

(Scattered Taste buds are also found on the PHARYNX, EPIGLOTTIS and LARYNX)

There are 4 types of taste bud – structurally similar. Although all are able to respond to more than one of the basic taste stimuli, each type responds most strongly to one of them – SWEET, SOUR, SALT or BITTER.

Substances in solution – e.g. in SALIVA or in serous secretion of Von Ebner's glands – enter pore → stimulate receptors for taste (Gustatory cells) → Nerve impulses relayed in nerves of taste → Taste Centres in Brain

Other tastes are probably due to combinations of these with smell or with ordinary skin sensations.

PATHWAYS and CENTRES for TASTE

The RECEPTORS for TASTE are linked by a chain of 3 Neurones with the RECEIVING CENTRES for TASTE in the CEREBRAL CORTEX.

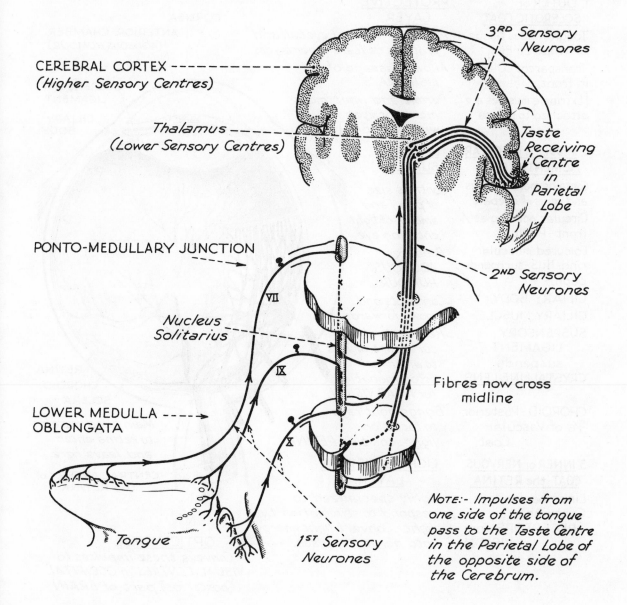

3RD Sensory Neurones

CEREBRAL CORTEX (Higher Sensory Centres)

Thalamus (Lower Sensory Centres)

Taste Receiving Centre in Parietal Lobe

PONTO-MEDULLARY JUNCTION

2ND Sensory Neurones

VII

Nucleus Solitarius

IX

Fibres now cross midline

LOWER MEDULLA OBLONGATA

X

Tongue

1ST Sensory Neurones

NOTE:- Impulses from one side of the tongue pass to the Taste Centre in the Parietal Lobe of the opposite side of the Cerebrum.

EYE

STRUCTURE

The Eyeball has three coats:-

1. OUTER or SCLEROTIC COAT

Tough fibrous tissue of "White of Eye".

Transparent CORNEA in front.

[Extrinsic muscles are attached to sclera.

2. MIDDLE or VASCULAR PIGMENTED COAT

Contains main arteries and veins of eyeball.

Circular opening at front – PUPIL.

Coloured muscular ring – IRIS – surrounds pupil.

CILIARY BODY.

CILIARY MUSCLE.

SUSPENSORY LIGAMENT suspends CRYSTALLINE LENS.

CHOROID – Posterior 5/6 of Vascular Coat.

3. INNER or NERVOUS COAT – the RETINA

Lines back of eye. Contains RECEPTORS for VISION

FUNCTION of PARTS

PROTECTIVE LAYER

Preserves shape of eyeball and protects delicate inner layers.

Allows passage of LIGHT RAYS.

Permit and limit movements of eyeball within ORBIT.]

LAYER of SUPPLY

Controls size of pupil and amount of light entering eye.

Produces AQUEOUS HUMOUR.

Contracts and moves forwards.

Relaxes to allow curvature of lens to alter for accommodation for NEAR VISION.

Brings light rays to focus on light-sensitive RETINA.

LIGHT-SENSITIVE LAYER

Highly specialized to respond to stimulation by light. Convert light energy into nerve impulses. ⟶

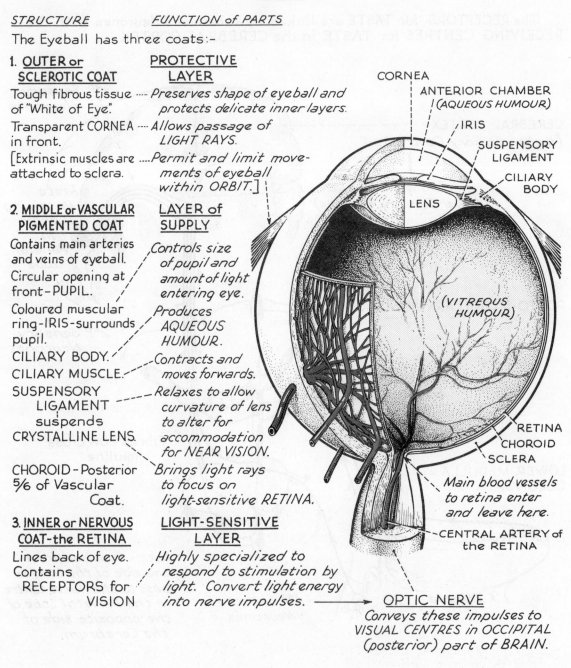

CORNEA

ANTERIOR CHAMBER (AQUEOUS HUMOUR)

IRIS

SUSPENSORY LIGAMENT

CILIARY BODY

LENS

(VITREOUS HUMOUR)

RETINA
CHOROID
SCLERA

Main blood vessels to retina enter and leave here.

CENTRAL ARTERY of the RETINA

OPTIC NERVE

Conveys these impulses to VISUAL CENTRES in OCCIPITAL (posterior) part of BRAIN.

PROTECTION of the EYE

The hidden posterior ⁴/₅ of the eyeball is encased in a bony socket – the ORBITAL CAVITY.   A thick layer of *Areolar* and *Adipose* tissue forms a cushion between bone and eyeball. The exposed anterior ¹/₅ of the eyeball is protected from injury by:-

The <u>EYELIDS</u> ------------------- close reflexly to protect eye from dust
Fringed with EYELASHES   and other foreign particles.

<u>CONJUNCTIVA</u> ------------------- smooth surfaces which glide over each
A delicate membrane lining eyelids   other when lids open and close.
and covering exposed surface of eye.

<u>LACRIMAL GLANDS</u> --------------- continuously secrete TEARS. These
                                   flow over, wash and lubricate surface
                                   of eye. They contain an ENZYME –
                                   LYSOZYME – which destroys bacteria.

<u>TARSAL GLANDS</u> ------------------- secrete a fluid to prevent lids
                                     from sticking together

<u>LACRIMAL DUCTS</u>
  drain tears
  from surface
  of eye

<u>LACRIMAL SAC</u>

<u>NASO-LACRIMAL</u>
<u>DUCT</u> ------------------- drains tears into back of nose.

--- CONJUNCTIVA

--- LACRIMAL GLANDS

--- OPENINGS OF
      TARSAL GLANDS

ORBICULARIS OCULI
MUSCLE

225

MUSCLES of EYE

The EYEBALLS are moved by SMALL MUSCLES which link the SCLEROTIC COAT to the BONY SOCKET.

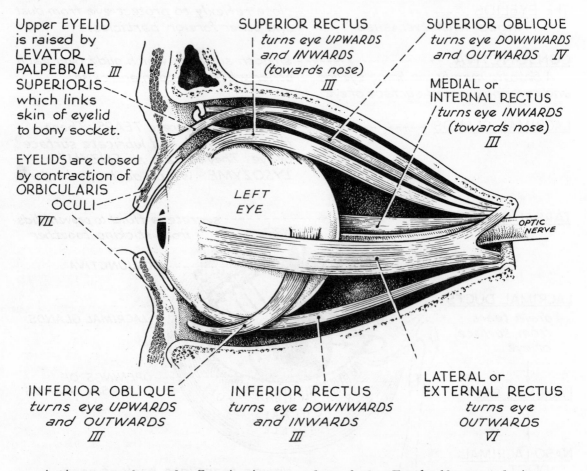

Upper EYELID is raised by LEVATOR PALPEBRAE SUPERIORIS *III* which links skin of eyelid to bony socket.

EYELIDS are closed by contraction of ORBICULARIS OCULI *VII*

SUPERIOR RECTUS *turns eye UPWARDS and INWARDS (towards nose)* *III*

SUPERIOR OBLIQUE *turns eye DOWNWARDS and OUTWARDS* *IV*

MEDIAL or INTERNAL RECTUS *turns eye INWARDS (towards nose)* *III*

LEFT EYE

OPTIC NERVE

INFERIOR OBLIQUE *turns eye UPWARDS and OUTWARDS* *III*

INFERIOR RECTUS *turns eye DOWNWARDS and INWARDS* *III*

LATERAL or EXTERNAL RECTUS *turns eye OUTWARDS* *VI*

Acting together, the Extrinsic muscles of the Eyeballs can bring about ROTATORY movements of the Eyes.
The Extrinsic muscles are supplied by motor fibres from Cranial Nerves III, IV and VI.

CONTROL of EYE MOVEMENTS

Both eyes normally move together so that images continue to fall on corresponding points of both retinae.

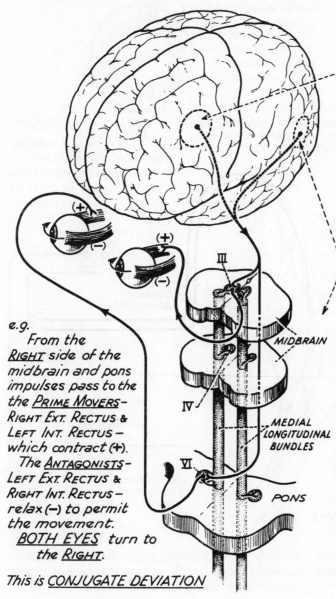

VOLUNTARY EYE MOVEMENTS are initiated in motor centres in FRONTAL LOBES.

Impulses from <u>one side</u> of the <u>Cerebral Cortex</u> turn <u>both eyes</u> to the <u>other side</u> of Visual Field.

REFLEX EYE MOVEMENTS *Two Groups* – (1) Those in response to Visual Stimuli. (2) Those in response to Non-Visual Stimuli. In control of these are :– CENTRES in OCCIPITAL LOBES: CENTRES in MIDBRAIN and PONS which give rise to CRANIAL NERVES III, IV and VI.

Impulses from <u>one side</u> of the <u>Midbrain and Pons</u> turn eyes to the <u>same side</u>.

These centres are <u>closely linked</u> with each other and with HIGHER and LOWER centres in the Central Nervous System, so that the eyes are moved reflexly in response to many stimuli, e.g. loud noises or proprioceptive messages from vestibular organs.

e.g.
From the <u>RIGHT</u> side of the midbrain and pons impulses pass to the the <u>PRIME MOVERS</u>– *RIGHT EXT. RECTUS & LEFT INT. RECTUS* – which contract (+).
The <u>ANTAGONISTS</u>– *LEFT EXT. RECTUS & RIGHT INT. RECTUS* – relax (–) to permit the movement.
<u>BOTH EYES</u> turn to the <u>RIGHT</u>.

This is <u>CONJUGATE DEVIATION</u>

Labels in diagram: III, IV, VI, MIDBRAIN, MEDIAL LONGITUDINAL BUNDLES, PONS

IRIS, LENS and CILIARY BODY

The <u>IRIS</u> is a muscular diaphragm with a central opening - the PUPIL.

The <u>LENS</u> is a transparent biconvex crystalline disc.

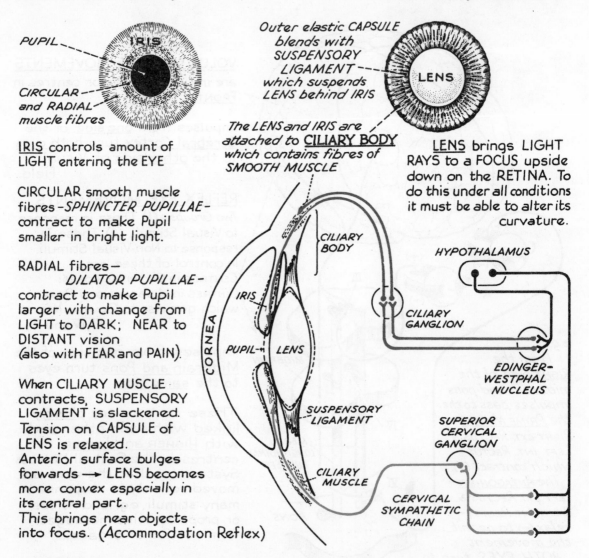

PUPIL

IRIS

CIRCULAR and RADIAL muscle fibres

Outer elastic CAPSULE blends with SUSPENSORY LIGAMENT which suspends LENS behind IRIS

LENS

The LENS and IRIS are attached to <u>CILIARY BODY</u> which contains fibres of SMOOTH MUSCLE

<u>IRIS</u> controls amount of LIGHT entering the EYE

CIRCULAR smooth muscle fibres –*SPHINCTER PUPILLAE*– contract to make Pupil smaller in bright light.

RADIAL fibres –
     DILATOR PUPILLAE–
contract to make Pupil larger with change from LIGHT to DARK; NEAR to DISTANT vision
(also with FEAR and PAIN).

When CILIARY MUSCLE contracts, SUSPENSORY LIGAMENT is slackened. Tension on CAPSULE of LENS is relaxed. Anterior surface bulges forwards ⟶ LENS becomes more convex especially in its central part.
This brings near objects into focus. (Accommodation Reflex)

<u>LENS</u> brings LIGHT RAYS to a FOCUS upside down on the RETINA. To do this under all conditions it must be able to alter its curvature.

CILIARY BODY

CORNEA

IRIS

PUPIL

LENS

SUSPENSORY LIGAMENT

CILIARY MUSCLE

HYPOTHALAMUS

CILIARY GANGLION

EDINGER-WESTPHAL NUCLEUS

SUPERIOR CERVICAL GANGLION

CERVICAL SYMPATHETIC CHAIN

These changes are brought about reflexly. The ingoing impulses travel in the Optic nerves. The outgoing motor impulses travel in Parasympathetic to Ciliary Body and Sphincter Pupillae and in Sympathetic to Dilator Pupillae.

ACTION of LENS

The normal lens brings light rays to a sharp focus upside down on the retina. It can do this whether we are looking at an object far away or one close at hand. The curvature increases reflexly to accommodate for near vision.

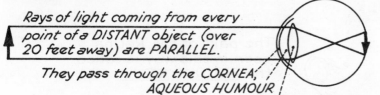

Rays of light coming from every point of a DISTANT object (over 20 feet away) are PARALLEL.

The conscious mind learns to interpret the image and project it to its true position in space.

They pass through the CORNEA, AQUEOUS HUMOUR

and LENS which refract them to a sharp focus — upside down and reversed from side to side — on the Retina.

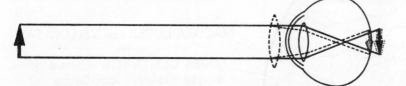

Rays of light coming from a NEAR object (less than 20 feet away) RADIATE from every point

A more convex lens is required to bring these rays to a sharp focus on the Retina.

If the EYEBALL is <u>too short</u>, rays from a distant object are brought into focus BEHIND the Retina when the ciliary muscle is relaxed.

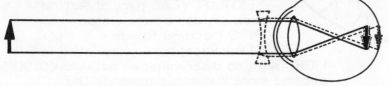

Vision is blurred.

This is longsightedness or HYPERMETROPIA.

The longsighted eye has to accommodate even for distant vision: i.e. Ciliary muscles contract to give a more convex lens and distant objects are then seen clearly. This limits amount of accommodating power left for near objects and the nearest point for sharp vision is then further away. It can be corrected by fitting spectacles with an additional convex lens.

If the EYEBALL is <u>too long</u>, rays from a distant object are brought into focus IN FRONT of the Retina.

This is shortsightedness or MYOPIA — only objects near the eye can be seen clearly.

It can be corrected by using a concave lens.

FUNDUS OCULI

Part of the RETINA can be seen by means of an instrument – the OPHTHALMOSCOPE – which shines a beam of light through the PUPIL of the eye on to the RETINA.

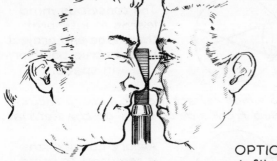

The part of the Retina seen in this way is called the <u>FUNDUS OCULI</u>.

OPTIC PAPILLA – or 'BLIND SPOT'– The nerve fibres from all parts of the retina converge on this area to leave the eyeball as the OPTIC NERVE. It has no RODS or CONES and therefore is not itself sensitive to light.

RETINAL BLOOD VESSELS enter or leave the eyeball here.

MACULA LUTEA – or 'YELLOW SPOT' – with

FOVEA CENTRALIS – area of acute vision – contains CONES only (the Receptors stimulated in BRIGHT and COLOURED LIGHT). When we look at an object the eyes are directed so that the image will fall on the fovea of each eye.

LEFT FUNDUS

EXTRAFOVEAL part of Retina – area of less acute vision – CONES become fewer – RODS (the Receptors stimulated in DIM LIGHT with no discrimination between COLOURS) become more numerous towards the peripheral part of the Retina.

RETINA

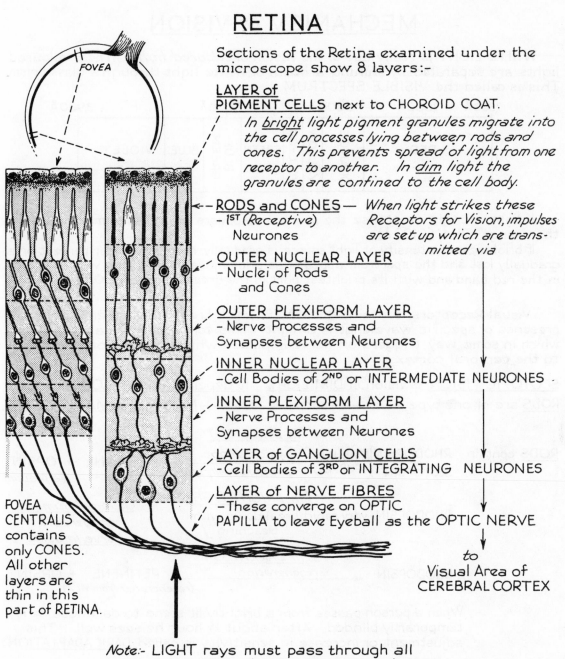

Sections of the Retina examined under the microscope show 8 layers:-

LAYER of
PIGMENT CELLS next to CHOROID COAT.

In bright light pigment granules migrate into the cell processes lying between rods and cones. This prevents spread of light from one receptor to another. In dim light the granules are confined to the cell body.

RODS and CONES —
- 1ST (Receptive) Neurones

When light strikes these Receptors for Vision, impulses are set up which are transmitted via

OUTER NUCLEAR LAYER
- Nuclei of Rods and Cones

OUTER PLEXIFORM LAYER
- Nerve Processes and Synapses between Neurones

INNER NUCLEAR LAYER
-Cell Bodies of 2ND or INTERMEDIATE NEURONES

INNER PLEXIFORM LAYER
- Nerve Processes and Synapses between Neurones

LAYER of GANGLION CELLS
- Cell Bodies of 3RD or INTEGRATING NEURONES

LAYER of NERVE FIBRES
-These converge on OPTIC PAPILLA to leave Eyeball as the OPTIC NERVE

to
Visual Area of
CEREBRAL CORTEX

FOVEA

FOVEA CENTRALIS contains only CONES. All other layers are thin in this part of RETINA.

Note:- LIGHT rays must pass through all these layers except the pigment cell layer to reach and stimulate RECEPTORS.

MECHANISM of VISION

WHITE LIGHT is really due to the fusion of *coloured lights*. These coloured lights are separated by shining a beam of white light through a glass prism. This is called the VISIBLE SPECTRUM.

7,800Å	7,000Å			6,000Å				5,000Å			3,900Å
RED		ORANGE	YELLOW	YELL.-GR.	GREEN	BLUE-GREEN	BLUE	VIOLET			

If light is bright or intense the spectrum appears brightest to man's eye in the orange band (6,100 Å).

If brightness or intensity of light source is gradually reduced, colour perception is gradually lost and the spectrum appears as a luminous band with a very dark area in the red band and with its brightest part in the green band (5,300Å).

Visual Receptors contain PIGMENTS which break down chemically in the presence of specific wavelengths of light. This forms a chemical stimulus which in some way triggers off nerve impulses which travel from the retina to the cerebral cortex.

SCOTOPIC VISION is vision in DIM LIGHT.   It depends on the RODS

RODS are of one type ——————and give——————MONOCHROMATIC VISION

In DIM LIGHT

RODS contain   RHODOPSIN ("VISUAL   *absorbs light* ────→ RETINENE + OPSIN
or   SCOTOPSIN   PURPLE")   *Pigment gradually splits*   (*a protein*)

↓

Chemical Stimulus
triggers off
Nerve Impulse

[In BRIGHT LIGHT
RHODOPSIN - very rapid breakdown with "bleaching".*]*

In DARKNESS
RHODOPSIN ←———— *regeneration* ————   RETINENE + OPSIN
(*regenerated from Vit.A*)

When a person passes from a brightly lit scene to darkness he is temporarily blinded. After about ½ hour he sees well. This adjustment or increase in sensitivity is called DARK ADAPTATION.

As light brightness or intensity increases rods lose their sensitivity and cease to respond.

232

MECHANISM of VISION

PHOTOPIC VISION is vision in <u>BRIGHT LIGHT</u>. <u>It depends on the CONES</u>

<u>CONES</u> are thought to be of 3 types —— giving <u>TRICHROMATIC VISION</u>.
Each type with a
different
photosensitive
<u>VISUAL</u>
<u>PIGMENT</u> with its own wavelength to which it is sensitive, which
it absorbs and by which it is broken down to form the
chemical stimulus.

Note:- The Photosensitive pigments in cones have not yet been isolated.

"<u>RED</u>" receptors absorb <u>YELLOW—ORANGE</u> light
"<u>GREEN</u>" receptors absorb <u>GREEN</u> light
"<u>BLUE</u>" receptors absorb <u>BLUE</u> light

All 3 types of "PHOTOPSINS" are probably stimulated in roughly equal
proportions when <u>WHITE light</u> falls on retina:
2 or more in varying proportions when <u>OTHER COLOURS of light</u> fall on
retina.

The various types of colour blindness could be explained in terms of the
absence or deficiency of one or more of these special receptors.

[A colour sensation has 3 qualities:-
HUE ——————— depends largely on wavelength.
SATURATION —— purity —
A "saturated" colour has no white light mixed with it.
An "unsaturated" colour has some white light mixed with it.
INTENSITY —— brightness — depends largely on "strength" of the light.]

As the intensity of light is reduced the Cones cease to respond and
the Rods take over.

When a person passes from darkness to bright light he is dazzled
but after a short time he sees well again.
This adjustment or decrease in sensitivity on exposure to bright light
is called <u>LIGHT ADAPTATION</u>.

VISUAL PATHWAYS to the BRAIN

The RECEPTORS for VISION are linked by a chain of Neurones with RECEIVING and INTEGRATING CENTRES in the OCCIPITAL LOBES of the CEREBRAL CORTEX.

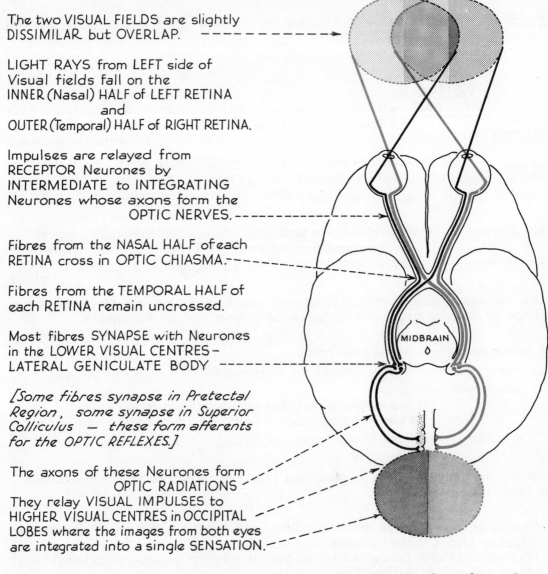

The two VISUAL FIELDS are slightly DISSIMILAR but OVERLAP. - - - - - - - - - -

LIGHT RAYS from LEFT side of Visual fields fall on the INNER (Nasal) HALF of LEFT RETINA
and
OUTER (Temporal) HALF of RIGHT RETINA.

Impulses are relayed from RECEPTOR Neurones by INTERMEDIATE to INTEGRATING Neurones whose axons form the OPTIC NERVES. - - - - - - - - - -

Fibres from the NASAL HALF of each RETINA cross in OPTIC CHIASMA. - - - - -

Fibres from the TEMPORAL HALF of each RETINA remain uncrossed.

Most fibres SYNAPSE with Neurones in the LOWER VISUAL CENTRES - LATERAL GENICULATE BODY - - - - - - - - - -

[Some fibres synapse in Pretectal Region, some synapse in Superior Colliculus — these form afferents for the OPTIC REFLEXES.]

The axons of these Neurones form OPTIC RADIATIONS
They relay VISUAL IMPULSES to HIGHER VISUAL CENTRES in OCCIPITAL LOBES where the images from both eyes are integrated into a single SENSATION. - -

MIDBRAIN

Note:- One side of the OCCIPITAL CORTEX receives impressions from the FIELD of VISION on the opposite side.

STEREOSCOPIC VISION

When we look at some object or scene the view seen by the RIGHT EYE is slightly different from the view seen by the LEFT EYE.

These two DISSIMILAR RETINAL IMAGES are fused in the Visual Centres of the Brain to give a 3-dimensional picture – an appreciation of DEPTH as well as of HEIGHT and WIDTH.

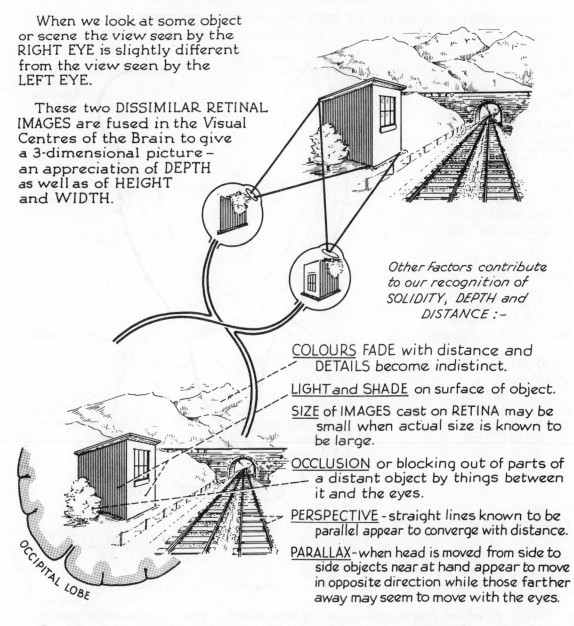

Other factors contribute to our recognition of SOLIDITY, DEPTH and DISTANCE :–

COLOURS FADE with distance and DETAILS become indistinct.

LIGHT and SHADE on surface of object.

SIZE of IMAGES cast on RETINA may be small when actual size is known to be large.

OCCLUSION or blocking out of parts of a distant object by things between it and the eyes.

PERSPECTIVE - straight lines known to be parallel appear to converge with distance.

PARALLAX - when head is moved from side to side objects near at hand appear to move in opposite direction while those farther away may seem to move with the eyes.

OCCIPITAL LOBE

By complex mental processes these points are interpreted in terms of Distance and Depth.

LIGHT REFLEX

When LIGHT falls on the RETINA the PUPILS CONTRACT.

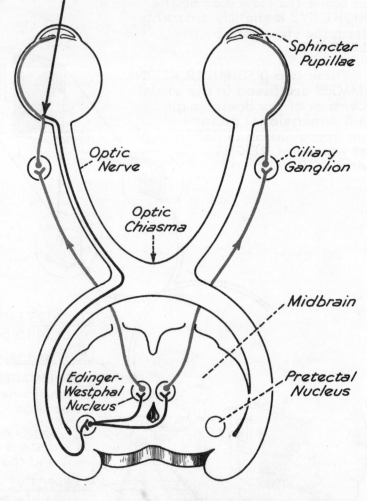

Impulses from
the Receptors in
the Retina pass
in the
OPTIC NERVE
to the
PRETECTAL
NUCLEUS
in the midbrain
of the same side

Impulses then
pass to the
EDINGER-WESTPHAL
NUCLEUS
of the 3ʀᴅ Cranial
Nerves of <u>both sides</u>

Impulses now travel
in efferent
Parasympathetic motor
nerves to cause
contraction of both
SPHINCTER
PUPILLAE

*Sphincter
Pupillae*

*Optic
Nerve*

*Ciliary
Ganglion*

*Optic
Chiasma*

Midbrain

*Edinger-
Westphal
Nucleus*

*Pretectal
Nucleus*

<u>DIRECT LIGHT REFLEX</u>
|
*When light shines into
one eye (as in diagram)
the PUPIL of that eye contracts* ———

<u>CONSENSUAL LIGHT
REFLEX</u>
|
*the PUPIL of the other eye also
contracts.*

*This cuts down the amount of light entering the eyes and protects
the Retinae from excessive stimulation. It also increases depth of
focus and improves the sharpness of the Retinal images.*

EAR

The ear has 3 separate parts, each with different rôles in the mechanism of HEARING:-

OUTER EAR

AURICLE or *PINNA*

CARTILAGE and SKIN

EXTERNAL AUDITORY MEATUS

MIDDLE EAR

Small chamber deep within the *TEMPORAL BONE*.
Contains 3 small bones – the *AUDITORY OSSICLES* – connected to form a small *LEVER*.

MALLEUS attached to drum and to *INCUS* linked to *STAPES* which fits into *OVAL WINDOW*.

STAPES
INCUS
MALLEUS
TEMPORAL BONE

TYMPANIC MEMBRANE (DRUM)

INNER EAR

Contains *ORGANS* of *EQUILIBRIUM* and *HEARING*.

SEMICIRCULAR CANALS (non-auditory part of inner ear – concerned with EQUILIBRIUM sense).

COCHLEA (in Spiral Canal) contains RECEPTORS for HEARING (Organ of Corti).

AUDITORY NERVE

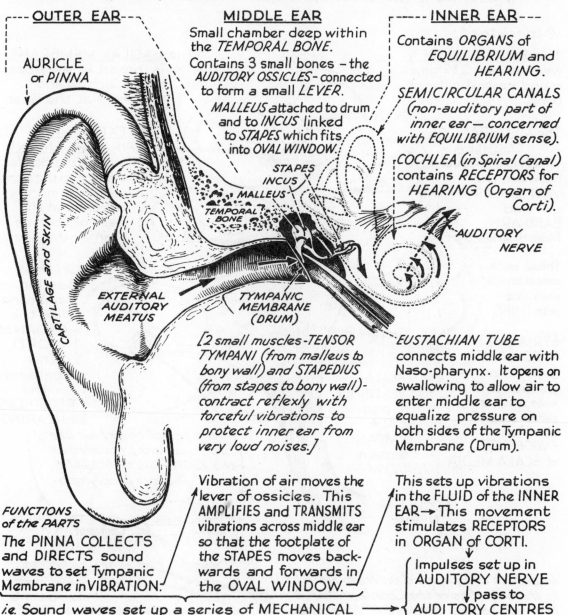

[2 small muscles - TENSOR TYMPANI *(from malleus to bony wall)* and STAPEDIUS *(from stapes to bony wall)* - contract reflexly with forceful vibrations to protect inner ear from very loud noises.]

EUSTACHIAN TUBE connects middle ear with Naso-pharynx. It opens on swallowing to allow air to enter middle ear to equalize pressure on both sides of the Tympanic Membrane (Drum).

This sets up vibrations in the FLUID of the INNER EAR → This movement stimulates RECEPTORS in ORGAN of CORTI.

FUNCTIONS of the PARTS

The PINNA COLLECTS and DIRECTS sound waves to set Tympanic Membrane in VIBRATION.

Vibration of air moves the lever of ossicles. This AMPLIFIES and TRANSMITS vibrations across middle ear so that the footplate of the STAPES moves backwards and forwards in the OVAL WINDOW.

Impulses set up in AUDITORY NERVE ↓ pass to AUDITORY CENTRES in BRAIN.

i.e. Sound waves set up a series of MECHANICAL STIMULI ⟶

COCHLEA

The Cochlea is the essential organ of HEARING.

It consists of :-

The BONY COCHLEA which spirals 2¾ times round central pillar of bone.

The MEMBRANOUS COCHLEA which is enclosed between the VESTIBULAR and BASILAR Membranes.

These spiral compartments are filled with FLUID

STRIA VASCULARIS (pigmented, granular cells with profuse blood supply) secretes ENDOLYMPH of SCALA MEDIA

SCALA VESTIBULI and SCALA TYMPANI contain PERILYMPH

COCHLEAR NERVE

APEX

BASE

The BASILAR MEMBRANE is broad at the Apex of the Cochlea and short at the Base.

The EXTERNAL SPIRAL LIGAMENT attaches the Basilar Membrane to the bony wall. It is more powerful at the Base of Cochlea than at the Apex.

On the Basilar membrane lies the ORGAN of CORTI. This contains the RECEPTORS for HEARING.

The tips of their hair-like processes are embedded in the TECTORIAL MEMBRANE which floats in Endolymph

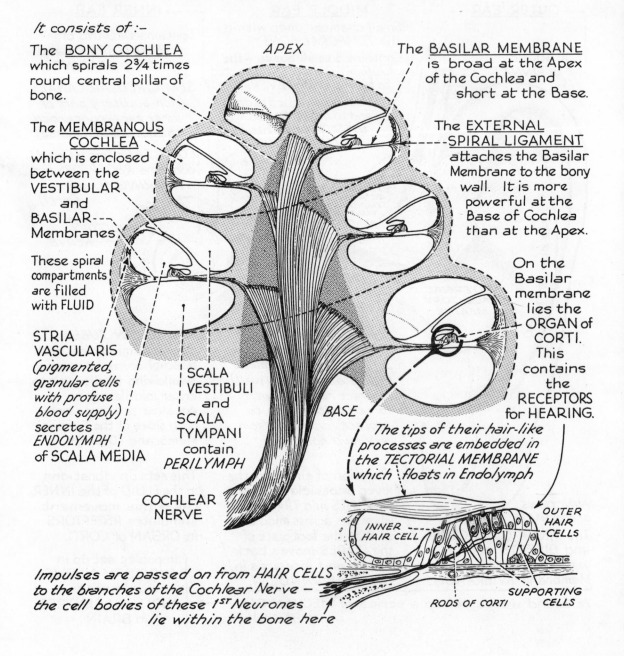

INNER HAIR CELL

OUTER HAIR CELLS

RODS OF CORTI

SUPPORTING CELLS

Impulses are passed on from HAIR CELLS to the branches of the Cochlear Nerve – the cell bodies of these 1ST Neurones lie within the bone here

MECHANISM of HEARING

This is most readily understood if the COCHLEA is imagined as straightened out:—

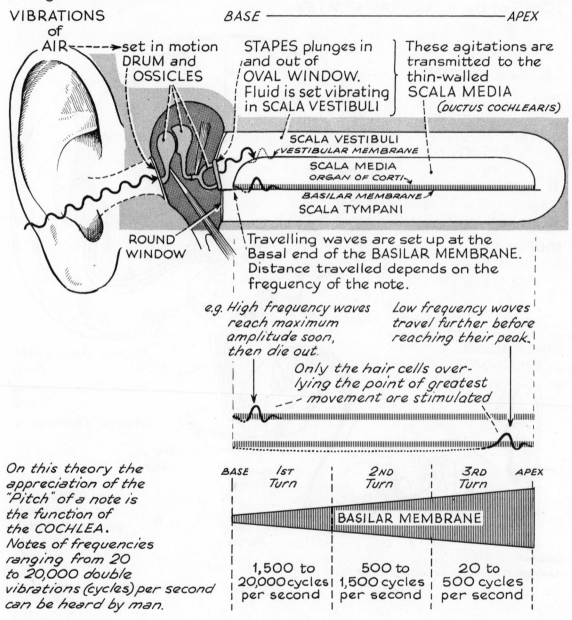

VIBRATIONS of AIR ---→ set in motion DRUM and OSSICLES

BASE ———————————— APEX

STAPES plunges in and out of OVAL WINDOW. Fluid is set vibrating in SCALA VESTIBULI

These agitations are transmitted to the thin-walled SCALA MEDIA *(DUCTUS COCHLEARIS)*

SCALA VESTIBULI
VESTIBULAR MEMBRANE
SCALA MEDIA
ORGAN OF CORTI
BASILAR MEMBRANE
SCALA TYMPANI

ROUND WINDOW

Travelling waves are set up at the Basal end of the BASILAR MEMBRANE. Distance travelled depends on the frequency of the note.

e.g. High frequency waves reach maximum amplitude soon, then die out.

Low frequency waves travel further before reaching their peak.

Only the hair cells over-lying the point of greatest movement are stimulated

On this theory the appreciation of the "Pitch" of a note is the function of the COCHLEA. Notes of frequencies ranging from 20 to 20,000 double vibrations (cycles) per second can be heard by man.

BASE	1ST Turn	2ND Turn	3RD Turn	APEX
	1,500 to 20,000 cycles per second	500 to 1,500 cycles per second	20 to 500 cycles per second	

BASILAR MEMBRANE

AUDITORY PATHWAYS to BRAIN

The RECEPTORS for HEARING are linked by a chain of NEURONES with the RECEIVING CENTRES for HEARING in the TEMPORAL LOBES of the CEREBRAL CORTEX.

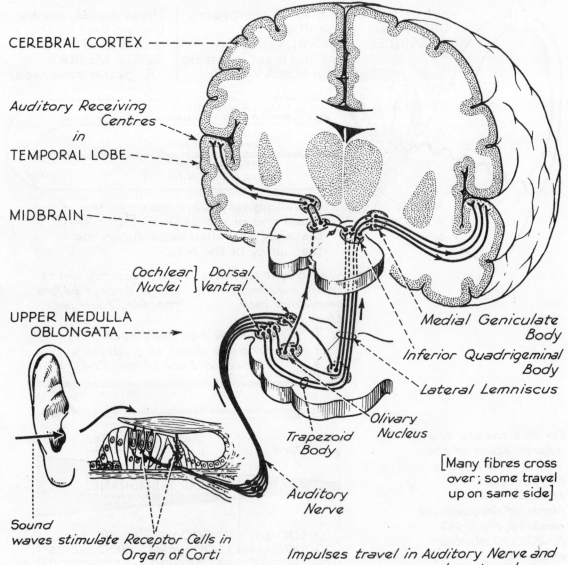

CEREBRAL CORTEX

Auditory Receiving
    Centres
        in
TEMPORAL LOBE

MIDBRAIN

Cochlear] Dorsal
Nuclei }Ventral

UPPER MEDULLA
OBLONGATA

Medial Geniculate
    Body

Inferior Quadrigeminal
    Body

Lateral Lemniscus

Olivary
Nucleus

Trapezoid
Body

Auditory
Nerve

[Many fibres cross
over; some travel
up on same side]

Sound
waves stimulate Receptor Cells in
Organ of Corti

Impulses travel in Auditory Nerve and
are relayed as shown.

SPECIAL PROPRIOCEPTORS

The "Special" Proprioceptors of the body are found in the Non-Auditory part of the Inner Ear — the Labyrinth. They are stimulated by movements or change of position of the head in space.

Three **SEMICIRCULAR CANALS** — *in each inner ear — one in each of three planes of space* — contain Receptors.

These receptors, situated in the Ampulla of each canal, are stimulated mechanically by the *starting or stopping of rotatory movements* of the head in space.

Membranous tubes containing Endolymph

One end of each canal has a swelling- the AMPULLA

Both ends of each canal open into the UTRICLE

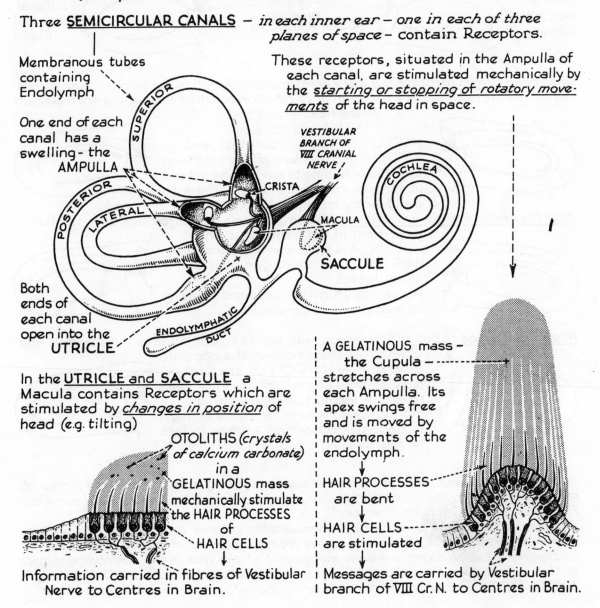

VESTIBULAR BRANCH OF VIII CRANIAL NERVE

SUPERIOR

POSTERIOR

LATERAL

CRISTA

COCHLEA

MACULA

SACCULE

ENDOLYMPHATIC DUCT

In the **UTRICLE** and **SACCULE** a Macula contains Receptors which are stimulated by *changes in position* of head (e.g. tilting)

OTOLITHS *(crystals of calcium carbonate)* in a GELATINOUS mass mechanically stimulate the HAIR PROCESSES of HAIR CELLS

Information carried in fibres of Vestibular Nerve to Centres in Brain.

A GELATINOUS mass — the Cupula — stretches across each Ampulla. Its apex swings free and is moved by movements of the endolymph.

HAIR PROCESSES are bent

↓

HAIR CELLS are stimulated

↓

Messages are carried by Vestibular branch of VIII Cr. N. to Centres in Brain.

ORGAN of EQUILIBRIUM: MECHANISM of ACTION

<u>At Rest</u>

RIGHT
Lateral
Horizontal
Canal

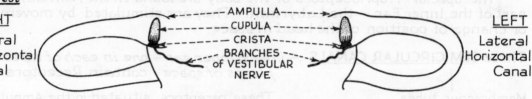

AMPULLA
CUPULA
CRISTA
BRANCHES
of VESTIBULAR
NERVE

LEFT
Lateral
Horizontal
Canal

<u>When head starts to rotate</u> (e.g. to the right) the endolymph in the semicircular canals which lie at right angles to the axis of rotation tends to lag behind the movement of the head — and the

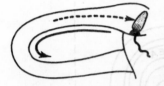

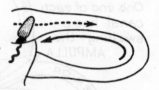

CUPULA is displaced,
HAIR CELLS are stimulated and
INGOING impulses form afferent pathways
for Reflexes leading to alterations in tone
of muscles in neck, trunk and limbs to
avoid body losing balance.

<u>After the initial inertia is overcome</u> the endolymph no longer lags behind the movement of the head — and the

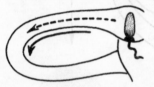

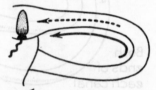

CUPULA is no longer displaced.
HAIR CELLS are no longer bent and
stimulated.
Nerve fibres no longer send signals
to medulla and cerebellum.

<u>When head stops rotating</u> the endolymph tends to continue to rotate and the

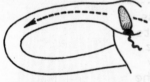

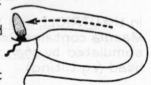

CUPULA is displaced in the opposite
direction → HAIR CELLS are bent →
Nerve fibres signal rotation of head
to left → Individual feels for a
moment as though he is rotating
in opposite direction — when in fact
he has ceased to rotate.

Rotating the head round a horizontal axis — i.e. tilting it backwards and forwards such as happens in the pitch and roll of a ship — stimulates the vertical canals. This can lead to "motion sickness".

Stimulation of the Semicircular canals also causes movements of the eyes to avoid too much displacement of the image being cast on the retinae.

During rotation there is a slow movement of the eyes in the direction opposite to that of rotation, then a quick return to the normal position. This is NYSTAGMUS. It continues for a short time after movement has ceased.

VESTIBULAR PATHWAYS to BRAIN

The Special Proprioceptive end-organs in the LABYRINTH are linked through the VESTIBULAR NUCLEI with RECEIVING and INTEGRATING CENTRES in the CEREBELLUM, and with MOTOR CENTRES in the MIDBRAIN and SPINAL CORD through which they initiate reflex muscular movements of EYES, HEAD and NECK, TRUNK and LIMB muscles to adjust balance and posture.

CEREBRAL CORTEX - - - - - - →

MIDBRAIN - - - - - - - - - -

CEREBELLUM - - - - - - - - - -

?

Vestibular { Superior
Nuclei { Medial
{ Lateral - - - - - - - -

UPPER MEDULLA OBLONGATA - - - - -

Medial Geniculate Body

Vestibular Nerve Ganglion - - - →

To Nuclei for Eye muscles

Lateral Vestibulo-spinal Tract

Medial Longitudinal Bundles

SPINAL CORD ←- - - - -

Semicircular Canals

GENERAL PROPRIOCEPTORS

Proprioceptors are the sense organs stimulated by MOVEMENT of the body itself. They make us aware of the movement or position of the body in space and of the various parts of the body to each other. They are important as ingoing afferent pathways in reflexes for adjusting Posture and Tone.

General Proprioceptors are found in SKELETAL MUSCLES, TENDONS and JOINTS.

GOLGI ORGAN — in tendons —
stimulated by tension which occurs
when muscle is STRETCHED and
when it is CONTRACTED

BONE

MUSCLE SPINDLE - - - - - - -
—in skeletal muscle-
stimulated when muscle
is STRETCHED or
SHORTENED

PACINIAN CORPUSCLES
similar to those in
the skin are found
in Deep Connective
Tissue and around
Joints. They are
stimulated by
PRESSURE of surrounding
structures when joints
are moved.

Muscle spindles
themselves contain
specialized muscle
fibres. These intrafusal
fibres are supplied with fine
motor nerves which are under
the control of higher centres.

PROPRIOCEPTOR PATHWAYS to BRAIN

General Proprioceptor end-organs may be linked with centres in
(a) CEREBELLUM or *(b)* PARIETAL LOBE of CEREBRAL CORTEX by a chain
of 3 Neurones.

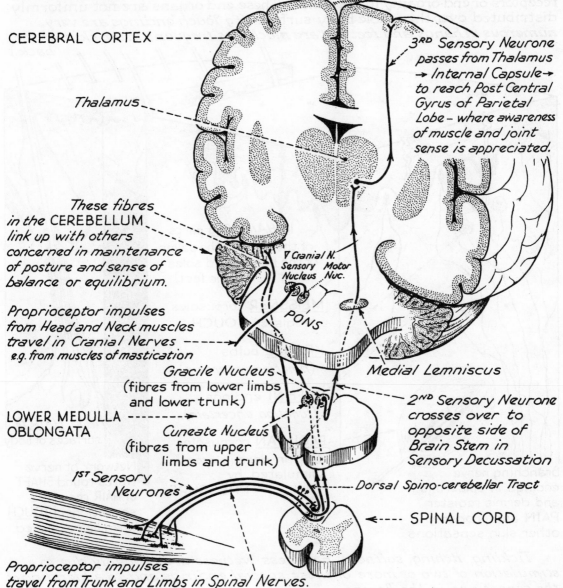

CEREBRAL CORTEX ------

*3RD Sensory Neurone
passes from Thalamus
→ Internal Capsule →
to reach Post Central
Gyrus of Parietal
Lobe - where awareness
of muscle and joint
sense is appreciated.*

Thalamus ---

*These fibres
in the CEREBELLUM
link up with others
concerned in maintenance
of posture and sense of
balance or equilibrium.*

V Cranial N.
Sensory Motor
Nucleus Nuc.

PONS

*Proprioceptor impulses
from Head and Neck muscles
travel in Cranial Nerves* ----
e.g. from muscles of mastication

Gracile Nucleus
(fibres from lower limbs
and lower trunk)

Medial Lemniscus

LOWER MEDULLA
OBLONGATA ------→

Cuneate Nucleus
(fibres from upper
limbs and trunk)

*2ND Sensory Neurone
crosses over to
opposite side of
Brain Stem in
Sensory Decussation*

1ST *Sensory
Neurones*

--- *Dorsal Spino-cerebellar Tract*

←---- SPINAL CORD

*Proprioceptor impulses
travel from Trunk and Limbs in Spinal Nerves.*

CUTANEOUS SENSATION

There are 5 basic skin sensations — TOUCH, PRESSURE, PAIN, WARMTH and COLD. There is much controversy as to how these are registered. In some areas they appear to be served by special nerve endings (sensory receptors or end-organs) in the SKIN. These end-organs are not uniformly distributed over the whole body surface. (*Eg. Touch "endings" are very numerous in hands and feet but are much less frequent in the skin of the back.*)

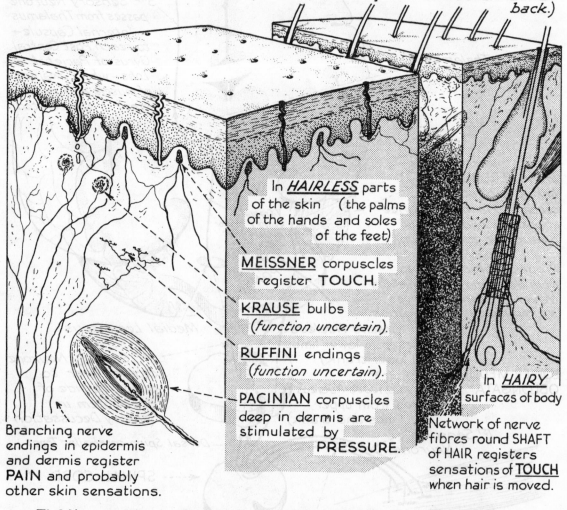

In *HAIRLESS* parts of the skin (the palms of the hands and soles of the feet)

MEISSNER corpuscles register TOUCH.

KRAUSE bulbs (*function uncertain*).

RUFFINI endings (*function uncertain*).

PACINIAN corpuscles deep in dermis are stimulated by PRESSURE.

Branching nerve endings in epidermis and dermis register PAIN and probably other skin sensations.

In *HAIRY* surfaces of body

Network of nerve fibres round SHAFT of HAIR registers sensations of TOUCH when hair is moved.

Tickling, itching, softness, hardness, wetness are probably due to stimulation of two or more of these special endings and to a blending of the sensations in the Brain.

246

SENSORY PATHWAYS from SKIN of FACE

The Receptors or nerve endings for ORDINARY SKIN SENSATIONS are linked by 3 Neurones with Receiving Centres in the PARIETAL LOBES.

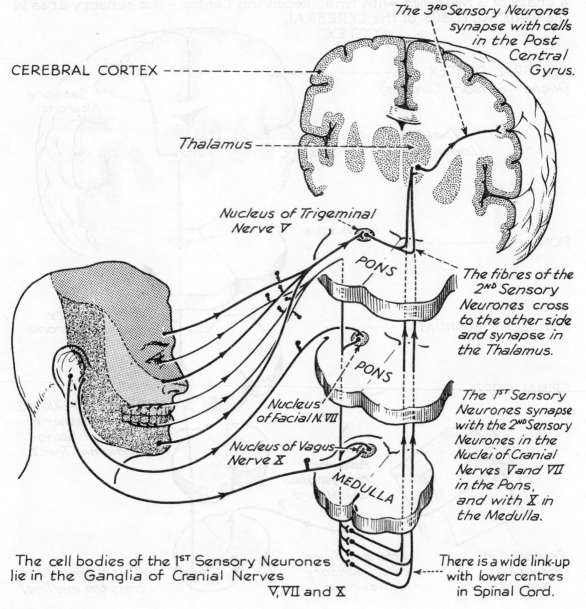

The 3RD Sensory Neurones synapse with cells in the Post Central Gyrus.

CEREBRAL CORTEX - - - - - - - - - - -

Thalamus - - - -

Nucleus of Trigeminal Nerve V

PONS

The fibres of the 2ND Sensory Neurones cross to the other side and synapse in the Thalamus.

PONS

Nucleus of Facial N. VII

The 1ST Sensory Neurones synapse with the 2ND Sensory Neurones in the Nuclei of Cranial Nerves V and VII in the Pons, and with X in the Medulla.

Nucleus of Vagus Nerve X

MEDULLA

The cell bodies of the 1ST Sensory Neurones lie in the Ganglia of Cranial Nerves V, VII and X

There is a wide link-up with lower centres in Spinal Cord.

PAIN and TEMPERATURE PATHWAYS from TRUNK and LIMBS

The nerve endings registering Pain and Warmth or Cold are linked by a chain of 3 Neurones with final Receiving Centre – the sensory area in the PARIETAL LOBES of the CEREBRAL CORTEX.

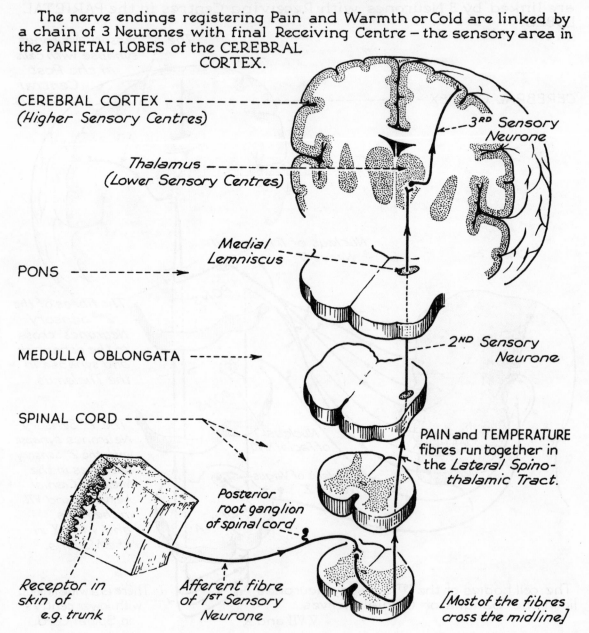

CEREBRAL CORTEX
(Higher Sensory Centres)

3RD Sensory Neurone

Thalamus
(Lower Sensory Centres)

Medial Lemniscus

PONS

2ND Sensory Neurone

MEDULLA OBLONGATA

SPINAL CORD

PAIN and TEMPERATURE fibres run together in the Lateral Spino-thalamic Tract.

Posterior root ganglion of spinal cord

Receptor in skin of e.g. trunk

Afferent fibre of 1ST Sensory Neurone

[Most of the fibres cross the midline.]

248

TOUCH and PRESSURE PATHWAYS from TRUNK and LIMBS

Touch and Pressure endings are linked by a chain of 3 Neurones with the PARIETAL LOBES.

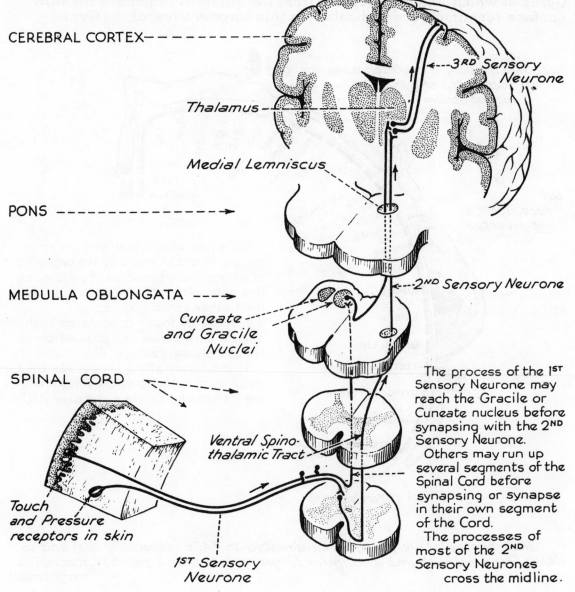

CEREBRAL CORTEX - - - - - - - - - - -

3RD Sensory Neurone

Thalamus -

Medial Lemniscus

PONS - - - - - - - - - - - - - - →

2ND Sensory Neurone

MEDULLA OBLONGATA - - - →

Cuneate and Gracile Nuclei

SPINAL CORD

Ventral Spino-thalamic Tract

Touch and Pressure receptors in skin

1ST Sensory Neurone

The process of the 1ST Sensory Neurone may reach the Gracile or Cuneate nucleus before synapsing with the 2ND Sensory Neurone.

Others may run up several segments of the Spinal Cord before synapsing or synapse in their own segment of the Cord.

The processes of most of the 2ND Sensory Neurones cross the midline.

SENSORY CORTEX

The 3ᴿᴰ Sensory Neurones (conveying information from the <u>opposite side</u> of the body) synapse with cells in the POST-CENTRAL GYRUS of the PARIETAL LOBE of the CEREBRAL CORTEX. The exact points on this Gyrus at which impulses coming from the different regions of the skin surface terminate are indicated on this coronal view of the Gyrus

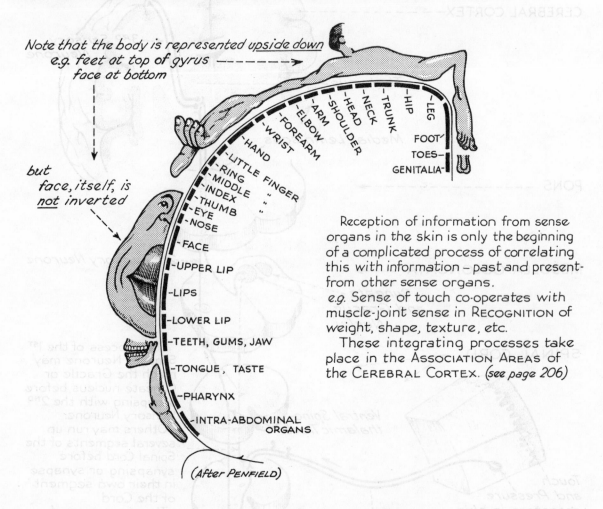

Note that the body is represented <u>upside down</u>
e.g. feet at top of gyrus ------->
face at bottom

but
face, itself, is
<u>not</u> inverted

LEG
HIP
TRUNK
NECK
HEAD
SHOULDER
ARM
ELBOW
FOREARM
WRIST
HAND
LITTLE FINGER
RING "
MIDDLE "
INDEX "
THUMB "
EYE
NOSE
FACE
UPPER LIP
LIPS
LOWER LIP
TEETH, GUMS, JAW
TONGUE, TASTE
PHARYNX
INTRA-ABDOMINAL ORGANS

FOOT
TOES
GENITALIA

Reception of information from sense organs in the skin is only the beginning of a complicated process of correlating this with information – past and present – from other sense organs.
e.g. Sense of touch co-operates with muscle-joint sense in RECOGNITION of weight, shape, texture, etc.

These integrating processes take place in the ASSOCIATION AREAS of the CEREBRAL CORTEX. (see page 206)

(After PENFIELD)

Note the relatively large area devoted to FACE (especially lips) and to HAND (especially thumb and index finger) while trunk representation is very small.

MOTOR CORTEX

The MOTOR NERVE CELLS which send out impulses to initiate VOLUNTARY MOVEMENT of SKELETAL MUSCLES lie in the PRE-CENTRAL GYRUS of each FRONTAL LOBE in the CEREBRAL CORTEX.

One cerebral hemisphere controls the muscles on the opposite side of the body.

The exact point in the GYRUS where neurones controlling any one part of the body are situated is indicated in this coronal view of the Gyrus.

Note that the body is represented upside down on motor cortex.
e.g. feet and legs at top,
    face at bottom.

Facial representation is probably
        bilateral
and is not, itself,
        inverted

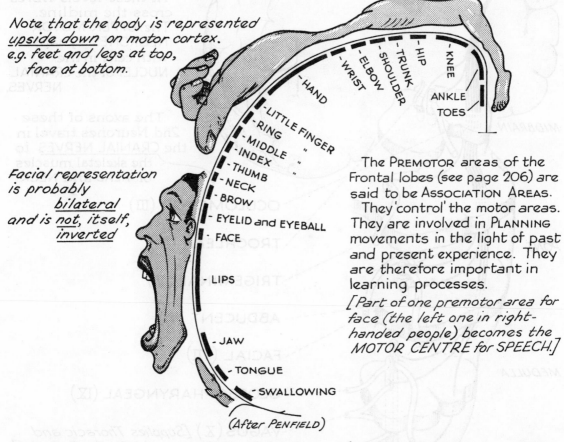

WRIST — ELBOW — SHOULDER — TRUNK — HIP — KNEE — ANKLE — TOES — HAND — LITTLE FINGER — RING " — MIDDLE " — INDEX " — THUMB — NECK — BROW — EYELID and EYEBALL — FACE — LIPS — JAW — TONGUE — SWALLOWING

(After PENFIELD)

The PREMOTOR areas of the Frontal lobes (see page 206) are said to be ASSOCIATION AREAS.

They 'control' the motor areas. They are involved in PLANNING movements in the light of past and present experience. They are therefore important in learning processes.

[Part of one premotor area for face (the left one in right-handed people) becomes the MOTOR CENTRE for SPEECH.]

Note the large area of the Motor Cortex (and therefore the very large number of neurones) devoted to control of voluntary movements of the HANDS. This enables them to perform complicated movements and to acquire highly intricate skills: similarly with muscles involved in ARTICULATION.

MOTOR PATHWAYS to HEAD and NECK

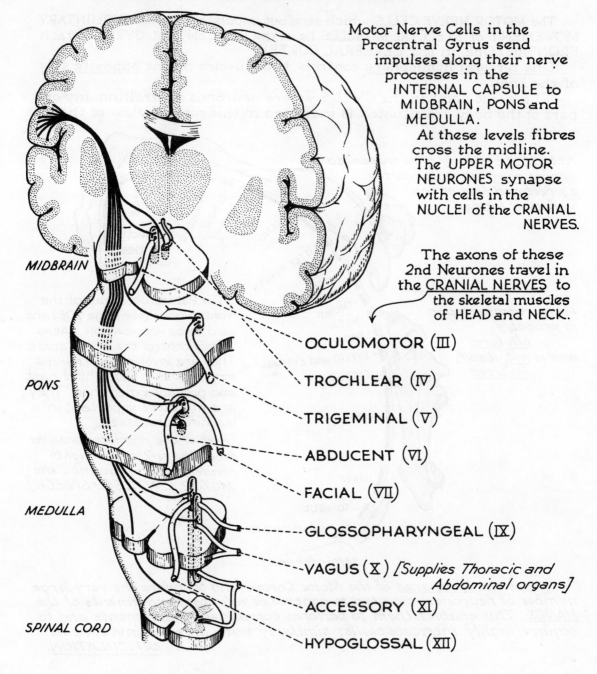

Motor Nerve Cells in the Precentral Gyrus send impulses along their nerve processes in the INTERNAL CAPSULE to MIDBRAIN, PONS and MEDULLA.

At these levels fibres cross the midline. The UPPER MOTOR NEURONES synapse with cells in the NUCLEI of the CRANIAL NERVES.

The axons of these 2nd Neurones travel in the CRANIAL NERVES to the skeletal muscles of HEAD and NECK.

MIDBRAIN

PONS

MEDULLA

SPINAL CORD

OCULOMOTOR (III)

TROCHLEAR (IV)

TRIGEMINAL (V)

ABDUCENT (VI)

FACIAL (VII)

GLOSSOPHARYNGEAL (IX)

VAGUS (X) [Supplies Thoracic and Abdominal organs]

ACCESSORY (XI)

HYPOGLOSSAL (XII)

MOTOR PATHWAYS to TRUNK and LIMBS

The CONTROLLING CENTRES in the MOTOR CORTEX are linked by 2 Neurones with the EFFECTOR ORGANS – the VOLUNTARY MUSCLES.

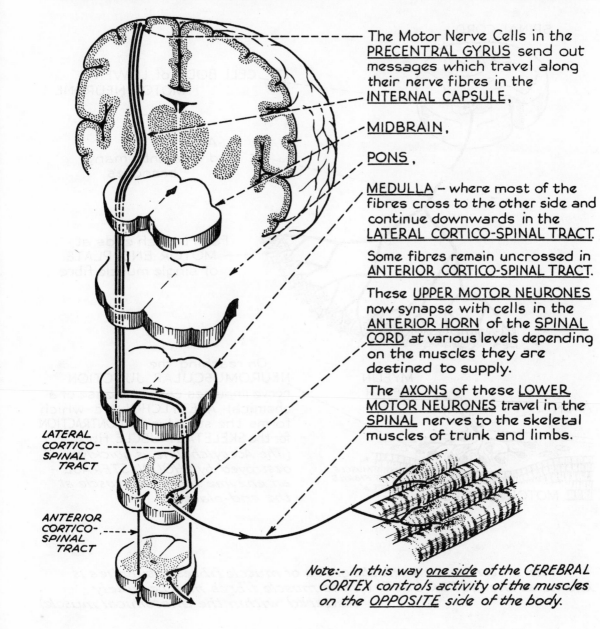

The Motor Nerve Cells in the <u>PRECENTRAL GYRUS</u> send out messages which travel along their nerve fibres in the <u>INTERNAL CAPSULE</u>,

<u>MIDBRAIN</u>,

<u>PONS</u>,

<u>MEDULLA</u> – where most of the fibres cross to the other side and continue downwards in the <u>LATERAL CORTICO-SPINAL TRACT</u>.

Some fibres remain uncrossed in <u>ANTERIOR CORTICO-SPINAL TRACT</u>.

These <u>UPPER MOTOR NEURONES</u> now synapse with cells in the <u>ANTERIOR HORN</u> of the <u>SPINAL CORD</u> at various levels depending on the muscles they are destined to supply.

The <u>AXONS</u> of these <u>LOWER MOTOR NEURONES</u> travel in the <u>SPINAL</u> nerves to the skeletal muscles of trunk and limbs.

LATERAL CORTICO-SPINAL TRACT

ANTERIOR CORTICO-SPINAL TRACT

Note:- In this way <u>one side</u> of the CEREBRAL CORTEX controls activity of the muscles on the <u>OPPOSITE</u> side of the body.

MOTOR UNIT

The AXON of the LOWER MOTOR NEURONE divides into many branches.
Each branch ends at the MOTOR END-PLATE of a single muscle fibre.

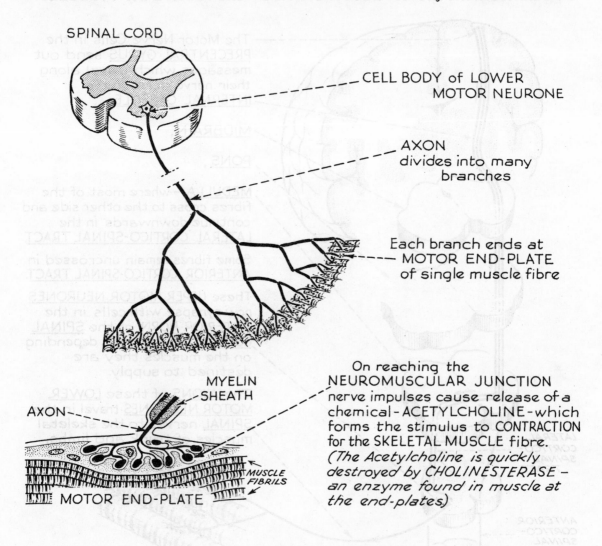

SPINAL CORD

CELL BODY of LOWER
MOTOR NEURONE

AXON
divides into many
branches

Each branch ends at
MOTOR END-PLATE
of single muscle fibre

MYELIN
SHEATH

AXON

MUSCLE
FIBRILS

MOTOR END-PLATE

On reaching the
NEUROMUSCULAR JUNCTION
nerve impulses cause release of a
chemical - ACETYLCHOLINE - which
forms the stimulus to CONTRACTION
for the SKELETAL MUSCLE fibre.
(The Acetylcholine is quickly
destroyed by CHOLINESTERASE –
an enzyme found in muscle at
the end-plates.)

*The motor nerve with the group of muscle fibres it supplies is
known as the MOTOR UNIT. (These muscle fibres may be widely
scattered within the anatomical muscle)*

FINAL COMMON PATHWAY

Each MOTOR NEURONE in the ANTERIOR HORNS of the SPINAL CORD serves as the PATHWAY for motor impulses initiated in — HIGHER MOTOR CENTRES in CEREBRUM *(of opposite side)* and travelling in —

① *CORTICO-SPINAL TRACT*

The Motor Neurone also serves as the PATHWAY for co-ordinating corrective (restraining or facilitating) impulses discharged by— LOWER MOTOR CENTRES in EXTRAPYRAMIDAL SYSTEM *(of same or opposite side)* and travelling in—

② *RUBRO-SPINAL TRACT*
   from Red Nucleus *(opposite side)*
③ *DORSAL VESTIBULO-SPINAL TRACT*
   from Dorsal Vestibular Nucleus *(same side)*
④ *OLIVO-SPINAL TRACT*
   from Olivary Nucleus *(same side)*
⑤ *RETICULO-SPINAL TRACT*
   from Reticular Nuclei *(same side)*
⑥ *VENTRAL VESTIBULO-SPINAL TRACT*
   from Vestibular Nuclei *(opposite side)*
⑦ *TECTO-SPINAL TRACT*
   from Tectum *(opposite side)*

The Motor Neurone also receives relays of afferent impulses from other reflex centres in –
   SPINAL CORD

⑧ *For REFLEXES of <u>same</u> segment of cord and from <u>same</u> side of cord.*
⑨ *For REFLEXES of <u>same</u> segment but <u>other</u> side of cord and body.*
⑩ *For REFLEXES from <u>other</u> segments of cord but <u>same</u> side of cord.*
⑪ *For REFLEXES from <u>other</u> segments and <u>other</u> side of cord.*

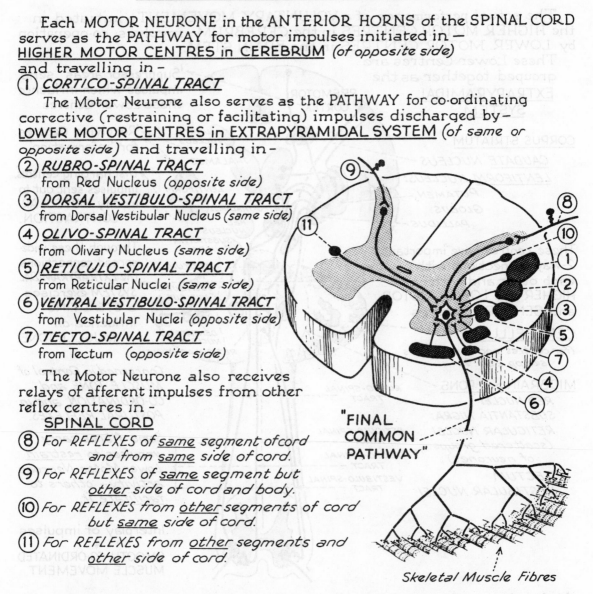

"FINAL COMMON PATHWAY"

Skeletal Muscle Fibres

If several sources compete for the "FINAL COMMON PATHWAY" at any one time, allied ones may reinforce each other: if incompatible, those which are initiated by PAINFUL stimuli (i.e. PROTECTIVE REFLEXES) take precedence.

EXTRAPYRAMIDAL SYSTEM

The actual performance of a VOLUNTARY MOVEMENT – initiated in the HIGHER MOTOR CENTRES in the CEREBRUM – involves co-operation by LOWER MOTOR CENTRES in the BRAIN STEM.

These Lower Centres are grouped together as the EXTRAPYRAMIDAL SYSTEM

CORPUS STRIATUM

   CAUDATE NUCLEUS

   LENTIFORM NUCLEUS:

      PUTAMEN,
      GLOBUS
      PALLIDUS

These centres have important 2-way connections with each other and with HIGHER and LOWER MOTOR and SENSORY Centres

CEREBELLUM
(for centres and connections see pp. 258-9)

MIDBRAIN and PONS

   RED NUCLEI
   SUBSTANTIA NIGRA
   RETICULAR NUCLEI
   (scattered groups of neurones)
   TECTUM
   VESTIBULAR NUCLEI

PREMOTOR CORTEX

CAUDATE NUCLEUS

THALAMUS

LENTIFORM NUCLEUS

SUB- & HYPO-THALAMUS

RED NUCLEUS

SUBSTANTIA NIGRA

TECTUM

VESTIBULAR NUCLEUS

DENTATE NUCLEUS

RETICULAR NUCLEI

RUBRO-SPINAL TRACT

RETICULO-SPINAL TRACT

TECTO-SPINAL TRACT

VESTIBULO-SPINAL TRACT

'Suppressor areas' send impulses which influence these Lower Centres

'Receiving Centres' for 'information'

'Discharging Centres'. Impulses sent out to modify MUSCULAR CONTRACTION.

Through these wide-spread connections the EXTRAPYRAMIDAL SYSTEM exercises a STEADYING influence on MUSCLE TONE.

Concerned in Control of MUSCLE TONE and CO-ORDINATION, and in POSTURAL REFLEXES.

Some fibres carry impulses to restrain Lower Motor Neurone activity – others to facilitate it.

Interplay of impulses
↓
SMOOTH CO-ORDINATED MUSCLE MOVEMENT

Little is known about the methods for regulating this interplay of impulses converging on the MOTOR UNIT during a voluntary action.

Where part of the Extrapyramidal System is destroyed by disease varying types of RIGIDITY, TREMOR and UNCO-ORDINATED MUSCLE MOVEMENT result.

CEREBELLUM

The CEREBELLUM has 2 HEMISPHERES.
Each Hemisphere has 3 LOBES. These differ in development and function.

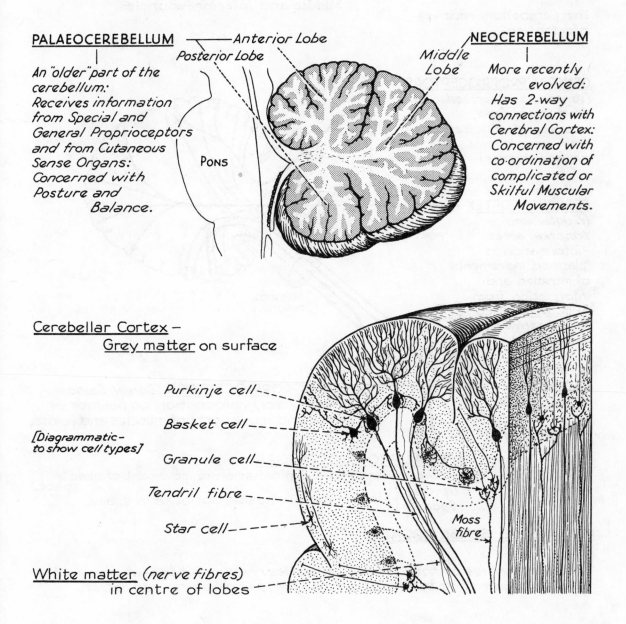

PALAEOCEREBELLUM

—Anterior Lobe
Posterior Lobe

An "older" part of the cerebellum:
Receives information from Special and General Proprioceptors and from Cutaneous Sense Organs:
Concerned with Posture and
　　Balance.

PONS

NEOCEREBELLUM

Middle Lobe

More recently evolved:
Has 2-way connections with Cerebral Cortex:
Concerned with co-ordination of complicated or Skilful Muscular Movements.

Cerebellar Cortex —
　Grey matter on surface

[Diagrammatic - to show cell types]

Purkinje cell----

Basket cell----

Granule cell----

Tendril fibre----

Star cell----

Moss fibre

White matter (nerve fibres)
　　in centre of lobes ----

257

CEREBELLUM

INGOING PATHWAYS

Each Cerebellar hemisphere is linked with the rest of the Nervous System through 3 bundles of nerve fibres — the Superior, Middle and Inferior Peduncles.

The Cerebellum receives information....

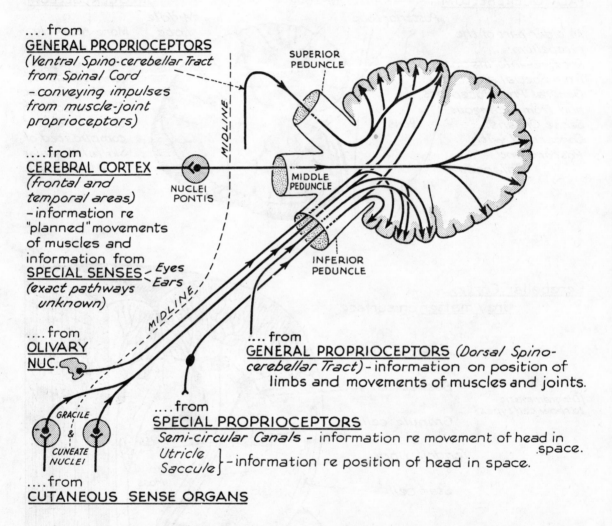

....from
GENERAL PROPRIOCEPTORS
(Ventral Spino-cerebellar Tract from Spinal Cord — conveying impulses from muscle-joint proprioceptors)

....from
CEREBRAL CORTEX
(frontal and temporal areas) — information re "planned" movements of muscles and information from **SPECIAL SENSES** *Eyes Ears (exact pathways unknown)*

SUPERIOR PEDUNCLE

MIDLINE

NUCLEI PONTIS

MIDDLE PEDUNCLE

INFERIOR PEDUNCLE

MIDLINE

....from
OLIVARY NUC.

GRACILE & CUNEATE NUCLEI

....from
SPECIAL PROPRIOCEPTORS
Semi-circular Canals — information re movement of head in space.
Utricle Saccule } — information re position of head in space.

....from
GENERAL PROPRIOCEPTORS *(Dorsal Spino-cerebellar Tract)* — information on position of limbs and movements of muscles and joints.

....from
CUTANEOUS SENSE ORGANS

The activities of the Cerebellum are carried out beneath the level of consciousness. The information reaching it gives no sensation.

CEREBELLUM

OUTGOING PATHWAYS

In addition to the grey matter on the surface of the cerebellar cortex, there are masses of grey matter — called CEREBELLAR NUCLEI — deep within the cerebellum.

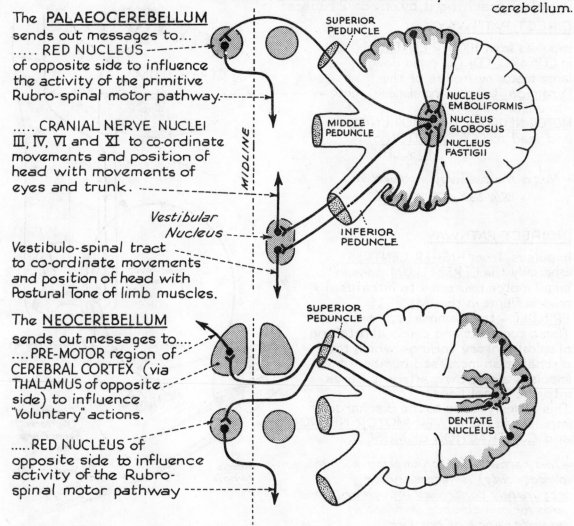

The **PALAEOCEREBELLUM** sends out messages to...
..... RED NUCLEUS of opposite side to influence the activity of the primitive Rubro-spinal motor pathway.

..... CRANIAL NERVE NUCLEI III, IV, VI and XI to co-ordinate movements and position of head with movements of eyes and trunk.

Vestibular Nucleus

Vestibulo-spinal tract to co-ordinate movements and position of head with Postural Tone of limb muscles.

The **NEOCEREBELLUM** sends out messages to....
....PRE-MOTOR region of CEREBRAL CORTEX (via THALAMUS of opposite side) to influence "Voluntary" actions.

.....RED NUCLEUS of opposite side to influence activity of the Rubro-spinal motor pathway

SUPERIOR PEDUNCLE

MIDDLE PEDUNCLE

NUCLEUS EMBOLIFORMIS

NUCLEUS GLOBOSUS

NUCLEUS FASTIGII

INFERIOR PEDUNCLE

MIDLINE

SUPERIOR PEDUNCLE

DENTATE NUCLEUS

The cerebral cortex initiates purposeful movements. During such movements proprioceptors are continually supplying information to the Cerebellum about the changing positions of muscles and joints. The Cerebellum is then responsible for collating this information and sending out impulses to bring about the smooth co-ordinated movements of different groups of muscles.

CONTROL of MUSCLE MOVEMENT

The muscle spindle is the key structure in the complex self-regulating mechanism for the control of movement of skeletal muscle.

In VOLUNTARY MOVEMENT a muscle can be made to contract by impulses reaching it by one of 2 routes:-

DIRECT PATHWAY

Impulses from HIGHER CENTRES in CEREBRAL CORTEX pass down large motor neurones of the Pyramidal tract — the alpha (α) route — to excite the LOWER MOTOR NEURONES → MOTOR UNITS — and lead to CONTRACTION of MUSCLE.

–This is the pathway involved in _voluntary movements_

INDIRECT PATHWAY

Impulses from HIGHER CENTRES especially the CEREBELLUM pass in small motor neurones to intrafusal muscle fibres in the MUSCLE SPINDLE — the gamma (γ) route — These contract and cause stretching of spiral sensory endings which then discharge an increased number of impulses along their afferent fibres into the Spinal Cord.
This reflexly leads to the discharge of impulses in the LOWER MOTOR NEURONE and CONTRACTION of MUSCLE.

–This pathway (in conjunction with the direct route) is necessary for _accurately controlled movements_ and also for movements involved in _maintenance of posture._

CEREBELLUM

Group Ia afferent fibre

α or lower motor neurone

γ motor neurone

Motor Unit

α efferent

Fusimotor (γ efferent) nerve fibre to intrafusal muscle fibre

Muscle spindle

It is probable that the CEREBRAL CORTEX decides _WHAT_ is to be done and the CEREBELLUM 'decides' _HOW_ it is to be done.

LOCOMOTOR SYSTEM

BONES and MUSCLES —— concerned with MOVEMENT of the body.

SKELETON — RIGID FRAMEWORK gives SHAPE and SUPPORT to body.

is JOINTED to permit MOVEMENT

FLAT BONES protect delicate organs.

LONG BONES act as levers.

SHORT BONES confer strength.

CLAVICLE

SCAPULA

HUMERUS

RIBS

VERTEBRAL COLUMN

ILEUM

ULNA

RADIUS

CARPALS

META-CARPALS

PHALANGES

FEMUR

PATELLA

FIBULA

TIBIA

TARSUS

METATARSALS

PHALANGES

The VERTEBRAL COLUMN is BASIS of SKELETON

33 VERTEBRAE

A series of bony rings united by intervertebral discs of cartilage

Hollow canal encloses and protects SPINAL CORD

7 CERVICAL	Atlas	Supports skull: permits nodding.
	Axis	Allows rotation of head.
		Allow bending and twisting movements of neck.
12 THORACIC		Support ribs. Allow rotation, forward bending and some sideways movement of trunk.
5 LUMBAR		Allow backward bending, sideways movement and some rotation of trunk.
5 SACRAL	Fused together	Transmit weight of body to pelvic girdle and legs.
4 COCCYGEAL	Fused together	

All bones give attachment to muscles.

SKELETAL MUSCLES

Bones are moved at joints by the CONTRACTION and RELAXATION of MUSCLES attached to them.

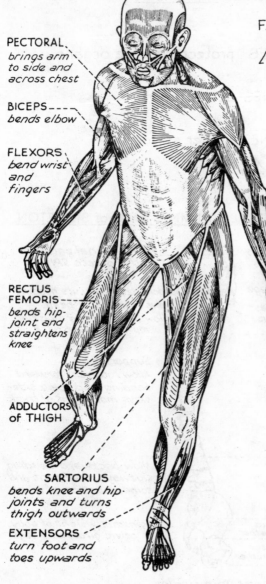

PECTORAL,
brings arm
to side and
across chest

BICEPS
bends elbow

FLEXORS,
bend wrist
and
fingers

RECTUS
FEMORIS
bends hip-
joint and
straightens
knee

ADDUCTORS
of THIGH

SARTORIUS
bends knee and hip-
joints and turns
thigh outwards

EXTENSORS
turn foot and
toes upwards

FACIAL muscles are involved in varying facial *EXPRESSION; SPEECH; MASTICATION* [Some muscles link bone to skin.]

The deep muscles of the THORAX, linking the ribs, contract and relax in *RESPIRATION.*

The muscles of the ABDOMEN are arranged in sheets and *PROTECT* delicate abdominal organs. They also contract to compress abdominal contents and aid in *MICTURITION, DEFAECATION, VOMITING* (and in the processes of *CHILDBIRTH* in the female).

In the LEGS are found the most powerful muscles of the body – especially those acting on the hip-joint.

Muscles which bend a limb at a joint are called **FLEXORS**.

Muscles which straighten a limb at a joint are called **EXTENSORS**.

Muscles which move a limb (or other part) away from the midline are called **ABDUCTORS**.

Muscles which move a limb (or other part) towards the midline are called **ADDUCTORS**.

262

SKELETAL MUSCLES

EXTENSORS - - - - - - - - - - -
straighten wrist and fingers

TRICEPS - - - - - - - - - - - - -
straightens elbow

DELTOID - - - - - - - - - - - -
raises arm

TRAPEZIUS - - - - - - - - -
*raises shoulder and
pulls head back*

LATISSIMUS DORSI - - - - -
*draws arm backwards
and turns it inwards
(It also draws downwards
an upstretched arm)*

The muscles of the
BACK play a large part in
maintaining erect posture

GLUTEALS
*straighten hip-joint
and move leg
outwards*

HAMSTRINGS
*bend knee and
straighten hip joint*

GASTROCNEMIUS
*bends knee and
turns foot downwards*

ACHILLES TENDON

FLEXORS
*turn foot and
toes downwards*

Some muscles
work together to
ROTATE a limb
or other part of
the body.

MUSCULAR MOVEMENTS

The long bones particularly form a light framework of LEVERS.
The skeletal muscles attached to them contract to operate these levers.

When a muscle contracts it shortens.

This brings its two ends closer together.

Since the two ends are attached to different bones by TENDONS, one or other of the bones must move.

Two bones meet or articulate at a JOINT.

Joint surfaces are covered with a layer of smooth CARTILAGE.

To avoid friction when the two surfaces move on one another a SYNOVIAL MEMBRANE secretes a LUBRICATING FLUID.

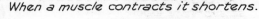

The muscle which contracts to move the joint is called the PRIME MOVER or AGONIST *(the Biceps in flexion of elbow)*

To allow the movement to take place, however, other muscles near the joint must co-operate :-
The oppositely acting muscles gradually relax — these are called the ANTAGONISTS and exercise a "braking" control on the movement. *(e.g. the EXTENSORS – chiefly Triceps - in flexion of the elbow.)*
Other muscles steady the bone giving "origin" to the Prime Mover so that only the "insertion" will move— these muscles are called FIXATORS.
Still other muscles help to steady, for most efficient movement, the joint being moved — called SYNERGISTS.

When the elbow is straightened the reverse occurs :-
TRICEPS, the Prime Mover, CONTRACTS;   BICEPS, the Antagonist, RELAXES.

264

RECIPROCAL INNERVATION

The co-ordinated group action of muscles is made possible by the many synaptic connections between CONNECTOR NEURONES of the INGOING or PROPRIOCEPTIVE NEURONES of one muscle group and the OUTGOING or MOTOR NEURONES of the functionally opposite group of muscles.

This is shown diagrammatically for the alternating contractions and relaxations of the Flexors and Extensors of one knee in Walking:-

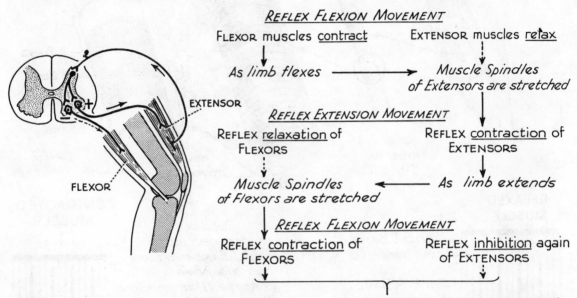

REFLEX FLEXION MOVEMENT

FLEXOR muscles <u>contract</u>   EXTENSOR muscles <u>relax</u>

As limb flexes ⟶ *Muscle Spindles of Extensors are stretched*

REFLEX EXTENSION MOVEMENT

REFLEX <u>relaxation</u> of FLEXORS   REFLEX <u>contraction</u> of EXTENSORS

Muscle Spindles of Flexors are stretched ⟵ *As limb extends*

REFLEX FLEXION MOVEMENT

REFLEX <u>contraction</u> of FLEXORS   REFLEX <u>inhibition</u> again of EXTENSORS

EXTENSOR

FLEXOR

The rhythmic alternating reflex movements of walking

i.e. The FINAL COMMON PATHWAY *is commandeered alternately by "exciting" or "inhibiting" messages.*

RECIPROCAL INNERVATION is probably due to division (within the SPINAL CORD) of the CONNECTOR NEURONES for each AFFERENT NERVE — *e.g.* one branch excites (+) extensor motor neurone; the other branch inhibits (–) the flexor motor neurone.

Thus:-
When limb is flexed-

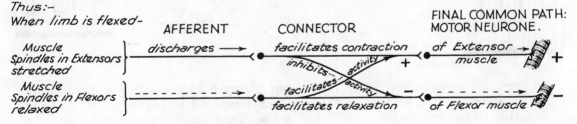

AFFERENT   CONNECTOR   FINAL COMMON PATH: MOTOR NEURONE.

Muscle Spindles in Extensors stretched   *discharges* ⟶   *facilitates contraction*   *of Extensor muscle*   +

inhibits activity   +

facilitates activity

Muscle Spindles in Flexors relaxed   *facilitates relaxation*   –   *of Flexor muscle*   –

RECIPROCAL INNERVATION helps make possible self-regulation of the rhythmical movements.

SKELETAL MUSCLE
and the MECHANISM of CONTRACTION

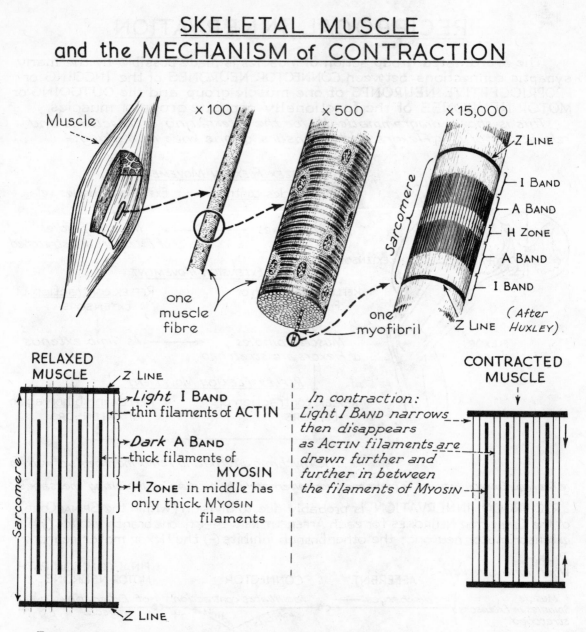

Muscle

×100   ×500   ×15,000

Z LINE
I BAND
A BAND
H ZONE
A BAND
I BAND
Z LINE

Sarcomere

one muscle fibre

one myofibril

(*After* HUXLEY)

RELAXED MUSCLE

Z LINE

Light I BAND
— thin filaments of ACTIN

Dark A BAND
— thick filaments of MYOSIN

H ZONE in middle has only thick MYOSIN filaments

Sarcomere

Z LINE

CONTRACTED MUSCLE

In contraction:
Light I BAND *narrows then disappears as* ACTIN *filaments are drawn further and further in between the filaments of* MYOSIN

Energy for contraction is derived from glucose and fat in the mitochondrion (*page 6*). The energy is transported from the mitochondrion to the contractile filaments in ATP. ATP splits readily into ADP and phosphate, releasing its trapped energy where needed (*page 42*).

AUTONOMIC NERVOUS SYSTEM

and

CHEMICAL TRANSMISSION
at NERVE ENDINGS

AUTONOMIC NERVOUS SYSTEM

The Autonomic Nervous System is concerned with maintaining a <u>Stable Internal Environment</u>. It governs functions which are normally carried out below the level of consciousness.

It has 2 separate parts – <u>PARASYMPATHETIC</u> and <u>SYMPATHETIC</u>. Both have their Highest Centres in the Brain.

Each <u>LOWER MOTOR PATHWAY</u> has <u>2 NEURONES</u> to <u>ORGANS</u> supplied

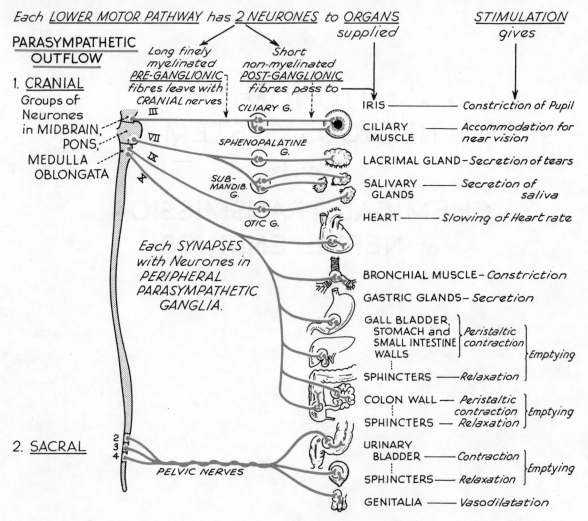

PARASYMPATHETIC OUTFLOW

1. <u>CRANIAL</u>
Groups of Neurones in MIDBRAIN, PONS, MEDULLA OBLONGATA

Each SYNAPSES with Neurones in PERIPHERAL PARASYMPATHETIC GANGLIA.

2. <u>SACRAL</u>

PELVIC NERVES

Long finely myelinated <u>PRE-GANGLIONIC</u> *fibres leave with CRANIAL nerves*

Short non-myelinated <u>POST-GANGLIONIC</u> *fibres pass to*

CILIARY G.
SPHENOPALATINE G.
SUB-MANDIB. G.
OTIC G.

III
VII
IX
X

2
3
4

STIMULATION gives

Organ	Effect
IRIS	*Constriction of Pupil*
CILIARY MUSCLE	*Accommodation for near vision*
LACRIMAL GLAND	*Secretion of tears*
SALIVARY GLANDS	*Secretion of saliva*
HEART	*Slowing of Heart rate*
BRONCHIAL MUSCLE	*Constriction*
GASTRIC GLANDS	*Secretion*
GALL BLADDER, STOMACH and SMALL INTESTINE WALLS	*Peristaltic contraction* } *Emptying*
SPHINCTERS	*Relaxation*
COLON WALL	*Peristaltic contraction* } *Emptying*
SPHINCTERS	*Relaxation*
URINARY BLADDER	*Contraction* } *Emptying*
SPHINCTERS	*Relaxation*
GENITALIA	*Vasodilatation*

General stimulation of the Parasympathetic promotes vegetative functions of the body. Stimulation of the Parasympathetic and inhibition of Sympathetic have the same overall effects.

AUTONOMIC NERVOUS SYSTEM

The Parasympathetic and Sympathetic systems normally act in balanced reciprocal fashion. The activity of an organ at any one time is the result of the two opposing influences.

Each *LOWER MOTOR PATHWAY* has *2 NEURONES* to *ORGANS* supplied *STIMULATION* gives

SYMPATHETIC OUTFLOW

Short finely myelinated *PRE-GANGLIONIC* fibres leave with *MOTOR ROOTS* of *SPINAL NERVES*. These *SYNAPSE* with Neurones in the *PARAVERTEBRAL GANGLIA* or the *PREVERTEBRAL GANGLIA* or directly with Cells in *SUPRARENAL MEDULLAE*.

Long *POST-GANGLIONIC* fibres pass to

SWEAT GLANDS ———— *Secretion*

SMOOTH MUSCLE in Skin — *Contraction*

BLOOD VESSELS in Skin —— *Constriction*

IRIS ———————— *Dilatation of Pupil*

BLOOD VESSELS of Head — *Vasoconstriction*

BLOOD VESSELS of Skeletal Muscle in upper limbs — *Relaxation*

HEART ————— *Heart Rate quickens*

BRONCHIAL MUSCLE ——— *Relaxation*

STOMACH WALL ——— *Relaxation*

SPHINCTERS ———— *Constriction*

BLOOD VESSELS of Abdomen ——— *Vasoconstriction*

LIVER — *Mobilization of Liver GLYCOGEN*

SUPRARENAL MEDULLAE — *Secretion of ADRENALINE*

SMALL INTESTINE and COLON WALLS ——— *Relaxation*

SPHINCTERS ———— *Constriction*

URINARY BLADDER —— *Relaxation* WALL

SPHINCTER ———— *Constriction*

GENITALIA ———— *Vasoconstriction*

PARAVERTEBRAL GANGLIA

COELIAC G.

SUP. MESENTERIC G.

T1

From Neurones in LATERAL HORNS of THORACIC and LUMBAR parts of SPINAL CORD

L2

INF. MESENTERIC GANGLION

PREVERTEBRAL GANGLIA

General stimulation of Sympathetic results in mobilization of resources to prepare body to meet emergencies. Stimulation of the Sympathetic and inhibition of the Parasympathetic have the same overall effects.

AUTONOMIC REFLEX

AUTONOMIC CENTRES in the BRAIN and SPINAL CORD receive SENSORY INFLOWS from the VISCERA. *(less is known about their exact pathways than about MOTOR OUTFLOWS.)* Some of the SENSORY NEURONES convey information about events in the viscera to HIGHER AUTONOMIC CENTRES which send impulses to modify the activity of ⎯⎯⎯⎯⎯⎯⎯⎯⎯

Both VISCERAL and SOMATIC AFFERENTS serve as <u>AFFERENT PATHWAYS</u> for <u>AUTONOMIC REFLEXES</u> by means of which much of the nervous regulation of vegetative functions is carried out below the level of consciousness.

e.g. The Simplest Autonomic Reflex arc:-

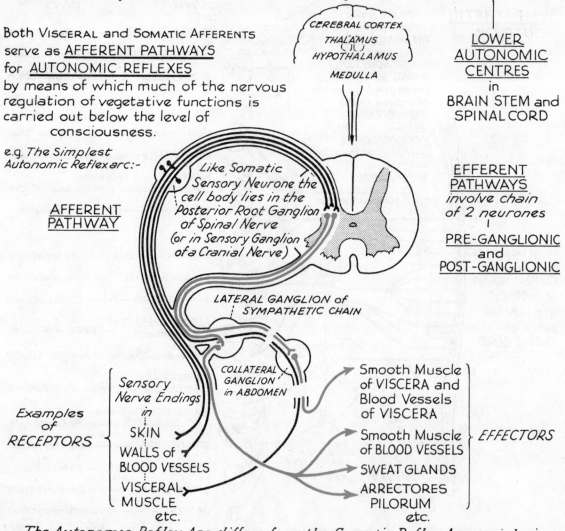

CEREBRAL CORTEX
THALAMUS
HYPOTHALAMUS
MEDULLA

LOWER AUTONOMIC CENTRES in BRAIN STEM and SPINAL CORD

AFFERENT PATHWAY

Like Somatic Sensory Neurone the cell body lies in the Posterior Root Ganglion of Spinal Nerve (or in Sensory Ganglion of a Cranial Nerve)

EFFERENT PATHWAYS *involve chain of 2 neurones*
|
PRE-GANGLIONIC and POST-GANGLIONIC

LATERAL GANGLION of SYMPATHETIC CHAIN

COLLATERAL GANGLION in ABDOMEN

Examples of RECEPTORS
Sensory Nerve Endings in SKIN
WALLS of BLOOD VESSELS
VISCERAL MUSCLE
etc.

Smooth Muscle of VISCERA and Blood Vessels of VISCERA
Smooth Muscle of BLOOD VESSELS
SWEAT GLANDS
ARRECTORES PILORUM
etc.

EFFECTORS

The Autonomic Reflex Arc differs from the Somatic Reflex Arc mainly in that it has 2 Efferent neurones. Transmission of impulse from Afferent to Efferent probably involves one or more connecting neurones.

CHEMICAL TRANSMISSION at NERVE ENDINGS

When an impulse passes along a nerve, sodium enters the fibre and potassium leaves it. When this process reaches the synapse between neurone and neurone or between neurone and effector organ a chemical substance is liberated. This bridges the gap and forms the stimulus to alter the membrane potential and initiate in the next neurone (or in the effector) the wave of depolarization and altered membrane permeability with ion exchange which becomes the transmitted nerve impulse.

Some nerves when stimulated liberate an acetylcholine-like substance. These are called CHOLINERGIC. Others liberate nor-adrenaline. These are called ADRENERGIC.

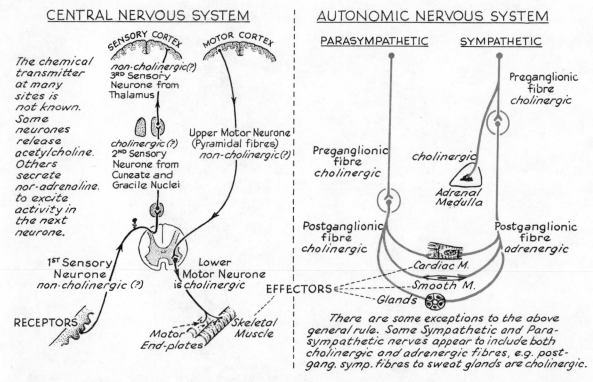

CENTRAL NERVOUS SYSTEM

SENSORY CORTEX   MOTOR CORTEX

The chemical transmitter at many sites is not known. Some neurones release acetylcholine. Others secrete nor-adrenaline. to excite activity in the next neurone.

non-cholinergic(?)
3RD Sensory Neurone from Thalamus

cholinergic (?)
2ND Sensory Neurone from Cuneate and Gracile Nuclei

Upper Motor Neurone (Pyramidal fibres) non-cholinergic(?)

1ST Sensory Neurone non-cholinergic (?)

Lower Motor Neurone is cholinergic

RECEPTORS

Motor End-plates   Skeletal Muscle

AUTONOMIC NERVOUS SYSTEM

PARASYMPATHETIC   SYMPATHETIC

Preganglionic fibre cholinergic

Preganglionic fibre cholinergic

cholinergic

Adrenal Medulla

Postganglionic fibre cholinergic

Postganglionic fibre adrenergic

Cardiac M.
Smooth M.
Glands

EFFECTORS

There are some exceptions to the above general rule. Some Sympathetic and Parasympathetic nerves appear to include both cholinergic and adrenergic fibres, e.g. post-gang. symp. fibres to sweat glands are cholinergic.

There are thought to be other chemical substances secreted by neurones which may act as Synaptic Transmitters to regulate certain aspects of cerebral activity.

Many neurones within the C.N.S. release inhibitory, _not_ excitatory, transmitters. One such substance found in the cerebral cortex is GABA (gamma aminobutyric acid). Another is the amino acid, glycine.

Inhibitory neurones block unimportant signals and thus permit onward transmission of selected information e.g. to centres which otherwise would be subjected to continuous bombardment by impulses from receptors all over the body.

INDEX

Abdomen
  muscles of, 262
  reflex vasoconstriction, 96
  reflex vasodilatation, 97
Abducens nerve (VI), **210**, 226, 227, 252
Absorption
  calcium and phosphorus
    —role of parathormone, **154,** 155, 156
  large intestine, 72
  small intestine, 39, 40, 41, **70, 71**
  stomach, 58
  striated border epithelium, **9**
  vitamin B$_{12}$, 106
Accessory nerve (XI), **210**, 252
Accommodation for near vision, 224, 228, 229
Acetylcholine
  chemical transmission, 271
    autonomic nervous system, 271
    gut wall, 74
    neuromuscular junction, 254
Acetylcoenzyme A, 41, **42**
Achilles Tendon, 215
Acoustic (auditory) nerve (VIII), **210, 237,** 240
Acromegaly, 165
Actin, 266
Adaptation
  in eye, 232, 233
  in receptors, 220
Addison's disease, 158
Adenosine diphosphate (ADP), 42, 266
Adenosine triphosphate (ATP), **42,** 266
Adipose tissue, **12**
  deposition in growth, 47
  in breast, 197
Adolescence
  energy requirements in, 48, 49
  growth, **47**
  puberty in female, 184
  puberty in male, 179
  reproductive organs, 186, 197
Adrenal cortex, **157**
  diseases of, in man, 158, 159, 166
  effect of anterior pituitary on, 163
  in carbohydrate metabolism, 40
  in water balance, 144, 145
  vitamin C in, **37**
Adrenaline, 160, **161**
  body temperature control, 45
  role in intermediate carbohydrate
    metabolism, **40**
  vasoconstrictor effects, 96
  vasodilatation effects, 97
Adrenal medulla, 157, **160**
  actions of adrenaline, **161**
  effect on metabolism, 40
  innervation, 271
  role in body temperature control, 45
Adrenergic nerves
  chemical transmission, **271**
  in gut, 74
Adrenocorticotrophic hormone (ACTH),
  157, **163,** 166
Afterbirth, **195**

Agglutination, **108, 109,** 110
Air
  capacity of lungs, 123
  composition of respired air, 124
  conducting passages, **118**
Albumin
  digestion of, 38
  human blood, 104
Aldosterone, 138, **144, 145,** 157, 159, **172**
Alimentary tract, **52**
  basic structural pattern, 67
  heat loss, food and faeces, 43, 44, 45
  innervation, 74
  nervous control, 75
  progress of food along, 53
Alkaline phosphate
  conservation by kidney, 143
Allergy, 11
Alveolar air, 119, 124, **125**
  partial pressure of oxygen, 126, 129
Alveoli, **119,** 121, 124, 125
Amino acids, **28**
  absorption and fate, 38, 39, 70, 71
  dietary sources of, 35
  reabsorption from glomerular filtrate,
    137
Ammonia
  deamination in liver, 39
  excretion in faeces, 39
  production by kidneys, 39, **142,** 148
Amniotic cavity, **19,** 192, 194
Amoeba, **3,** 4
Ampulla
  of semicircular canals, 241, 242
  of Vater, 62
Amylase, 38, 40, **62, 66**
Anaemia, 107
Anal canal, 52, 53, **72**
  defaecation, 73
  innervation, 75
Androgens
  of suprarenal cortex, **157,** 159, 166
Angiotensin(ogen), 144, 172
Ankle jerk, 215
Antibodies, 11, **112,** 113, 174
  blood group antibodies, 108, 109, 110
Antidiuretic hormone (ADH), **138, 144,**
  145, 168, **170, 171, 172**
Antigens, 112
  blood group antigens, 108, 109, 110
Antihaemophilic globulin, 105
Aorta, **81,** 82, 83, **94**
  baroreceptors, **89, 96,** 97, 132
  blood flow in, 100
  blood pressure in, 92
  chemoreceptors, **96, 97,** 131, 132
  diameter with blood flow, 100
  valve, 84
Aortic bodies
  chemoreceptors
    in regulation of arteriolar tone, 96,
      97
    in regulation of respiration, **131, 132**
Aortic sinus
  baroreceptors

  in cardiac reflexes, 89
  in regulation of arteriolar tone, 96, 97
  in regulation of respiration, 132
Apocrine glands
  mammary, 198
Appendix, 53, 71, 72
Apperception, 220
Aqueous humour, **224,** 229
Arachnoid, 114
Arcuate arteries of kidney, 134
Areola of breast, 197, 198
Arrectores pilorum (smooth muscle of
  skin), 45, 161
  in autonomic reflex, 270
Arterioles
  blood pressure in, 92
  histology, 91
  nervous regulation of, 95
  of kidney, 134, **135,** 136, 172
  of pulmonary circulation, 101
  reflex and chemical regulation of tone,
    96, 97
Artery, 78
  blood flow in, 100
  blood pressure in, **92, 93**
  histology, 91
  mesenteric, 70
  pulmonary, 101, **119**
  renal, 134, 135
  retina, central artery of, 224
  splenic, 113
  superior hypophyseal, 163
  supra-optic, 168
  umbilical, 191, 193
Artificial respiration, 122
Ascorbic acid (see vitamin C)
Association areas of cerebral cortex, **206**
  functions of, 250, 251
Atria of heart, 79, 80, 81, 82
Atrio-ventricular node, groove, 80, 81,
  85, 86
Auditory
  cortical representation, 240
  nerve (VIII), **210,** 237, 240
  ossicles, 237, 239
  pathways, 240
  receptors, 17, 238, 240
  stimuli, 220, 239
Auerbach's plexus, 67
  oesophagus, 56, 57
  small intestine, 66, 68, 70
Auricles (or atria) of heart, 79, **80,** 81, 82
  conduction of heart beat, **83, 85,** 90
  innervation, 87
Auriculo-ventricular groove, 80, 81
  node
    conduction of heart beat, 85, 86
Autonomic nervous system, 20, 21, **268,**
  269
  autonomic reflex, 270
  chemical transmission in, 271
  in control of gastro-intestinal move-
    ments, 60, 65, 73, **74, 75**
    gastro-intestinal secretions, 59, **63,**
      68

Autonomic nervous system—*continued*
    micturition, 147
    respiration, 132
    suprarenal medulla, 160
  in eye reflexes, 228, 236
  in regulation of arteriolar tone, 95, 96, 97
    heart action, 87, 88, 89
  outflows, **268, 269**
  salivary centres, 55
Axon, **15, 16, 17,** 212
  reflex, 97

Baby
  birth, 194
  development, 19, 188 to 194
  energy expenditure, 49
  growth, 46
Bacteria
  defence against, 11, 112
  nitrogen cycle, 32
  of intestine, 37, 72
  synthesis of vitamins K and B, 36, 37
Bainbridge reflex, 88
Balance, sense of, (see Equilibrium)
Baroreceptors
  in cardiac reflexes, 88, 89
  in respiratory reflexes, 132
  in vasomotor reflexes, 96, 97
Basal ganglia, 207, (256)
Basal metabolism, 48, 49
  anterior pituitary activity and, 165, 167
  thyroid activity and, 151, 152, 153
Basilar membrane of cochlea, 238, 239
Basophil cells
  of anterior pituitary, 163, 166
  of blood, 11, 107
Beri-beri, 37
Betz cells, 16, 206
  functions of, 251, 252, 253, 255, **260**
Bicarbonate
  carriage in blood, 102, 104, 127, 128, 129
  conservation by kidney, 143
Biceps, 262, 264
Bile, 64
  duct, 53, 62, 64
  expulsion of, 65
  pigments, 64, 72, 104
  salts, 64, 72
Bilirubin, 64, 104
Biliverdin, 64
Binocular vision, (234), 235
Biotin, 37
Bipolar cells, 17
  auditory, 238, 240
  retinal, 231
  vestibular, 243
Birth
  body proportions at, 46
  change in circulation at, 193
  energy requirements, 49
  normal, 194
Bladder (urinary), 134, **146,** 147, 178
  innervation, 147, 268, 269
Blastocyst, 19, 188, 189
Blind spot, 230

Blindness, colour, 233
Blood, 78, **104**
  acid base balance, 142, 143
  cells, 11, 106, 107, 112, 113
  cholesterol, (152)
  circulation, 78, 79
    foetal, 193
    kidney, 134, 135
    placental, 191
    portal, 71
    pulmonary, 101, (119)
    splenic, 113
  coagulation (11), **105,** (161)
  exchange of water and electrolytes, 98, 102
  flow
    pulmonary, 101
    systemic, 100
  groups, 108, 109, 110, 111
  haemopoiesis, 106, 107
  heat dispersal, 43, 44, 45
  plasma, 104
    clearance, 136, 137, 138, 139, 140
    proteins, 104, 105, 112
  platelets, 11, 107
  pressure
    and adrenaline, 161
    and angiotensin, 172
    and noradrenaline, 160
    and posterior pituitary, 168
    arterial, 92 to 97
    capillaries, 98
    in underactivity of suprarenal cortex, 158
    in underactivity of anterior pituitary, 167
    kidney glomeruli, 136
    pulmonary, 101
  regulation of water balance, 103, 144, 145
  reservoir, spleen, 113
  sugar, 40, 41, (71), (104)
    adrenaline and, 161
    anterior pituitary and, 167
    role of glucagon, 172
    role of insulin, 172
    role of suprarenal cortex, 158, 159, 166
  transport
    foodstuffs, 71
    haemopoietic factor, 106
    heat, 43, 44, 45
    hormones
      endocrine glands, 150-173
      gastro-intestinal, 59, 60, 63, 65, 68, 70
      reproductive system, male, 176-179; female, 180-201
    respiratory gases, 117, 119, 125-129, 131
    waste products, (excretion), in
      kidney, 134-140
      in faeces, 39
  volume, **90,** 102, (158), 172
Blood vessels, 78, 79, **91**
  blood pressure in, 92, 93, 101
  capillaries, 91, 98, 101

  control of, 95 to 97
  development, 19, 190
  effects of adrenaline, 161
  elastic arteries, 94
  flow in, 100, 101
  in heat dispersal, 43, 44, 45
  innervation of, 95, 268, 269, 270
  role of vitamin C, 37
  structure of, 9, 12, **91**
  veins, 91, 98
Body-building compounds, 33, 35, 50
Body cavities, (9)
  development of, 19
  pericardial, 80
  peritoneum, 67
  pleural, 119
Body fluids
  composition, 25, 102
    control of, 98, 102, 142, 143
    role of ADH, 168, 170, 171
    role of kidney, 137, 142, 143
    role of suprarenal cortex, 157, 158, 159
  water balance, 103, 144, 145
Body temperature, **43, 44, 45**
  effect of thyroid, 152, 153
  effect on respiration, 131
  in panhypopituitarism, 167
Bones, **13,** 35, 36, **261-264**
  action of growth hormone, 163-167
  action of parathormone, 154-156
  action of thyrocalcitonin, 151
  chemical constituents of, 25, 35
  marrow, 11
    haemopoiesis, 106, 107
  vitamin D deficiency, effect of, 36
Bowman's
  capsule, 135-140
  glands (nose), 221
Brain, 20, 204, **206**
  cells of, 16
  centres and pathways (see Cerebral cortex)
  cerebrospinal fluid, 114
  control of respiratory movements, 130-132
  coronal section, 209
  cranial nerves, 210
  development of, 205
  horizontal section, 207
  links with cerebellum, 258, 259
  links with extrapyramidal system, 256
  link-up of neurones, 217-219
  membranes of, 114
  motor cortex, **251**
  role in control of muscle movement, **260**
  sensory cortex, **250**
  ventricles of, 114
  vertical section, 208
Brain stem, (207), (208), (209), 217, 256
Breast (see Mammary gland)
Breathing
  mechanism of, 121
Broad ligament (uterine), 181
Bronchi, 117-119

Bronchioles, 118, 119
  action of adrenaline on, 161
  muscle, innervation of, 268, 269
  stretch receptors in, 132
Brünner's glands, 66, 68
Buffer
  nerves (heart), 89
  salts in urine, 142, 143
Bulbo-urethral glands, 178
Bundle of His, 85

Caecum, 53, 69, 71, 72
  control of movements, 73, 75, 268, 269
Calcium, 25, 50
  absorption, 71, 154-156
  content of food, 35
  excretion, 148
  in blood clotting, 105
  metabolism and parathormone, 36, 154-156
  metabolism and thyrocalcitonin, 151
Calorie
  intake, 43
  requirements, 48, 49
  values of foods, 33
Capillaries, 78, 79
  and tissue fluid exchange, 98, 102
  blood pressure in, 92, 101
  cross-sectional area/blood flow in
    pulmonary, 101, (119)
    systemic, 100
  kidney, 135-141
  nervous control, 95 to 97
  structure, 91
  vitamin C and, 37
Carbamino compound, 128, 129
Carbohydrate, 25, 26
  absorption, 70, 71
  body fat and, 41
  caloric value of, 33
  carbon cycle, 30, 31
  dietary sources, 34, 50
  digestion, 38, 54, 62, 66
  metabolism, 40
    adrenaline and, 161
    islets of Langerhans and, 173
    pituitary and, 163-167
    suprarenal cortex and, 157
  release of energy from, 42
  requirements, 48, 50
  specific dynamic action of, 48
Carbon cycle, (27), 30, 31
Carbon dioxide
  in blood
    carriage and transfer, 128, 129
    exchange of respiratory gases, 125-127
    in regulation of arteriolar tone, 96, 97
      blood pressure, 96
      respiration, 131, 132
  in carbon cycle, 30, 31
  in respired air, 125
Carbonic anhydrase
  in blood (R.B.C.), 128, 129
  kidney, 142, 143
Carboxypeptidase, 62
Cardiac

cycle, 83, 84
  murmurs, 84
  muscle, 80, 81
    chemical transmission, 271
    histology, 14
  nerves, 87
  output, 90, 172
    to kidneys, 134, 136, 172
  reflexes, 88, 89
  sphincter (stomach), 53, 56-58
    control of, 75
Carotene, 36
Carotid body
  and blood pressure, 96, 97
  in regulation of respiration, 131, 132
  in vasomotor reflexes, 96, 97
Carotid sinus, 89, 132
  and blood pressure, 96, 97
  and respiration, 132
  and vasomotor reflexes, 96, 97
  nerves, 210
Cartilage, 13
  at joints, 264
  in ear, 237
  in trachea, 118
Cauda equina, 211
Caudate nucleus, 207, (209)
  links with other centres, 217, 256
Cell, 3, 6
  Betz, 16, 206
  chemical constituents, 25-29
  differentiation, 8
  division
    meiosis, 18
    mitosis, 6, 7
  enzyme systems, 42
  fluid in, 102, 144, 145
  Leydig, 177
  oxidations, 42
  control of, role of thyroid, 151-153
  pancreas α and β, 172
  permeability, 6, 102
  Purkinje, 16, 257
  Schwann, 15, 16
  Sertoli, 177
  (see also Tissues)
Central artery of retina, 224, 230
Central canal of spinal cord, (114), 211
Central fissure of cerebrum, 206, 209
Centriole, 6, 7, 18
Centrosome, 6, 7, 18
Cerebellum, 208-210, 217, 257-259
  development of, 205
  histology, 16, 257
  in control of muscle movements, 260
  links with extrapyramidal system, 256
  proprioceptor pathways, 245
  vestibular pathways and centres, 243
Cerebral cortex
  association areas, 206
  centres and pathways
    auditory, 240
    autonomic, 268-270
    eye movements, 227
    motor to head and neck, 252
    motor to trunk and limbs, 253
    olfactory, 221

pain and temperature from trunk and limbs, 248
    proprioceptor, 245
    sensory, 247
    taste, 223
    touch and pressure, 249
    vestibular, 243
    vision, 234
  cranial nerves, 210
  development of, 205
  'edifice', 217
  histology, 16, 206
  links with cerebellum, 258, 259
  links with extrapyramidal system, 256
  motor areas, 251
  role in control of muscle movement, 260
  sensory areas, 250
  structure, 206
    coronal section, 209
    horizontal section, 207
    vertical section, 208
Cerebrospinal fluid, 114
  in development, 205
Cervical nerves, 211
Cervix, 194
  (see also under Uterus)
Chemical senses, 220
  smell, 221
  taste, 54, 222, 223
Chemical transmission, 212, 271
  in gut, 74
Chemoreceptors
  and arteriolar tone, 96, 97
  in regulation of respiration, 131, 132
Chest (see Thorax)
Chiasma, optic, (210), 234
Childbirth, 194
  effect of oxytocin, 169
Children
  anterior pituitary in, 163-165
  effect of thyroid deficiency in, 152
  energy requirements in, 48, 49
  growth, 46, 47
    effect of vitamins on, 36, 37
  mammary glands in, 197
  overactivity of suprarenal cortex in, 159
  reproductive system in, 176, 180
  thymus in, 174
  uterus in, 186
  vitamin D deficiency in, 36
Chloride shift, 128, 129
Chlorophyll, 30
Cholecystokinin, 65
Choleretic action, 64
Cholesterol
  in bile, 64
  in blood, 104
  in myxoedema, 152
Cholic acid, 64
Choline, 37
Cholinergic nerves, 271
Cholinesterase, 254
Chordae tendineae, (80), 81
Chorionic
  gonadotrophins (see Placenta)

Chorionic—*continued*
  membranes, 192
  villi, 190, 191
Choroid
  coat of eye, 224
  plexuses, 114
Christmas factor, 105
Chromaffin granules
  suprarenal medullae, 160
Chromatin, 6, 7
  in meiosis, 18
Chromophils of anterior pituitary, 163
Chromophobes of anterior pituitary, 163
Chromosomes, 6, 19
  DNA in, 29
  in meiosis, 18
  in mitosis, 7
  nucleotides in, 29
Chyme, 60
Chymotrypsin, 62
Cilia, 5, 9, 10
  in trachea, 118
  in uterine tube, 181
Ciliary
  body, 224, **228**
  ganglion, 228, 236, 268
  muscle, 224, 229
  nerves, 210, 268
Circulation
  course of, 79
  foetal
    changes at birth, 193
    placental, 191
  portal, 71
  pulmonary, 101
  splenic, 113
Citric acid cycle, 41, **42**
Claustrum, 207, 209
Clearance tests
  diodone, 141
  inulin, 139
  urea, 140
Cleavage of zygote, 19, 188
Clot retraction, 105
Clotting (see Blood coagulation)
Coccygeal nerves, 211
Cochlea, 237, **238**, (241)
  centres and nerve pathways, 210, **240**
  mechanism of hearing, 239
Coeliac ganglion, (75), 269
Coenzyme systems, 42
  vitamin B complex in, 37
Cold
  as skin sensation, 246
  effect on arteriolar tone, 96
  effect on respiration, 132
  effects on skin, 45
  pathways from skin, 247, 248
Collagen fibres, 12
Collateral ganglion in abdomen, 270
Collecting tubules (ducts), 135, 137, 138
Colloid
  osmotic pressure, 98
    (see under Osmotic pressure)
Colon (ascending, transverse, descending
  and pelvic), 53, **72**
  absorption in, 71

innervation, 75, 268, 269
  movements in, 73
Colour blindness, 233
Coma, diabetic, 173
Conditioned reflexes, 219
  expulsion of bile from gall bladder, 65
  salivary, **55**
  secretion of gastric juice, 59
  secretion of pancreatic juice, 63
Conducting arteries, 91, 92, 101
Conduction
  in heat loss, 43-45
  in nerve, 15, **213**
  of heart beat, 85, 86
Cones (retina), 231
  mechanism of vision, 233
Conjugate deviation (eyes), 227
Conjunctiva, 225
Connective tissues, 11-13
  development of, 19
  role of vitamin C, 37
Consensual light reflex, 236
Contractile vacuole, 3, 4, 5
Contractility, 4, 8, 14, 22
Contraction, muscle, 262-266
  control of, 260
Convection
  in heat loss from skin, 43-45
Convoluted tubules, 135-139, 142-145
  action of ADH, 168, 170, 171
  action of insulin, 172
  action of parathormone, 154-156
  action of suprarenal corticoids, 157-159
Cornea, 224, 228, 229
Coronary blood vessels
  innervation, 268, 269
  vasodilatation, 95, 97
Corpora quadrigemina, 208, (240)
  development of, 205
Corpus albicans, 182, 185
Corpus callosum, 208
Corpus luteum, 182, 183 to 185, 187, 189,
  190, 192, 196, 197 to 199, 201
  control of progesterone secretion by
    anterior pituitary, 163, 184, 201
Corpus striatum, 207, 209, 256
  links with other centres, 217, 256
Cortex, cerebral, (see under Cerebral
  cortex and Brain)
Corti, organ of, 237-240
Corticoids, suprarenal, 157-159
  and anterior pituitary, 163, 166
Corticotrophin, 157, 158, 163
  overproduction of, 166
Coughing, respiration in, 132
Counter-current system/kidney, 138
Cranial nerves, 210, 252
Creatinine
  blood, 104
  urinary, 137, 148
Cretin, 152
Crista (of semicircular canals), 241, 242
Crypts of Lieberkühn, 66
  control of secretion, 68
Crystalline lens, 224, 228, 229
Cubical epithelium, 9
Cumulus oophorus, 182

Cuneate nucleus
  chemical transmission in, 271
  links with cerebellum, 258
  proprioceptor pathways, 245
  touch and pressure pathways from
    trunk and limbs, 249
Cupula of semicircular canals, 241, 242
Cushing's syndrome, 159, 166
Cutaneous sensation
  centres, 206, 250
  nature of stimuli, 220
  pathways, 210, 247-249
  receptors, 17, 20, **246**
    links with cerebellum, 258
Cystic duct, 64
Cytochrome system, 42
Cytoplasm, 3, 5, **6**, 7

Dark adaptation of eye, 228, 232
Dead space air, 124
Deamination, 39, (142)
Decidua, 190
  after childbirth, 195
Defaecation, 72, **73**, (75)
Defence
  digestive tract in, 58, 66, 67, 72
  lymphatic tissue, 112, 113
  white blood cells, 11, 104
Dehydration, tissues,
  causes, **145**
Dendrites, **15**, 16, 17, 212
Dense fibrous tissue, 12
Dentate nucleus (Cerebellum), 259
Deoxyribonucleic acid (DNA), 6, **29**
Depot fat, 40, 41
Depth perception, **235**
  action of lens in, 229
Development of the individual, 19
  growth, 46
  intrauterine, 188-190
Diabetes
  insipidus, 171
  mellitus, 173
Diaphragm, 117, **120, 121,** 122
  innervation, 130
  in vomiting, 61
Diarrhoea, 144, 172
Diastole, **83**
  blood pressure in, 92, 93
  elastic arteries in, 91, 94
  heart sounds in, 84
Diencephalon, 205
Diet
  balanced, 33, **50**
  body-building requirements, 33, 35
  effect on haemoglobin regeneration,
    106
  energy requirements, 33, 34, 48, 49
  influence on growth, 46
  protective requirements, 33, 36, 37
Digestion, 21, 38, 39-41, 52
  in intestine, 62, 66
  in mouth, 54
  in stomach, 58
Diglycerides, 41
Dihydroxycholecalciferol (1:25 DHCC)
  150, 154-156

Diiodotyrosine, 151
Dilator pupillae, **228**
  action of adrenaline, 161
Diodone
  renal clearance, 141
Dipeptides, 28
Disaccharide, 26
Dissociation curves, 126, 127
Diuresis, 144, (171)
Donor (blood), 108, (109), (110)
Dorsal vestibulo-spinal tract, (243), **255,**
  256
Drum membrane, ear, 237, 239
Ductus arteriosus, 193
Ductus venosus, 193
Duocrinin, 68
Duodenum, 53, 62, **66**
  absorption in, 70
    vitamin B$_{12}$, 106
  bile in, 64, 65
  digestion in, 66
  emptying, 60
  hormones of, 60, 63, 65, 68, 70
  movements of, 69
  secretion, 66
    control of, 68
Dura mater, 114
Dwarfing
  in pituitary deficiency, 164
  in thyroid deficiency, 152

Ear, 237, 238
  mechanism of hearing, 239
  nature of stimuli, 220
  pathways to brain, 240
  receptors, 17, 238, 240
Ectoderm, 19
  development of nervous system, 205
Edinger-Westphal nucleus, 228, 236
Effector organs
  autonomic nervous system, 268-270
  central nervous system, 20, 253
  chemical transmission to, 271
  in reflex action, 214-218
Elastic arteries, **91**
  function of, **94**
  in pulmonary circulation, 101
Elastic fibres, 12
  in arteries, 91
  in cartilage, 13
  in respiratory system, 13, 118, 119,
  (121)
Electrocardiogram, 86
Electrolytes, 102
  control of distribution, **98**
    in blood, 104
    in urine, 141
  control of excretion
    role of anterior pituitary, 166, 167
    role of kidney, 134, 136, 137, 142-145
    role of parathyroid, 154-156
    role of posterior pituitary, 170, 171
    role of suprarenal cortex, 157-159
  distribution in body, 102
Electromagnetic waves, 220
Embryo
  development, 19, 188, 205

implantation, 19, 189, 190
  placenta, 191
Emotion
  and ADH, 170
  and adrenaline, 160, 161
  and lactation, 198
  and respiration, 132
  (see also Stress)
Endocardium
  histology, 80
Endocrine glands, 150-174
Endoderm, 19
Endolymph
  of ear, 238, 239
  of semicircular canals, 241, 242
Endolymphatic duct, 241
Endometrium, 181, 185-190, 195, 196, 201
Endothelium, 9
  heart valves, 81
  of blood vessels, 91
  permeability, 102
Energy
  balance, 43
  foods, 34, 50
  release, 33, **42**
  requirements, 33, 48-50
  role of thyroid, 151-153
  sources, 30
  uses, 43-47, 266
Enterogastrone, 59, 60
Enterokinase, 62, 66
Environment
  influence on growth, 46
  internal, stability of,
    acid-base balance, 142, 143
    role of endocrines, 150, 154-156, 157-
    159, 168, 170, 171
    role of kidneys, 134
    water balance, 103, 144, 145
    stimuli from, 20, 220
Enzymes, 9, 38-42
  carbonic anhydrase
    in kidney, 142, 143
    in red blood cells, 128, 129
  glutaminase, 142
  intrinsic factor, 106
  lysozyme, 225
  mouth, 38, 40, 54
  pancreas, 38, 41, 62
  role of vitamin B complex, 37
  small intestine, 38-40, 66
  stomach, 38-40, 58, 106
Eosinophils
  anterior pituitary, 163-165
  blood, 11, 107
  parathyroid, 154
Epididymis, 176, 177
Epithelia, 9, 10
  development of, 19
  (for sites see under Organs)
Equilibrium
  mechanism of action, 220, **242**
  organs of, 237, 241
  vestibular pathways to brain, 210, **243,**
  245
Erepsin, 66
Ergastoplasm, 6

Erythrocytes (red blood corpuscles), 11
  agglutinogens, 108-111
  carriage and transfer of O$_2$ and CO$_2$,
  128, 129
  destruction of, 113
  electrolytic composition, 102
  formation, 106, 107
  functions, 104
Erythropoietin, 106, 150
Eustachian tube, 221, 237
Eve's method of artificial respiration, 122
Excretion, 4, 8, 21, 22
  digestive system, 52
    in bile, 64
    large intestine, 72
    salivary glands, 54
  kidneys, 134-147
    diodone clearance, 141
    inulin, 139
    urea, 140
  respiratory system, 117-132
Exercise
  blood elements in, 107
  body temperature in, 43-45
  cardiac output in, 90
  energy requirements for, 48, 49
  pulmonary ventilation in, 101, 123
  urine formation in, 148
Exophthalmos, 153
Expiration
  expired air, 123-125
  in artificial respiration, 122
  mechanism of, 121
  nervous control of, 130
Expiratory
  capacity and reserve, 123
  centres, 130
  muscles, 121
Extensor reflexes
  knee jerk, 215
  reciprocal innervation, 265
External auditory meatus, 237
External genitalia
  female, 180, 184, 200
    innervation, 268, 269
  male, 176-179
    reflex vasodilatation, 95, 97
External rectus of eye, 226
  in conjugate deviation, 227
  innervation, 210
External respiration, 117, 125
External spiral ligament of cochlea, 238
Exteroceptors, 220
Extracellular fluid, 102
  aldosterone and, 157, 172
  distribution, control of, 98
Extrapyramidal system, 217, **256**
  pathways, 255
Extrinsic factor, 106
Extrinsic muscles of eye, 226
  innervation, 210
  reflex movements of, 227
Extrinsic thromboplastin, 105
Eyelids, 225, 226
Eyes, 224
  accommodation, 228, 229
  autonomic fibres to, 228, 268, 269

Eyes—*continued*
conjugate deviation, 227
control of movements, 227
cortical centres, 227, 234
effects of adrenaline on, 161
extrinsic muscles of, 226
fundus oculi, 230
innervation, 210
light reflex, 236
mechanism of vision, 231, 232, 233
nature of stimulus, 220
nerve pathways from, 234
protective mechanisms, 225
stereoscopic vision, 235
visual purple, 36

Facial muscles, 262
nerve supply, 210, 252
Facial nerve (VII), 210
motor pathways to head and neck, 252
pathway for taste, 223
sensory pathways from face, 247
Factors V, VIII, IX, X, 105
VII, 36, 105
Facultative reabsorption, 171, 172
Faeces, 21, **72**
bile pigments in, 64
calcium in, 154
expulsion, 73
fluid loss in (or in water balance), 103, 145
heat loss in, 43
Fallopian tubes (uterine tubes), 180, 181, 185
at maturity, 187
at menopause, 200
at puberty, 184
fertilization in, 188
Fat, 25, 27, 30
absorption of, 64, 70, 71
caloric value of, 33
dietary, 34, 50
specific dynamic action, 48
digestion of, 38, 58, **62**, 64, 66
from carbohydrate, 40
heat insulation, 44, 45
in growth, 47
metabolism, 41
endocrines affecting, 152, 153, 157, 159, 163, 165, 166, 173
transport, 71
Fat-soluble vitamins, 36
absorption, 71
Fatty acids, 27, 41
absorption and transport, 70, 71
role in emulsion of fats, 64
Female reproductive system and sex hormones (see under Reproductive system)
Fertilization, 19, 188, 201
Fibrin, (104), 105
Fibrinogen, 104, **105**
Fibrous tissue, 12
'Fight or Flight' mechanism, 161
Filtration (see under Kidney)
Final common pathway, **255**
in reciprocal innervations, 265

Fixators, 264
Flavoproteins, 42
Flexion, 264
reciprocal innervation, 265
Fluids of body (see under Body fluids)
Foetal circulation, 193
Foetus
birth weight, 46
circulation in, 193
development of, 19, 188-192
haemolytic disease, 110, 111
haemopoiesis in, 113
placental circulation, 191
Rh incompatibility, 110, 111
Folic acid, 37
in haemopoiesis, 106
Follicle-stimulating hormone (FSH), 163, 179, 184-187, 189, 190, 196-201
Food
absorption of, 70, 71
balanced diet, 50
body-building, 33, **35**
digestion of, 38, 54, 58, 62, 66
energy balance, 43-46
energy-giving, 33, **34**
energy requirements, 33, 48, 49
metabolism, 39-41
oxidation of, 42
progress along alimentary canal or digestive tract, 53
protective, 33, 36, 37
specific dynamic action, 48
Food vacuole, 3-5
Foramen ovale, 193
Forebrain, 208, 209
development of, 205
Fourth ventricle, 114
Fovea centralis, 230, 231
Fröhlich's dwarf, 164
Frontal lobes, 206-208, 251, (253), (256)
centres for eye movements, 227
Frontal sinus, 221
Fructose, 34, 40
Functional residual capacity (lungs), 123
Fundus
of eye, 230
of stomach, 53, 58
movements of, 60
Fusimotor fibres, 260

GABA, 271
Galactogogue action of oxytocin, 169
Galactose, 26, 40
Gall bladder, 52, 53, **64**
expulsion of bile, control of, 65
parasympathetic innervation, 268
Gametes, maturation of, 18
(see also under Ovum and Spermatozoa)
Gamma aminobutyric acid, 271
Gamma globulin
in plasma, 104
Gases, in blood (see under Blood)
Gastric glands, 58, 59
parasympathetic innervation, 268
Gastric hormones, 59
Gastric juice, 58

control of secretion, 59
Gastrin, 59
Gastro-colic reflex, 73
Gastro-intestinal hormones, 59, 60, 63, 65, 68, 70, 150
Gastro-intestinal secretion
in water balance, 103
Gastro-intestinal tract, 52, 53
absorption—calcium and phosphorus—
role of parathormone, 154-156
actions of adrenaline on, 161
in Addison's disease, 158
Genes, **6**, 7, (29)
Germinal epithelium, 182
Giantism, 165
Glands
Brünner, 66, (68)
bulbo-urethral, 178
chemical transmission at, 271
endocrine
islets of Langerhans, 173
ovary, 182, 183
parathyroid, 154
pituitary, 162, 163, 168
suprarenal cortex, 157
medulla, 160
testis, 177, 179
thymus, 174
thyroid, 151-153
endometrial, 181, 186, 187, 189
gastric, 58
intestinal, 66
lacrimal, 225
liver, 64
mammary, **197**, 199
mucous, trachea, 118
pancreatic, 62
prostate, 178
salivary, 9, 54
sweat, 43-45
tarsal, 225
Globulin, 104, 112
Globus pallidus, 207
links with other centres, 217, 256
Glomerulus, kidney, 134, **135-140**
volume of filtrate, 171
Glossopharyngeal nerve (IX), 210
carotid sinus nerve, 89, 96, 97, (131), 132
motor pathways to head and neck, 252
pathway for taste, 223
Glottis, 118
in sneezing, 132
Glucagon, **173**
Glucocorticoids, (40), 157-159, 166
Glucose, 26
absorption and transport, 71
conversion to fat, 40
formation from protein, 157-159
kidney reabsorption, 137
metabolism, 40
oxidation of, 42
role of glucagon and insulin, 173
Glutaminase, in kidney, 142
Glutamine, in kidney, 142
Glycerides, 27
absorption in small intestine, 70, (71)

Glycerides—*continued*
  digestion in intestine, 41, 66
  role in emulsifying fats, 64
Glycerol, 27, 41, 70, 71
Glycine, 271
Glycogen
  action of adrenaline on, 161
  and suprarenal cortical hormones, 157-159
  and thyroid, 152, 153
  in liver, 34, **39-41**
  in muscles, 34, **40**
  in vaginal epithelium, 181
  Krebs citric-acid cycle, 42
  role of glucagon and insulin, 173
Goblet cells, **9**, 72
  in trachea, 118
Golgi body, 6
Golgi organ (in tendons), 244
Gonadotrophins, 163, 201
  at menopause, 196, **200**
  at puberty, 179, 184
  in regulation of female reproductive cycle, 184-201
  lack of, 167
  placental, 191
Gonads, 176, 180
  at puberty, 179, 184
  role of anterior pituitary in control of, 163
Graafian follicle, **182**, 184-187, 189, 196, 197, 199
  role of anterior pituitary, 201
Gracile nucleus
  chemical transmission in, 271
  links with cerebellum, 258
  proprioceptor pathways, 245
  touch and pressure pathways from trunk and limbs, 249
Granular leucocytes, 11, 107
Grey matter
  brain, 206, 207
  spinal cord, 211
Groups, blood, 108-111
Growth, 4, 22, 31, 42, **46, 47**
  at puberty, 179, 184
  dietary requirements for, 35, 48, 49
  hormone, 39, 163-165, 167, 173
  role of endocrines in control of, 150
    anterior pituitary, 46, 163-166
    gonads, 179, 184
    suprarenal cortex, 157-159
    thyroid, 151
  role of vitamins, 36, 37
Gustatory cells, 222
Gyrus, 250, 251

Haemocytoblast, 107
Haemoglobin
  appearance in red blood corpuscle, 107
  breakdown products, 64
  carbamino compound, 128, 129
  carriage of respiratory gases, 126-129
  iron, 35
Haemolysis, 108-110
Haemolytic disease of newborn, 110
Haemopoiesis, 107

  in lymph nodes, 112
  in spleen, 113
  nutritional factors for, **106**
Haemopoietic factor, 106
Haemostasis, 105
Hageman factor, 105
Hair
  adrenaline and, 161
  body temperature regulation, 45
  in panhypopituitarism, 167
  in thyroid deficiency, 152
  touch sensation, **246**
  (see also under Hirsutism and Secondary sex characteristics)
Hassall's corpuscles, 174
Haversian canal (in bone), 13
Health, 22
  role of vitamins, 36, 37
Hearing
  auditory receptors, 17, 238, 240
  centres for, 206, 240
  mechanism of, **239**
  nature of stimulus, 220
  nerve pathways, 210, 240
  organ of, **237, 238**
Heart, 78-90
  blood pressure, 92
  cardiac cycle, 83
  cardiac muscle, 14
  cardiac output, 90
  cardiac reflexes, 88, 89
  effects of adrenaline on, 161
  effects of thyroid on, 152, 153
  electrocardiogram, 86
  foetal, 193
  histology, 9, 14, 80, 85
  innervation, 87, 210, 268, 269
  in panhypopituitarism, 167
  origin and conduction of heart beat, 85
  rate, 87
  sounds, 84
  valves, 12, 81, 83, 84
  venous return to, 99
Heat
  balance, **43-45**
  effect of thyroid, 152, 153
  effect on blood vessels, 97
  effect on respiration, 131, 132
  release of, 42
  skin sensation, 246
Height, 46, 47
  (see also under Growth)
Henle's loop (kidney), 135, 137
Heparin, 11
Heredity, 7
  influence on growth, 46
  inheritance of Rh blood group factor, 111
  interchange of hereditary material in meiosis, 18
  nucleic acids, 29
Hering-Breuer reflex, 132
Heterozygous, 111
High altitude
  acclimatization to (red blood corpuscles in), 107
  effect on respiration, 131

Hindbrain, 208, 209
  development of, 205
Hirsutism
  in adrenogenital syndrome, 159
  in Cushing's syndrome, 166
Homozygous, 111
Hormones, 150
  (see under Endocrines, Gastro-intestinal and Reproduction)
Humidity
  and control of body temperature, 44
Humoral theory of transmission of nerve impulses, 212, **271**
Hyaline cartilage, 13
Hyaloplasm, 6
Hydrochloric acid, in gastric juice, 58, 59
Hydrogen
  carriers, 42
  ions, kidney, 142, 143
Hydrostatic pressure
  of blood, glomeruli (kidneys), 136
    systemic capillaries, 98
  of cerebrospinal fluid, 114
  tissue fluids, 98
Hyperinsulinism, 173
Hypermetropia, 229
Hypoglossal nerve (XII), 210, 252
Hypophysis, 162
  (see also under Pituitary)
Hypothalamus, 150, **168**, 177, 208
  autonomic centres in, 270
  development of, 205
  effect on anterior pituitary, 163
  in control of suprarenal cortex, 157
  in control of thyroid, 151
  in eye reflexes, 228
  influence on heart action, 87
  in release of oxytocin, 168, 169, 195, 196
  in suckling reflexes, 195, 196, 198
  links with extrapyramidal system, 256
  osmoreceptors in, 144, 145, 168, 170, 171, 172
  releasing factors, 151, 157, 163, (197), 198, 199, 201
  role in body temperature regulation, 44, 45
  role in control of blood vessels, 95
  role in control of suprarenal medullae, 160
  role in regulation of water balance, 144, 145
  thyrotrophin-releasing factor (TRF), 151

Ileo-caecal valve, 53, 66, 69, 72
  innervation and control, 73, 75
Ileum, 53, 66
Immunity, 11, 174
  antibody formation, 112, 174
  gamma globulin in, 112
Implantation, 19, 189, 190, 201
Incisura angularis, 60
Incus, centres for smell, 221
  middle ear, 237
Infancy (see under Baby and Growth)
Inferior mesenteric ganglion, (75), 269

Inferior oblique muscle of eye, 226
  innervation, 210
Inferior quadrigeminal body, 240
Inferior rectus muscle of eye, 226
  innervation, 210
Infundibular stalk, 162
Inheritance (see under Heredity)
Inhibition
  at motor neurones, 265
  of gastric motility and secretion, 59, 60
Inner ear, cochlea, 237-239, 241
  special proprioceptors, 241
Inositol, 37
Insensible perspiration, 43
Inspiration
  air in, 125
  in artificial respiration, 122
  mechanism of, 121
  nervous control of, 130, 131, 132
Inspiratory capacity, 123
Insula, 209
Insulation of skin, 43-45
Insulin, 40, 41, 173
Intercalated discs, 14
Intercellular cement, 9, 11
  role of vitamin C, 37
Intercostal muscles, 117, 120, 121, 130
Intermediate metabolism, 39-41
Internal capsule, 207, 209, 245, 247-249,
  253, 256
Internal rectus muscle of eye, 226
  innervation of, 210, 227
Internal respiration, 117, 125
Interoceptors, 20, 220
Interstitial cells
  of ovary, 185
  of testis, 177, 179
Interstitial cell stimulating hormone
  (ICSH), (163), 179
Interstitial fluid (tissue fluids), 98, 102
Intervertebral discs, 13, 261
Intestinal
  absorption, 70, 71
    role of parathormone, 154
  bacteria, 36, 37
  digestion, 66
    functions of bile, 64
  enzymes, 38, 66
  epithelium, 9
  hormones, 59, 60, 63, 65, 68, 70
  innervation, 74, 75, 268, 269
  juice, 66, 68
  motility, 69
  phase of gastric secretion, 59
  tract, general structure of, 53, 67
  villi, 70
Intracellular water, 102, (103)
Intrafusal muscle fibres, 244, 260
Intrapleural pressure, 120, 121
Intrathoracic pressure, 99, 121
Intrinsic factor
  formation in stomach, 58
  function, 106
Intrinsic thromboplastin, 105
Inulin clearance, 139
Involuntary muscle, 14
Involution of uterus, 196

Iodine
  dietary, 35, 50
  thyroid hormone, 151, 153
Iris, 224, 228
  effect of adrenaline on, 161
  in light reflexes, 236
  nerve supply, 210, 268, 269
Iron, 35, 50
  carriage of, 104
  in haemopoiesis, 106
  storage in liver, 64
Irritability (see Phenomena of life)
Islets of Langerhans, 62, 173
  role in carbohydrate metabolism, 40

Jejunum, 54, 66
  (see also Small intestine)
Joints
  muscular movements at, 264, 265
  proprioceptors, 244
Juxtaglomerular apparatus, 144, 145, 172

Ketogenic amino-acids, 41
Ketone bodies, 41
  in diabetes mellitus, 173
Kidney, 21
  ammonia formation in, 142, 143
  anatomy and histology, 9, 10, 134, 135
  and acid base balance, 142, 143
  and insulin, 173
  and parathyroid, 154-156
  and posterior pituitary, 170, 171
  and suprarenal cortex, 157-159
  and water balance, 103, 138, 144, 145
  clearance tests, 139-141
  concentration of filtrate, 137, 138
  counter-current theory / mechanism,
    138
  deamination of amino-acids, 39
  enzymes in, 142, 143
  excretion of ketone bodies, 41, 173
  filtration of blood, 136
Knee jerk, 215
Krause bulbs, in skin, 246
Krebs citric-acid cycle, 42
Kupffer cells (liver), 64

Labour, 194
Labyrinth, 241, 242
  pathways to brain, 243
Lacrimal ducts, 225
Lacrimal glands, 225
  parasympathetic innervation, 268
Lacrimal sac, 225
Lactase, intestinal, 66
Lactation, 198
  caloric requirements for, 49
  role of anterior pituitary in, 163
Lacteals, 70
Lactic acid, in muscle, 40
Lactogenic hormone, 163
Lactose, 34
Large intestine, 52, 53, 72
  absorption in, 71
  bacteria in, 39, 64, 72
  deamination in, 39
  innervation, 75, 268, 269

movements, 73
Larynx, 117, 118
  nerves to, 210
Lateral cortico-spinal tract, 253
Lateral fissure of cerebrum, 206
Lateral ganglion of sympathetic chain, 270
Lateral geniculate body, 234
Lateral horns of spinal cord, 211
  sympathetic outflow, 269
Lateral lemniscus, 240
Lateral spino-thalamic tract, 248
Lateral vestibulo-spinal tract, 243
Lecithin, 64
Lens of eye, 224, 228, 229
Lentiform nucleus, 207, 256
Leucocytes, 11, 107
Leucocytosis, 107
Leucopenia, 107
Leucopoiesis, 106, 107
Levator palpebrae superioris, 226
Leydig, cells of, 177, (179)
Ligaments, histology of, 12
Light adaptation, 233
Light reflex, 236
Limbs
  muscles, 262-264
    innervation, 211
Lipase
  gastric, 58
  intestinal, 66
  pancreatic, 41, 62
    activation of, 64
Liquor folliculi, 182
Liver, 52, 53, 64
  action of insulin, 173
  and adrenaline, 161
  and anterior pituitary, 163
  and glucocorticoids, 157-159
  bile formation and excretion, 64
  development of, 19
  glycogen in, food value of, 34-37
  in foetal circulation, 193
  innervation, 269
  intermediate metabolism, 39-41
  storage of vitamin B_{12}, 106
  transport of food from gut to, 71
Locomotor system, 22, 261-266
Longitudinal fissure of cerebrum, 206, 208
Loop of Henle, 135, 137, 138
Loose fibrous tissue, 12
Lorain dwarf, 164
Lower motor neurones, 253, 254, 255, 260
  chemical transmission in, 271
  in reciprocal innervation, 265
Lower visual centres, 234
Lumbar enlargement of spinal cord, 211
Lumbar nerves, 211
Lungs, 117, 119, 120
  capacity of, 123
  composition of respired air, 124
  exchange of respiratory gases, 126-129
  heat loss, 43
  Hering-Breuer reflex, 132
  movement of respiratory gases, 125
  nerves to, 210
  stretch receptors in, 132
  water loss from, 103

Luteinizing hormone (LH), 163, **179, 184,** 187-201
Luteotrophin (luteotrophic hormone), 184
Lymph, 102, **112**
  absorption of glycerides, 38, (41)
  in haemopoiesis, 107, 174
  nodes, **112**
  reticular tissue of, 11
Lymphatics, **112**
  small intestine, 70, 71
  transport of absorbed foodstuffs, 41, 71
Lymphocytes, 11, 112
  formation of, **107,** 112, 113
  of gastro-intestinal tract, 66, 67, 72
  of thymus, 174
  role in immunity, 11
Lymphoid tissue
  growth of, 46
  in digestive tract, 67
    in large intestine, 72
    Peyer's patch in ileum, 66
  in spleen, 113
  in thymus, 174
Lysosome, 6
Lysozyme, 225

Macrophage cells, 112
  (see also Reticulo-endothelial cells)
Macula
  lutea (retina), 230
  of utricle, 241
Magnesium, 25, 102
Male
  hormones, suprarenal, 157-159
    testis, 176-179
  organs, 176-179
Malleus, 237, (239)
Maltase
  intestinal, 40, 66
  pancreatic, 62
Maltose, 54, 62, 66
Mammary glands, 180, 197-199
  action of anterior pituitary on, 163
  action of oxytocin, 168, 169
  after menopause, 199, 200
  at puberty, 184, 197
  hormonal control of changes in menstrual cycle, 185
  suckling reflexes, 195, 196, 198
Manganese, 25, 106
Mass peristalsis, 72, 73
Master tissues, 20, (203-271)
Mastication, 54
Maturation
  of ovum, (18), **188**
  of sperm, 18
Maxillary sinus, 221
Meal, rate of passage, 53
Medial geniculate body, 240, 243
Medial lemniscus, 245, 248, 249
Medial longitudinal bundles, 227, 243
Medulla oblongata, 208, 209, (210), 211
  centres
    autonomic, 268 to 270
    cardiac, 87-89
    expulsion of bile, 65

gastric secretion, 59
    pancreatic secretion, 63
    respiratory, 130, 131, 132
    salivation, 55
    swallowing, 57
    vasomotor, 95-97
    vomiting, 61
  development of, 205
  pathways
    hearing, 240
    links with cerebellum, 258
    links with other centres, 217
    motor, 252, 253
    pain and temperature, 248
    parasympathetic outflows from, 268
    pressure and touch, 249
    proprioceptor, 245
    sensory from head and neck, 247
    taste, 223
    vestibular, 243
  role in control of suprarenal medullae, 160
Megakaryoblast, 107
Megakaryocyte, 107
Meiosis, 18
  in oogenesis, 182
  in spermatogenesis, 177
Meissner corpuscles, in skin, 17, 246
Meissner's nerve plexus, in gut, 67, 70
Melanophores, 158
Membrana granulosa, 182
Mendelian inheritance, rhesus factor, 111
Menopause, **200**
  mammary glands, 199
  uterus, after, 196
Menstrual cycle, **181,** 185, 187, **201**
  at puberty, 184
  cessation at menopause, 196, 200
  endometrial changes in, 187
  ovarian hormones in, 185
  ovary in, **182**
  restoration after childbirth, 196
Mesencephalon, 205
Mesentery, of gut, 67, 70
Mesoderm, 19
Metabolism, 4, 8, 22, 42, 43, 45
  basal, 48, 49
  carbohydrate, 40
  excretion of waste products of, 134, 137, 142, 143
  fat, 41
  heat production in, 43, 44, 45
  protein, 39
  role of endocrines in control of, 150
    adrenaline, 161
    anterior pituitary, 163-167
    insulin, 172
    parathyroid, 154-156
    suprarenal, 157-159
    thyroid, 151-153
  specific dynamic action of protein, 48
  vitamins and, 36, 37
  water of, 103
Metaphase, 7, 18
Metencephalon, 205
Micturition, 147
Midbrain, 208, 209

centres
    for eye reflexes, 227, 234, 236
    parasympathetic, 268
    vestibular, 243
  development of, 205
  pathways
    for hearing, 240
    links with extrapyramidal system, 256
    motor, 252, 253
    vestibular, 243
Middle ear, 237, (239)
Milk
  dietary, 34, 35
  expression of, 169
  (see also under Lactation)
Mineralocorticoids, 157-159
  and pituitary, 163, 166
Minerals, 25
  absorption in large intestine, 72, 73
  absorption in small intestine, 70, 71
  dietary, 35, 50
  excretion (see under Kidney)
Mitochondria, **6,** 42, 266
Mitosis, **7**
  in haemopoiesis, 107
  in spermatogenesis, **177**
Mitral valve, 81, 83, 84
Monoblast, 107
Monochromatic vision, 232
Monocyte, 11, 107
Monoglycerides, 27, 41, (66), (70)
Monoiodotyrosine, 151
Monosaccharide, 26
Moss fibre, of cerebellum, 257
Motion sickness, 61, 242
Motor cortex, 206, 251
  chemical transmission in, 271
  in control of skeletal muscle movement, 260
  pathways from, 252, 253, 256, 258
Motor end-plate, 15, **254**
  chemical transmission at, 271
Motor nerve cells, 15, 16
  in cerebral cortex, 251-253, 260
  spinal cord, 211, 214-218, 254-256, 260
Motor pathways
  in control of voluntary movements, **260**
  to head and neck, **252**
  trunk and limbs, **253**
Motor unit, **254,** 255, **260**
Mouth, **54**
  lining, 10
Mouth-to-Mouth Resuscitation, 122
Mucin (mucus), 9, 10
  in gall bladder, 64
  intestine, 68, 72
  mouth, 54
  oesophagus, 56
  stomach, 58
  trachea, 118
Mucosa
  in digestive tract, 67
  large intestine, 72
  oesophagus, 56
  small intestine, 66

Mucosa—continued
    stomach, 58
    muscularis mucosae, 67
Mucous glands, 9
  in trachea, 118
  oesophagus, 56
  salivary glands, 54
Multipolar neurones, (15), 16
Muscles
  action of insulin, 173
  chemical transmission in, 271
  contraction of, 264, 266
  development, 19
  effect of adrenaline on, 161
  eye, 224-226
  heat production in, 43, 44, 45
  histology, 12, 14, 266
  in diet, 34, 35
  -joint sense, 220, 244
    pathways, 245, 258
  of mastication, nerve supply to, 210
  reciprocal innervation of, 265
  release of energy in, 42, 266
  respiratory, 120 121
    innervation, 130
  skeletal, 262, 263, 266
  spindle, 244
    in reciprocal innervation, 265
    role in control of muscle movement, 260
    stretch reflexes, 215
  tone, 44, 45
    role of extrapyramidal system, 256
    role of proprioceptors, 244
  utilization of glucose by, 40
Muscular arteries, 91
  blood pressure in, 92
  nervous regulation of, 95
Muscularis mucosae, 67
  in oesophagus, 56
  'villus pump', 70
Muscular veins, 91
  blood pressure in, 92
  'tone' in, 99
Myelencephalon, 205
Myelin, 15, 16
  and conduction of nerve impulse, 213, 254
Myeloblast, 107
Myocardium, 80
Myoepithelial cells, mammary glands, 169
Myofibril, 14, 266
Myogenic movement, 69, 73, 75
Myometrium, 181, 186, 187, 190, 192, 194-196
Myopia, 229
Myosin, 266
Myxoedema, 152

Nasal passages, 117, 221
Naso-lacrimal duct, 221, 225
Negative intra-thoracic pressure, 120, 121
Neocerebellum, 257, 259
Nephrons, 135
  clearance of diodone, 140
  inulin, 138

  urea, 139
Nerves
  adrenergic, 271
  cholinergic, 271
  cranial, 210
  histology, 15-17
  nature of impulse, 212, 213
  spinal, 211
Nervous tissues, 15, 16, 17, 20, 204
  development of, 19, 46, 205
Neurilemma, 15, 16
Neurogenic movement, 69, 73, 75
Neuroglia, 15, 16
Neurohormones
  antidiuretic hormone, 168, 170, 171
  oxytocin, 168
Neuromuscular excitability
  role of blood calcium and parathyroid, 154, 155
Neuromuscular junction, 254
  chemical transmission at, 271
Neuromuscular transmission, 254, 271
Neurones, 15, 16, 17
  arrangement of, 219
  nature of nerve impulse, 213
  synapse, 212
Neurosecretion
  hypothalamus, 168, 170, 171, 172
Neutral (unsplit) fat, (triglycerides), 27
  absorption of, 41, 70, 71
  digestion of, 66
    role of bile, 64
Neutrophil, 11, 107
Newborn
  circulation, 193
  energy requirements of, 49
  haemolytic disease of, 110
  haemopoiesis in, 106
  thymus in, 173
  urine in, 141
Nicotinamide nucleotides, 42
Nicotinic acid, 37
Nitrogen, 25, 28, 29
  cycle, 32
  retention, role of growth hormone, 163-165
Nodal tissue (heart), 85, 86
Node of Ranvier, 15, 16
  in conduction of nerve impulse, 213
Non-granular leucocytes, 11
  formation of, 107, (112), (113)
Noradrenaline, 160
  in chemical transmission, 271
  in gut, 74
Normoblast, 107
Nose, 117, 210, 220, 221
Nucleases, 62
Nucleic acids, 29
  digestion of, 62, 66
  in leucopoiesis, 106
Nucleolus, 6, 7
Nucleoplasm, 3, 6
Nucleoproteins, 29
Nucleotides, 29
Nucleus
  caudate, 207, 256
  cochlear, 240

  cranial nerve nuclei, 245, 247, 252
  cuneate, 245, 249
  dentate, 259
  division of, 7, 18
  Edinger-Westphal, 228, 236
  emboliformis, 259
  fastigii, 259
  globosus, 259
  gracile, 245, 249
  histology, 6, 7
  lentiform, 207, 256
  nucleic acids, 29
  olivary, 240, 258
  red, 256, 259
  solitarius, 223
  supra-optic, 144, 145, 168-171
  vestibular, 243, 256
Nutrition, 4, 8, 22, 33, 34-37
  factors in haemopoiesis, 106
  influence on growth, 46
Nystagmus, 242

Obesity
  in Cushing's syndrome, 159, 166
  in Fröhlich's syndrome, 164
Obligatory reabsorption, 171
Occipital lobes, 206-208
  centres for eye reflexes, 227
  centres for vision, 234, 235
Oculomotor nerve (III), 210, (226), (227), 252
  parasympathetic outflow to ciliary muscle, (228), 268
Oddi, sphincter of, 62
Oedema
  in overactivity of suprarenal cortex, 159
Oesophagus, 52, 53, 56, (154)
  innervation, 75
  lining, 10, 56
  swallowing, 57
Oestrogen, 180, 182, 185, 186, 187, 189-192, 196, 199, 200, 201
  action on mammary glands, 196-198
  after childbirth, 195
  at menopause, 196, 200
  at puberty, 184
  effect on anterior pituitary, 163
  in male, 157, 179
  suprarenal cortex, 157-159
Olfactory
  bulb, (210), 221
    development of, 205
  centres, 221
  epithelium, 221
  nerve, 210, 221
  receptors, 221
Olivary nucleus, 240, 258
Olivo-spinal tract, 255
Oocyte, 182
Oogenesis, 182
  hormonal control of, 184
Ophthalmoscope, 230
Optic
  chiasma, (210), 234, (235), 236
  nerve, 210, 224, 226, 230, 231, 234, 236
  papilla, 230, 231

Optic—*continued*
  radiations, 234
  reflexes, (228), 234
Orbicularis oculi, 225, 226
Organ of Corti, **237**, 238, 240
  mechanism of hearing, 239
  pathways to brain, **240**
Organelles, 6
Osmoreceptors, **144, 145,** 168-171, 172
Osmotic pressure
  adjustment in stomach, 58
  in regulation of water balance, 138, 144, 145
    role of ADH, 170, 172
    role of corticoids, 157-159, 172
  plasma proteins and, 98, 114, 120, 136
  filtration force,
    kidney glomeruli, 136
    systemic capillaries, 98
    pulmonary capillaries, 101
Osteitis fibrosa cystica, 156
Osteomalacia, 36
Otic ganglion, 268
Otoliths, 241
Oval window, 237, 239
Ovarian cycle, **182, 185,** 201
  at menopause, 200
  at puberty, 184
  in pregnancy, 183
Ovarian hormones, 185, 201
  at menopause, 200
  at puberty, 184
  in pregnancy, 189, 190
Ovary, 150, 180, 181, **182, 183**
  action of anterior pituitary on, 163, 201
  at menopause, 200
  at puberty, 184, 186
  hormones of, **185**
  influence of suprarenals, 157
  in menstrual cycle, 187, 201
Ovulation, **182, 185,** 187, **201**
Ovum, **182**
  fertilization and maturation of, 18, 188
Oxidation (tissue), 40, 41, **42**
  heat production, 43-45
  role of thyroid in, 151-153
Oxygen, 25-31
  carriage and transfer, 11, **128, 129**
  dissociation from haemoglobin, **126**
  effect on uptake and release of CO_2 by blood, 127
  exchange of respiratory gases, 125
  in regulation of arteriolar tone, 96, 97
  in regulation of respiration, 131, 132
  in release of energy, 40, 41, **42**
  in respired air, 124, 125
  role in adrenaline secretion, 160
  transfer in placenta, 191
  uptake by blood in lungs, **90**
Oxyntic cells, 58
Oxytocin, 168, 169, 194-196, 198

Pacemaker (S.A. node), **85,** (86), (87)
Pacinian corpuscles, 17
  in joints, 244
  in skin, 246
Pain, 207, 208, 220

effect on respiration, 132
pathways from skin, 248
skin sensation, 246
Palaeocerebellum, 257, 259
Palate, 53
  soft palate in swallowing, **57,** 210
Palmitic acid, 27
Pancreas, 52, 53, **62, 63**
  development of, 19
  islets of Langerhans, 173
Pancreozymin, 63
Paneth cells, 66
Panhypopituitarism, 167
Pantothenic acid, 37
Papillae of tongue, 222
Papillary muscles, (80), 81
Para-aminobenzoic acid, 37
Parafollicular cells, thyroid, 151
Parallax, 235
Paramecium, 5
Parasympathetic outflow, 21, **268**
  chemical transmission, 271
  in control of expulsion of bile, 65
    gastric secretion, 59
    secretion of pancreatic juice, 63
    secretion of small intestine, 68
    swallowing, 57
  to blood vessels, 95
    gall bladder, 65
    gut, 73-75
    heart, 87-89
    iris, 228, 236
    urinary bladder, 147
Parathyroid, 150, **154-156**
Paraverterbral ganglia, (75), (87-89), (95), **269**
Parietal lobes, **206,** 208, 209
  proprioceptor centres, 245
  sensory centres, 247-250
  taste centres, 223
Parotid salivary gland, 54
  innervation, 210
Pars distalis (anterior pituitary), 163
  development of, 162
Pars glandularis, 162, 163
Pars intermedia, 162
Pars nervosa, 168-171
  development of, 162
Pars tuberalis, 162
Parturition, 194, 195
Patellar tendon, 215
Pellagra, 37
Pelvic nerves, 73, 75, (95), (147), **268**
Penis, 176, 178, 179
Pepsin, 38, 39, **58**
Peptic cells, 58
Peptides, 39, 66
Peptones, 39, 58
Perception, 220
Pericardial sac, 80
Pericardium, 80
Perilymph (cochlear), 238, (239)
Peripheral resistance, 92
  pulmonary circulation, 101
Peristalsis
  in large intestine, 72, 73
    oesophagus, 56, 57

small intestine, **69**
stomach, 60
uterine tubes, 181, 188, 189
Pernicious anaemia, 37, (58)
Perspiration, 43-45
Peyer's patch, **66,** 174
pH (hydrogen-ion concentration)
  of blood, 104
  of urine, 141
Phagocytes, 11
  in lymph nodes, 112
  in spleen, 113
Pharynx, 52, 53, 56
  innervation, 210
  in swallowing, 57
Phenomena of life, **4,** 8, 22
Phosphate metabolism, 25, 30, 35
  energy rich bond, 42, 266
  excretion, 137, 143
  role of parathormone, 154-156
  vitamin D and, 36
Phosphatides, 27
Photopic vision, 233
Photopsin, 233
Photosynthesis, 30, 31
Physiological haemostasis, 105
Pia mater, 114
Pigment
  areola of breast, 198
  choroid coat of eye, 224
  in skin, Addison's disease, 158
  retina, 231-233
Pitch, appreciation of, 239
Pituicytes, 168
Pituitary, 150, 162-171
  anterior, **163-167**
    adrenocorticotrophic hormone (ACTH), corticotrophin, 157
    and insulin production, 173
    and the ovarian and endometrial cycles, 201
    and thyroid, 151-153
    at menopause, 200
    control of testes, 177, 179
    effect on mammary glands, 199
    gonadotrophic hormones, 47, **179, 184,** 189, 190, 192, 196
    growth hormone, **46,** 163-165
      role at puberty, 179, 184, 197
      role in fat metabolism, 41
      role in intermediate carbohydrate metabolism, 40
      role in lactation, 198
      role in protein metabolism, 39
  development of, **162,** 205
  posterior, 168-172
    antidiuretic hormone in water balance, 138, 144, 145, 170, 171, 172
    in pregnancy, 194
    in suckling reflexes, 195, 196, 198
    oxytocin after childbirth, 195, 196, 198
    role in lactation, 198
    vasopressor effect, 168
Placenta, 190, **191,** 192, 201
  after childbirth, 195
  effect on lactation, 198
  in foetal circulation, 193

Placenta—*continued*
progesterone, **183**
transfer of antibodies in haemolytic
disease of newborn, 110
Plasma, 104
agglutinins in, 108, 109
clearance of, 136-141
(see also under Plasma proteins)
Plasma cells, 112, 113
Plasma globulin, 104
(see also under Plasma proteins)
Plasma proteins, 39, **104**
as weak acids, 127
osmotic pressure, 98, 114, 120, 136
role in carriage and transfer of O_2, and
CO_2, **128, 129**
Plasma thromboplastin antecedent, 105
Platelets, 11, 104
formation of, 107
role in blood clotting, 105
Pleura, 119, 120
intrapleural pressure, 121
Pneumotaxic centres, 130
Polar bodies, 18, 188
Polycythaemia, 107
Polydipsia, 173
Polymorphonuclear granular leucocytes,
11
formation of, 106, **107**
Polypeptides, 28
Polyphagia, 173
Polysaccharide, 26
Polyuria, 156, 173
Pons, 208, 209, 256
centres
for eye reflexes, 227
pneumotaxic, 130
development of, 205
pathways through
motor, 252, 253, 256
proprioceptor, 245
sensory, 247-249
taste, 223
links with cerebellum, 258
parasympathetic outflow from, 210,
268
Portal canal, 64
Portal vein, 71
absorption amino-acids, 39
fat, 41
glucose, 40
Post-central gyrus, 247, 250
Posterior horns of spinal cord, 211
Posterior root ganglion, 17, **211**, 214,
248, 270
Post-ganglionic fibres, 268-270
Postural reflexes, 243, (245)
role of extrapyramidal system, 256
Posture
muscles maintaining, 263
role of cerebellum, 259, 260
extrapyramidal system, 256
proprioceptors, 244, 245
vestibular mechanism, 243
Potassium, 25, 50, **102**
depletion, 158
in blood, 104, 127, **128, 129**

intoxication, 159
role in passage of nerve impulse, **213,**
271
tubular excretion in urine, 137,157-159
Pre-central gyrus, 251-253
Pre-ganglionic fibres, 268-270
Pregnancy
anterior pituitary in, (163), 201
energy requirements in, 49
fertilization in uterine tube, 188
mammary glands in, 198
menstrual cycle ending in, 201
ovary in, 183
uterus in, 189-196
Premotor cortex, **206, 251**, 256, 259
Pressure
joints, 244
skin sensation, 17, 220, 246, 249
(see under Blood pressure, Hydrostatic pressure and Osmotic
pressure)
Pretectal nucleus of midbrain, 236
Prevertebral ganglia, 269
Prime movers, 264
Proerythroblast, 106, 107
Progesterone, 180, **182-185,** 187, 189,
190-192, 195-201
action on mammary glands, 197, 198
and anterior pituitary, 163, 201
suprarenal cortex, 157
Prolactin, 163, 198, 199
Prolactin-inhibiting factor, 198, 199
Prolymphocyte, 107
Promyelocyte, 107
Prophase, 7, 18
Proprioceptors, 20, 220, 241, 242, 244
links with cerebellum, 258
pathways to brain, 243, 245
role in reciprocal innervation, 265
Prosencephalon, 205
Prostate gland, 176, 178, 179
Protective foods, 36, 37
in balanced diet, 50
Protein, 25, 28-30, 39, 41, 42
absorption, 70, 71
and body temperature, regulation, 44,
45
and growth, 46, 47
and haemopoiesis, 106
caloric values, 45
carbon cycle, 31
dietary sources, 35, 50
specific dynamic action, 45, 48
digestion, 38, 58, 62, 66
metabolism, **39**
and anterior pituitary, 163-167
and glucocorticoids, 157-159
nitrogen cycle, 32
(see also Plasma proteins)
Proteoses, 58
Prothrombin, 36, **105**
Protoplasm, **3, 6**
carbohydrates in, 26, 34, 40
constituents of, 25-29, 34, 35
essential materials for building and repair, 33
fats in, 27, 34, 41

proteins in, 28, 35, 39
Pseudopodia, 3, 4
Ptyalin, 40, **54**
action in stomach, 60
Puberty, 179, 184
growth in, 47
haemopoiesis after, 106
mammary glands at, 197
thymus regression, 174
uterus at, 186
Pudendal nerves, 73, 75, 97, 147, (268)
Puerperium, 195
Pulmonary
artery, 81, 83, 119
circulation, 79, 101, 119
exchange of respiratory gases, 129
valve, 84
veins, 81, 101, 119
ventilation, 123
Pulse (radial), 94
Pupil of eye, 224, 228, 236, 268, 269
Purkinje
cell, 16, 257
tissue, 85
Putamen, 207, 256
Pyloric glands, 58
sphincter, 53, 58
innervation, 268, 269
movements, 60
control of, 75
Pyramidal tract, 208, 217, 260
chemical transmission in, 271
Pyridoxine, 37
Pyruvate, 42
Pyruvic acid, 40

Radiation
heat loss from skin, 43-45
optic, 234
Ranvier, node of, 15, 16
in conduction of nerve impulse, 213
Rathke's pouch, 162
Reabsorption, kidney, 137
regulation, 144, 145
(see also under Renal tubule)
Receptive relaxation stomach, 60
Receptors, 15, 17, 20, 220
cutaneous, 246
hearing, 237, 238
heart, 88
in autonomic reflexes, 270
in reflex action, 214-219
muscle spindle, 244
of saccule, 241
of semicircular canals, 241
of utricle, 241
olfactory, 221
stretch receptors in lungs, 132
(see also under Stretch receptors)
taste, 222, 223
urinary bladder, 147
vision, **231,** 232, 233
Reciprocal innervation, **265**
Rectum, 52, 53, **72**
defaecation, 73
innervation, 75
Red blood cells (corpuscles), 11

Red blood cells—*continued*
  blood group factors, 108-111
  carriage and transfer of O_2 and CO_2, 126-129
  destruction in spleen, 113
  enzymes in, 128, 129
  formation of, **107**
    nutritional requirements for, 25, 35-37, 106
  haemoglobin, 128, 129
  volume and numbers, 104
Red nucleus, 217, 256, 259
Reflex
  accommodation for near vision, 228, 229
  action, 214 to 218
  arc, 214-218
  autonomic, 270
  cardiac, 88, 89
  conditioned, **55**, 59, 60, 63, 65, 219
  control of gastric motility, 60
  control of muscle movement, 260
  defaecation, 73
  effects on anterior pituitary, 163
  effects on secretion of antidiuretic hormone, 170
  emptying of small intestine, 69
  expulsion of bile from gall bladder, 65
  eye movements, 227
  factors in regulation of respiration, 132
  gastro-colic, 73
  gastro-ileo-colic, 69
  Hering-Breuer, 132
  inhibition and facilitation of, 265
  light, 236
  micturition, 147
  muscles of middle ear, action of, 237
  pathways, 219
  postural, 243, 256
  reciprocal innervation in, 265
  salivation, 55
  secretion of gastric juice, 59
    intestinal juice, 68
    pancreatic juice, 63
  spinal, 216
  stretch, 215
  suckling, 195, 196, **198**
  swallowing, 57
  vasomotor, 96, 97
  vomiting, 61
Release of energy, 31, 40, 41, 42, 43, 266
  role of vitamins in, 36, 37
  thyroid in control of, 151-153
Releasing factors, 151, 157, 163, (197-199), 201
Renal artery, 134, 135
Renal corpuscle, 135-141
Renal tubules, **135**
  conservation of base, 137, 142, 143
  hormones, action of
    ADH, 168, 170, 171, 172
    corticoids, 157-159, 166, 172
    insulin, 173
    parathormone, 154-156
  reabsorption in regulation of water balance, 138, 144, 145, 172
  reabsorption of urea, 140

secretion of ammonia, 142
secretion of creatinine, 141
secretion of diodone, 141
secretion of hydrogen ions, 142, 143
Renal vein, 135
Renin, 144, 145, 150, 172
Rennin, 58, 62
Repair, 35, 42, 48, 49, 163
Reproduction, 4, 22, **176-201**
  organs
    development of, 47
    female, 180
    male, 176
  role of hormones in control of, 201
Residual volume (lungs), 123
Respiration, 4, 8, 22, 117-132
  action of adrenaline on, 161
  artificial, 122
  carbon cycle, 31
  inhibition in swallowing, 57
  lungs, capillary bed, 101
  muscles of, 117, 262
  rate, **123**
    in thyroid deficiency, 152
    in thyroid overactivity, 153
  respiratory
    centres, 130
      in vasomotor reflexes, 96
    epithelium, 10
    gases
      composition of, **124**
      interchange, 119
      movement of, 125
      volume of, 123
    movements, 117, 120, 121
      control of, 130-132
      role in movements of lymph, 112
    passages, (12), (13), **118**, 221
    surfaces, 119, 124
    system
      heat loss from, 43
      water loss from, 103
  uptake of oxygen by blood in lungs, 90
Rete testis, 177
Reticular nuclei (of brain stem), 256
Reticular tissue, 11
  in haemopoiesis, 107
  lymph node, 112
  spleen, 113
Reticulocyte, 107
Reticulo-endothelial system, 11
  in bone marrow, 107
  lymph nodes, 112
  spleen, 113
Reticulo-spinal tract, 255, 256
Retina, **224, 229, 231**
  blood vessels, 224, 230
  fundus oculi, 230
  light reflex, 236
  mechanism of vision, 232, 233
  nerve pathways from, 234
  visual purple, 36
Retinene, 232
Rhesus factor, 110, 111
Rhodopsin, 232
Rhombencephalon, 205
Riboflavine, 37

Ribonucleic acid (RNA), 6
Ribosomes, 6
Ribs, 117, 120, 121, 261
Rickets, 36
Rods of corti, 238
Rods of retina, 231, 232
Round window, ear, 239
Rubro-spinal tract, 255, 256, 259
Ruffini endings, in skin, 246
Rugae (stomach), 58

Saccule, 241, (243), 258
Sacral
  nerves, 211
  parasympathetic outflow, 75, 147, 268
Saliva, **54**, 58, 60
  control of secretion, 55
  glands forming, 9, 52-55
    control of blood supply, 95, 97
    nerves to, 210, 268
  role in taste, 222
  role in water balance, 103, **145**
Sarcolemma, 14
Sarcomere, 266
Sarcoplasm, 14
Scala, media, tympani, vestibuli, 238, 239
Schäfer's method of artificial respiration, 122
Schwann cell, 15, 16
Sclerotic coat of eye, 224, 226
Scotopic vision, 232
Scotopsin, 232
Scrotum, 176, 178, 179
Scurvy, 37
Secondary sex
  characteristics
    development at puberty, 47, 176-179, 180, 181, 184
    role of pituitary, 164, 167
    role of suprarenal cortex, 157-159
  organs
    female, 180, **181**, 184
      at menopause, 200
      cycle of events in menstrual cycle, **185**
      hormonal control of, 182
    male, 176, 177, **178**, **179**
Secretin, 63, 68
Segmentation, in small intestine, 69
Sella turcica, 162
Semicircular canals, 237, **241**, **243**
  links with cerebellum, 258
  mechanism of action, 242
  nerve pathways, 210
  pathways to brain, 243
Semilunar valves, 81, 83, 84
Seminal fluid, 178
Seminal vesicles, 176, 178
Seminiferous tubules, 177
Senility, 167
Sense organs, 220
Sensory cortex, 206, (245), (247), (249), **250**, 271
Sensory decussation, 208, 245, (247), 249
Sensory neurones, 15, 17
  (see also under Receptors)
Serosa, in digestive tract, 67

Serous acini, 9, 54, 62, 118
Sertoli, cells of, 177
Serum albumin, 104
Serum globulin, 104
Sex glands, (see Gonads)
Sex hormones
  at puberty, 47
  of suprarenal cortex, 157-159
    (see Oestrogen, Progesterone, Testosterone)
Simmond's disease, 167
Sino-auricular node, 85, 86
Skeletal growth, and thyroid deficiency, 152
Skeletal muscle, 14, 20, 262-266
  actions of adrenaline, 161
  actions of insulin, 173
  and suprarenal cortex, 157-159
  control of blood supply, 95, 97, 269
  control of movements, 260
  control of postural tone, 259
  heat production and tone in, 43-45
  in Addison's disease, 158
  in giantism and acromegaly, 165
  initiation of movement, 251
  innervation, 254, 255
  metabolism, 40, 43
  motor pathways to, 253
  muscular movements, 264, 266
  nerve endings in, 15, 16
  neuromuscular transmission, 254, 271
  oesophagus, 56
  proprioceptors in, 244
    pathways from, 245
  reciprocal innervation, 265
  reflex contractions, 214-216
  respiratory, 120, 121
  role in movement of lymph, 112
  role in venous return to heart, 99
  role of cerebellum in control of postural tone, 259
  role of extrapyramidal system, 256
  sphincters
    anal, 72, 73, 75
    bladder, 146, 147
  sympathetic innervation of blood vessels to, 269
Skeleton, 261
Skin, 44, 45, 246
  adipose tissue in, 12
  afferents in autonomic reflexes, 270
  control of blood supply, 95-97
  effects of adrenaline on, 161
  fluid loss from, 103
  heat loss from, 43-45
  in Addison's disease, 158
  in overactivity of anterior pituitary, 165
  in panhypopituitarism, 167
  in thyroid deficiency, 152
  nerve pathways from, 210, 247-249
  sensation, 246
  sense organs in, 44, 45, 220, 246
  stratified squamous epithelium, 10
  sweat glands, 43-45, 103
  vitamins A and D and, 36
Sleep

heat production during, 45
urine volume in, 141
Small intestine, 9, 21, 38-41, 66-71
  absorption, 70, 71
  innervation, 74, 75, 268, 269
  movements of, 69
  secretion, 66, 68
Smell, 220, 221
  centres for, 209
  olfactory nerve, 210
Smooth muscle, (visceral, involuntary), 14
  and adrenaline, 161
    oxytocin, 169
    vasopressin, 168
  chemical transmission, 271
  innervation, 16, 268-270
  in blood vessels, 91, 92, 95
    bronchioles, 118
    digestive tract, 56-74
    oesophagus, 56
    urinary passages, 146, 147, 178
    uterine tubes, 181
    uterus, 181, 186, 190, 192
    vas deferens, 178
Sodium, 25, 50
  and antidiuretic hormone, 170, 172
  and kidney, 137, 138
  and suprarenal cortex, 138, 144, 145, 157-159, 172
  in blood, 104
  in urine, 148
  role in passage of nerve impulse, 213, 271
Somatotrophin, (growth hormone), 39, 163-165, 167, 172
Sound waves, 220, 237, (239)
Special proprioceptors, 241-243
Specific dynamic action of food, 48
Speech, 118
  interruptions in respiration, 132
  motor area of cerebral cortex, (centre for), 206
  role of saliva in, 54
Spermatogenesis, 176, 177, 179
Spermatozoa, 176, 177, 188
  development of, 177
  maturation of, 18
  Rh factor inheritance, 111
Sphenopalatine ganglion, 268
Sphincters
  digestive tract, 53, 60, 66, 69, 72, 73, 75
  Oddi, 53, 62, 65
  pupillae, 228, 236
  urinary bladder, 146, 147
Sphygmomanometer, 93
Spinal cord, 20, 204, 209, 211
  anterior horn cell, 15, 16
  as reflex centre, 214 to 218
  cerebrospinal fluid, 114
  coverings of, 114
  development of, 205
  in autonomic reflexes, 270
  in control of muscle movement, 260
  in reciprocal innervation, 265
  links with cerebellum, 258
  links with extrapyramidal system, 256

nerve supply to respiratory muscles, 130
pathways
  final common, 254, 255
  motor, 252, 253
  proprioceptor, 245
  sensory, 247-249
  sacral parasympathetic outflow, 268
  sympathetic outflow, 269
  vestibular links, 243
Spinal nerves, 211
  posterior root ganglion, 17
Spinal reflexes, 216
  final common pathway, 255
  stretch, 215
Spleen, 113
  circulation, 71, 113
  haemopoiesis, 106, 107, 174
  reticular framework, 11
Stapedius muscle, 237
Stapes, 237, 239
Starch, 25, 26
  dietary, 34
  digestion of, 38, 54, 62, 66
  metabolism, 40, 42
    (see also under Carbohydrate)
Stercobilin, 64
Stercobilinogen, 64
Stereoscopic vision, 235
Sternum, 120, 121
  in haemopoiesis, 106
Steroids
  bile salts, 27, 64
  cholesterol, 27, 157
  of suprarenal cortex, 157
Stomach, 9, 21, 52, 53, 58-61
  absorption from, 71
  hormones of, 59, 150
  innervation, 75, 268, 269
  intrinsic factor formation, 106
  movements, 60
  secretion, 38-41, 58, 59
Straight arteries, (kidney), 134
Stratified squamous epithelium, 10
  in oesophagus, 56
  skin, (44), (45), 246
  vagina, 181, 185
Stress
  and ADH, 170
  and adrenaline, 160, 161
  and corticotrophin, 157
  and lactation, 198
  effect on anterior pituitary, 163
  effect on cardiovascular system, 87, 96, 97
  effect on respiration, 132
Stretch receptors
  in arch of aorta, 89, 97, 132
  bronchioles, 132
  carotid sinus, 89, 97, 132
  great veins, 88
  intestine, 69
  skeletal muscle and tendons, 244
Stretch reflexes, 215
Stria vascularis, 238
Striated border epithelium, 9
  in kidney tubules, 135, 136

Striated border epithelium—*continued*
    large intestine, 72
    small intestine, 66, 70
Striated muscle, 14
    (see Skeletal muscle)
Stroke volume (heart), 90
Stuart-Prower factor, 105
Subarachnoid space, 114
Sublingual salivary glands, 54
Submandibular ganglion, 268
Submandibular salivary glands, 54
Submucosa (gastro-intestinal tract), 66, 67
Substantia nigra, 256
Succus entericus, 66, 68
Suckling, 169, 195, 196, **198**
Sucrase, intestinal, 66
Sugars, 26
    absorption, 70, 71
    dietary, 34
    metabolism, 40-42
    (see also under Carbohydrate)
Superior cervical ganglion, 228
Superior mesenteric ganglion, (75), 269
Superior oblique muscle of eye, 210, 226
Superior rectus muscle of eye, 210, 226
Suppressor areas, in cerebral cortex, 256
Supraoptic nucleus, 144, 145, 168-171
Suprarenal
    cortex, **157-159**, 172
        and anterior pituitary, 163, 166
        and insulin, 173
        role in water balance (aldosterone), 138, 144, 145, 172
    medulla, 45, **160**, 161
        chemical transmission, 271
        innervation, 269
Suspensory ligament of lens, 224, 228
Swallowing, 56, 57
    respiration in, 132
    role of saliva, 54
Sweat
    fluid loss, 103, 145
    glands, 43-45
        chemical transmission, 271
        innervation, 269, 270
    in body temperature control, 43-45
Sympathetic ganglion cell, 16
Sympathetic nervous system, 21, 269
    chemical transmission in, 271
    in autonomic reflex, 270
    outflow, 269
        blood vessels, 95-97
        digestive tract, 74, 75
        heart, 87-89
        iris, 228
        suprarenal medullae, 160
        urinary bladder, 147
        veins, 99
    vasoconstrictor tone in body temperature regulation, 44, 45
Sympathomimetic amines, 271
Synapse, 212, 213, 214
    chemical transmission at, 271
Synaptic transmitters, 271
Synergists, 264
Synovial membrane, 264

Systemic circulation, 79
    blood flow, 100
Systole, 83
    blood pressure in, 92, 93
    elastic arteries in, 91, 94
    heart sounds in, 84

Tarsal glands, 225
Taste, 220, **222**
    pathways and centres, 210, 223
Tears, 225
Tectorial membrane of cochlea, 238, (240)
Tecto-spinal tract, 255, 256
Teeth, 25, 35, 36, 53, **54**
Telencephalon, 205
Telereceptors, **20**, (220)
Telophase, 7, 18
Temperature
    body, control of, 43-45
    cutaneous sensation, 246, 248
    effect on uptake and dissociation of O_2 from haemoglobin, 126, 128
    in regulation of respiration, 131
Temporal lobes, 206
    centres for hearing, 240
Tendon, 12, 264
    proprioceptors in, 244
Tensor tympani, 237
Testis, 150, 176, **177**, 178, 179
    action of anterior pituitary on, 163
    influence of suprarenals, 157
Testosterone, 176, 177, **179**
    effect on anterior pituitary, 163
Tetany, 155
Tetraiodothyronine, 151
Thalamus, 207-209
    centres
        autonomic, 270
        pain and temperature, 248
    chemical transmission in, 271
    development of, 205
    links with other centres, 217, 256, 259
    pathways
        proprioceptor, 245
        sensory, 247, 249
        taste, 223
Theca interna (ovary), 182
Thiamine, 37, 41
Thirst, 145, 156, 171, 173
Thoracic
    cord, 130, 211, 269
    duct, 71, 112
    nerves, 211
Thorax, 117, **120**, **121**, 122
Thrombin, 105
Thrombocytopenia, 107
Thromboplastin, 105
Thymocytes, 174
Thymus, 150, **174**
Thyrocalcitonin, 151
Thyroglobulin, 151
Thyroid, 150, **151-153**
    and anterior pituitary, 163
    development of, 19
    in haemopoiesis, 106
Thyrotrophin (TSH), 151-153, 163, 166
    releasing factor (TRF), 151

Thyroxine, 151, 163
Tidal volume, 123
Tissue fluids, 102, 103, 112
    exchange of water and electrolytes, 98, 138
    heat dispersal, 43
    (see also under Body fluids)
Tissue oxidation, 40-42
    role of thyroid, 151-153
Tissues, 9-17
    constituents of, 25-29
    development of, 19
    differential growth of, 46, 47
    exchange of O_2 and CO_2, 125-129
    heat production, 43-45
    metabolism of, 39-42
    water of metabolism of, 103
Tone
    blood vessels
        arterioles, 92, 96, 97
        veins, 99
    large intestine, 73
    skeletal muscle, 44, 45, 244, 256
    stomach, 60
Tongue, 53, **54**, 57, 210, 222, 223
Touch
    Meissner corpuscle, 17, 246
    pathways, 247, 249
Trachea, 117, **118**, 119, 151, 154
Transitional epithelium, 10, 146
Transmission of nerve impulse, 213
Transverse (central) fissure, 209
Trapezoid body, 240
Triceps, 264
Trichromatic vision, 233
Tricuspid valve, 81, 83, 84
Trigeminal nerve (V), 210, 247, 252
Triiodothyronine, 151
Trochlear nerve (IV), 210, 226, 227, 252
Trypsin, 38, 39, **62**
Trypsinogen, pancreatic, 62
Tympanic membrane, 237, 239
Tyrosine, 151

Ultra-violet rays, 36, 220
Umbilical blood vessels and cord, 191-**193**
Unicellular organisms, 3-5
Unipolar nerve cells, 17
    (see also Posterior root ganglion)
Unstriped muscle, 14
Upper motor neurones, 217, 252, 253, 260
    chemical transmission in, 271
Urea, 39
    blood, 104
    clearance, 140
    excretion, 136, 137
    urinary, 148
Ureter, 134, 135, **146**
Urethra, 134, **146**, 147, 178
Uric acid
    blood, 104
    excretion, 136, 137
    urinary, 148
Urinary bladder, 21, 134, **146**, 178
    actions of adrenaline on, 161
    innervation, 147, 268, 269

Urinary bladder—*continued*
lining, 10, 146
storage and expulsion of urine, 147
Urinary tract, 134-148
heat loss from, 43
water loss from, 103
Urine, 21, **148**
clearance of inulin, 139
control of Ca and P loss, 154-156
control of output, 103, 138, **144, 145,**
170, 171, 172
control of salt loss, 157-159, 166
diodone clearance, 141
excretion of urea, 140
expulsion, 134, 146, **147**
formation, 134-138
heat loss in, 43
in diabetes mellitus, 173
in maintenance of acid-base balance,
142, 143
in water balance, 103, 138, 144, 145
ketone bodies in, 41
storage, 134, 146, **147**
urea and ammonia in, 39, 142, 148
Urobilin, 64
Urobilinogen, 64
Urochrome, 148
Urogenital system (see Reproduction and
Excretion)
development of, 19
Uterine tube, 9, 181, **186-188**
hormonal control of changes in menstrual cycle, 185
Uterus, 180, **181, 184-196**
action of oxytocin, 168, 169
at and after menopause, 196, 200
at and after parturition, 194, 195
childhood, 186
hormonal control of changes in
menstrual cycle, 185, 201
implantation, 19, 190
involution of, 169, 196
maturity, 187
menstrual cycle, 187
placentation, 190
puberty, 184, 186
puerperium, 195
Utricle, 241
links with cerebellum, 258

Vagal, tone, 87
Vagina, 180, **181,** 184, 185, 200
Vagus nerve (X), **210,** 223, 247, **252, 268**
contraction of gall bladder, 65
gastric motility, 60
gastric secretion, 59
in regulation of heart action, 87-89
in regulation of respiration, 131, 132
in swallowing, 57
in vomiting, 61
outflow to
blood vessels, 95-97
gastro-intestinal tract, 75
heart, 88
relaxation of sphincter of Oddi, 65

secretion of pancreatic juice, 63
Valves (see Heart)
lymph, 112
vein, 99
Vas deferens, 176-178
Vasoconstriction, 95, **96,** 268, 269
role of adrenaline, 161
role of noradrenaline, 160
role of posterior pituitary, 168
Vasodilatation, 95, **97,** 268, 269
Vasopressin, 168
Vegetative functions, 21
regulation of, 268-270
Veins, 78, 79, **91**
area of cross-section, 100
baroreceptors in vasomotor reflexes,
96
portal, 71
pressure in, 92
pulmonary, 101, 119
renal, 135
stretch receptors in, 88
subdural sinuses, 114
umbilical, 191, 193
venous return to heart, 99
Venoconstriction, 99
Ventral spino-cerebellar tract, 245
Ventral spino-thalamic tract, 249
Ventral vestibulo-spinal tract, 255
Ventricles of brain, 207-209
development of, 205
Ventricles of heart, 80-83, 94
conduction of heart beat to, 85, 86
in cardiac reflexes, 89
output, 90, 101
Venules, 91
blood pressure in, 92
Vermiform appendix, 72
Vertebrae, 117, 120, 121, 261
Vestibular branch of VIII cranial nerve,
210, 241-243
membrane, 238, 239
nuclei, 217, 243, 256, 259
pathways to brain, 243
Vestibulo-spinal tract, 256
links with cerebellum, 259
Vibration sense, 220
Villi
chorionic, 190, 191
intestinal, 66, 70
Villikinin, 70
Virilism, 159, (166)
Viscera, 20, 21
abdominal, 52, 53
innervation of, 75, 268, 269
control of blood supply, 95:97
in giantism and acromegaly, 165
interoceptors, 20, 220
pelvic (see External genitalia, Urinary
bladder, Rectum and Large intestine)
innervation of, 268, 269
sensations, 220
smooth muscle of, 14
visceral reflex arc, 270
Vision, 224-236
centres for, 206, 234

colour, 233
fields of, 234
mechanism of, 232, 233
monochromatic, 232
nature of stimulus, 220
optic nerve, 210, 224
pathways for, 210, 234
photochemical mechanism, 232, 233
pigments in, 36, 232, 233
receptors for, 231
spectrum, 220, 232
stereoscopic, 235
Visual purple, 36, (232)
Vital capacity, 123
Vitamins, 36, 37, 50
absorption and transport, 70, 71
in haemopoiesis, 106
reabsorption, kidney, 137
rhodopsin regeneration, 232
Vitreous humour, 224
Vocal cords, 118
Voluntary muscle, 14, **262-266**
(see Skeletal muscle)
Vomiting, **61,** 144, 172
Von Ebner's glands, 222
Vulva, 169

Walking, reciprocal innervation in, 265
Warmth
skin sensation, 246, 248
body temperature, 43, 44, 45
Water
absorption and transport, 71-73
balance, **103**
internal secretions in, 103
regulation of, **144, 145**
role of parathyroid, 156
role of posterior pituitary, 138, 170,
171, 172
role of suprarenal cortex, 157-159,
172
daily requirements, 50
distribution in body, 102
excretion of, 134, 137, 141, 144, 145
filtration in kidney, 136
reabsorption in kidney tubules, 137,
138
Water-soluble vitamins, 37, 71
White blood corpuscles, **11,** 104, 106, **107,**
112
White fibro-cartilage, 13
Womb (see Uterus)
Work
energy requirements for, 48, 49
release of energy for, 42

Xerophthalmia, 36

Yolk sac, 19

Zona glomerulosa, 144, 157, 172
Zymogen, 62
Zygote, 19, 188